The Political Economy
of International Oil
and the Underdeveloped Countries

BY MICHAEL TANZER

TEMPLE SMITH   LONDON

First published in Great Britain 1970 by
Maurice Temple Smith Ltd
37 Great Russell Street, London w.c.1

SBN. 851170013

Reproduced photolitho in Great Britain by
J. W. Arrowsmith, Ltd, Winterstoke Road, Bristol 3

To the people of the underdeveloped countries

70/-

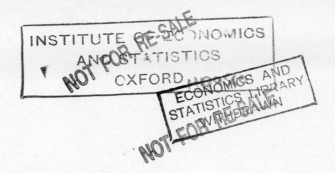
THE POLITICAL ECONOMY
OF INTERNATIONAL OIL
AND THE UNDERDEVELOPED COUNTRIES

Contents

Contents

Preface

THE LITERATURE on the oil industry is vast. Millions of words have been written on the history, politics, and economics of this giant and enormously important industry. The reader may well ask then, Why the need for another book? My answer is that no existing work deals comprehensively with the complex political and economic relationships between and within the underdeveloped and developed countries centering around that most vital commodity, oil.

This vacuum became clear to me in the course of my employment as an economist covering underdeveloped countries for a major international oil company. While I had a background in problems of economic development and could find ample materials on petroleum technology and general petroleum economics, I found a dearth of published work dealing systematically and realistically with the manifold impact of oil on the underdeveloped countries. The unfortunate tendency in the social sciences to divorce "economics" from "politics" has meant that those scholars who write on oil have too often tended to limit themselves to relatively isolated "economic" or "political" analyses; almost by definition such an approach would prevent comprehensive treatment of relationships which are so many-sided as to seem at times overwhelming. In addition, few academicians have ever had the practical experience of working within this industry; their consequent lack of first-hand knowledge can be a serious, albeit not insurmountable, handicap.

Several general intellectual debts should be mentioned. My academic training in economics, all received at Harvard and culminating in a Ph.D., provided me with some of the formal tools of analysis used in this study. Also, my ability to develop a broad-gauge approach to "economic" problems was strengthened by several years of teaching experience at Harvard—particularly in Professor Samuel Beer's pioneering social science course devoted to exploring applications of economic, political, sociological, and psychological theories to analysis of historical problems. Finally, my work as an economist for Esso Standard Eastern, the Asian-African affiliate of Standard Oil of New Jersey, gave me invaluable experience for this study. My on-the-job training involved a systematic familiarization with all phases of the international oil industry. In particular, the section in this volume on oil and India benefits from my various assignments on that country, including a stay of several months.

With regard to the actual carrying out of this study, my greatest debt is to Harry Magdoff, economist-writer and lecturer at the New School for Social Research. His continuing encouragement along with his careful and insightful criticisms of the various drafts of the manuscript have been invaluable to me. I also wish to express my appreciation to the Louis M. Rabinowitz Foundation for financial support which helped me to undertake and complete this work.

Finally, my deepest gratitude to my wife, Dr. Deborah Tanzer. She not only lent moral support but also made time between running a household and carrying out her own brilliant researches in psychology to provide extensive editing and proof-reading services. To her, as well as our sons, David and Kenneth, my thanks for bearing with my preoccupation with this study.

PART I

The Political Economy of International Oil:
Forces and Issues

CHAPTER ONE

Introduction

No COMMODITY plays a more vital role than oil in the economic life of the underdeveloped countries of the world. Obviously, for nations with enormous oil deposits, such as Venezuela, parts of the Middle East and North Africa, oil *is* the economic life. Oil exports provide not only a large share of national income but also most of these countries' foreign exchange earnings. Oil thus represents the great asset which could potentially provide all the capital necessary for economic development. In a real sense, these countries represent prime candidates for testing the validity of the theory of economic development through "unbalanced growth" which has been advocated by influential economists.[1]

The main focal point of this book, however, is the less obvious impact of oil on the vast majority of underdeveloped countries which are presently dependent upon imported oil. For these countries oil's vital impact can be felt in a variety of ways. First, energy fuels are indispensable for modern industry and agriculture. In fact any economist who like this author has studied the relationship between energy and the total economy in a variety of countries would be tempted to add Energy as the fourth factor of production to the classic ones of Land, Labor, and Capital. For just as capital without labor is useless, so too is sophisticated capital without energy. (Historically, the whole course of industrialization, starting with the waterwheel and steam power, can be seen as a joint capital-energy substitution process; it involves using sophisticated capital, which magnifies human power through the mechanism of inanimate energy, to replace primitive capital, which simply combines directly with human power.)

3

I. The Political Economy of International Oil

Besides being a vital motive power for physical transformations, energy is also a necessity for many modern industrial processes which involve chemical transformations. The manufacture of cement or steel, for example, both require great quantities of heat. In addition, for certain leading industries, energy is necessary as a basic "feedstock," i.e., where the energy source itself becomes part of the final product. In the manufacture of pig iron or petrochemicals, for instance, the hydrocarbons in coal and oil are incorporated bodily. Finally, energy is an irreplaceable element of final consumer demand, e.g., for household cooking, heating, transportation, and lighting.[2]

Existing literature has tended to stress the qualitative role of energy as a necessary element for physical production. In the words of one pioneering study prepared by the United Nations:

> Although in many cases the cost of energy, especially of electricity, represents but a minor percentage of total costs, energy exercises great influence because of its qualitative effects. It is the key element without which the production process cannot operate adequately, and the lack or shortage of energy may cause serious difficulties. It stands in the same position as other tangible or intangible factors of industrial production, the economic effect of which is more important than their net cost.[3]

This author has no quarrel with the notion of energy as an indispensable element in the production process. (The history of India's recent economic development, for example, reveals numerous examples of idle production capacity owing solely to lack of available energy.) At the same time, for example in formulating energy policy, it is also important to recognize the various costs associated with different policies. Despite the fact that energy costs amount to only about two to three percent of gross national product, its quantitative impact on key aspects of the economy is much greater.

First, for many underdeveloped countries the cost of importing energy is a major burden on their scarce foreign exchange; oil imports typically account for between 5 and 10 percent of all imports. Second, energy costs are much more significant in the "advanced" industries than in "backward" indus-

4

tries or in agriculture. Agriculture in underdeveloped countries typically requires virtually no energy fuels, relying as it does almost exclusively on human and animal power. At the other extreme, modern heavy industries, energy costs as a proportion of total costs have been estimated as follows: 25 percent in steel making, 20 percent in production of other metals, 10 to 15 percent in chemical production, 15 percent in railways, and 20 to 25 percent in motor transport. The light industries such as textiles lie in the middle; for this group energy costs probably account for somewhat less than 10 percent of total cost.[4] Clearly then, energy costs are relatively most important for those industries which form the foundation of an industrialized economy.

Another significant aspect of energy is the capital required to produce or process it. It is frequently said that the energy sector requires relatively large amounts of capital:

> In Latin America, energy contributes 2.3 per cent of the gross product and a much larger share towards capital formation. As a result, the energy sector has a high capital-output ratio. . . .
> . . . The high capital intensity required by energy production exerts strong pressure on the capital formation capacity of most Latin American countries.[5]

While this generalization may be true for the energy sector as a whole, it is still necessary to analyze the productivity of capital investment in developing different types of energy. This is particularly important in calculating whether it is better to rely on foreign energy (e.g., by importing refined petroleum products) or to develop the energy indigenously (e.g., by building refineries at home).

One aspect of oil's potential role in an underdeveloped country is worth special mention: oil's capacity to be a leading factor in the development process even without being exported. Most development economists would probably say that an indigenous petroleum sector is not well suited to playing a leading role, because crude oil production involves little "backward linkage" as compared with an industry like steel which tends to generate demand for the output of other industries.[6] It is our belief that such a view is mistaken in that it fails to take account of the enormous potential for capturing capital from an indigenous crude oil production center.

This potential arises from the fact that, of all the major commodities involved in international trade, crude oil has by far the largest gap between the average cost of production and the price. The existence of this gap lays a basis for large capital generation by an indigenous crude oil sector. It is particularly important because all of this potential capital is directly translatable into foreign exchange through import replacement (at least up to the point where indigenous crude oil production meets domestic consumption needs). The fact that the huge capital generation in various oil-exporting underdeveloped countries has generally failed to trigger economic development should not be seen as necessarily negating this principle; usually there are other forces operating to prevent the effective harnessing of this capital. One way in which an indigenous crude oil sector can lead in economic development is shown in the case of Mexico.

Finally, because energy supplies are a vital necessity to most developing economies, they play an important role in determining the ability of a country to maintain an independent course of political and economic development. Two aspects of the petroleum picture are particularly relevant here: (a) whether the ownership of the resources is foreign or indigenous; (b) whether the physical location of the energy resources is external or indigenous. As we shall show, the mere fact of foreign ownership automatically involves a country in the complex relationships of what we have termed "the political economy of international oil." On the other hand, if a country's petroleum resources are physically located abroad, regardless of the locus of ownership the country is vulnerable to political and economic pressures of various kinds. Like an iceberg, only the tip of the totality of these pressures is normally visible; the whole only comes to the surface in the extreme case of attempts to completely cut off a country's flow of petroleum, i.e., oil boycotts.

Vulnerability to oil boycotts is rarely if ever mentioned in the literature of energy and development economics. Nevertheless, it is clearly very much in the mind of many countries in developing their own energy policies. The United States, for example, has always tended to encourage domestic exploration and production of crude oil partly to insure that domestic supplies would be available in case of war. The Union of South Africa for years has been attempting to develop indigenous petroleum resources, both through exploration for crude oil and research on processing petroleum from oil shale. Cuba, the intended victim of an oil boycott in recent years, has energetically sought to find indigenous crude oil. It seems obvious that for any country seriously seeking to minimize its dependence upon others, indigenous energy resources are a vital necessity. There is no surer way to bring to its knees any but the most primitive economy than through cutting off its energy supplies.

This particular aspect of energy and petroleum policy is analyzed in some detail later, partly because of the problem's intrinsic interest, but also because the vantage point of this book is that of an underdeveloped country seeking to combine rapid development with the maximum possible economic and political independence. These two goals encompass the range of specific aims of most enlightened groups in underdeveloped countries.

This vantage point, we feel, is the best one for achieving one of the fundamental aims of this study: to assist government planners and others responsible for energy and oil policy in underdeveloped countries by providing a practical framework for decision making. At the same time, it is hoped that the present work will be useful for all concerned with the problems of economic development, by providing insight into a crucial sector of the economy. Finally, it is also hoped that this study will be of general interest insofar as it illuminates the present international rivalries and struggles for power in the underdeveloped world.

The basic methodology and structure of this work is designed to serve all three aims simultaneously. Part One provides a general conceptual framework for analyzing the major forces and issues affecting oil and the underdeveloped countries. Part Two widens and deepens this conceptual framework, through a detailed analysis of the history of the petroleum industry in In-

7

dia over the last two decades. Part Three broadens the scope of the framework through briefer case studies of various underdeveloped countries with different energy and petroleum policies. From comparison of these various experiences an attempt is made to derive some useful general conclusions.

CHAPTER TWO

Basic Economics of International Oil

SOME UNDERSTANDING of certain basic economic characteristics of the international oil industry is necessary for a full comprehension of oil's manifold impact on the underdeveloped countries. This chapter examines the following aspects: (1) world oil prices and costs; (2) size of the world oil trade and division of the profits. For this sketch of the theory and practice of oil pricing, the basic economic tools of analysis used are the classic ones of supply and demand theory.

The theoretical *upper limit* to the price of crude oil is fundamentally determined by the maximum demand for the various refined oil products which can be obtained from this crude oil. That demand, as measured by the maximum price that consumers would be willing to pay for specified amounts of oil products, reflects in turn the value to them of using these products. For example, to the owner of an electric power station which could run only on diesel oil, the value would be the value of electricity output less all his other costs. In the case of motor gasoline, the upper limit to price would be the monetary expression of the pleasure or value to a car owner of being able to operate his vehicle.[1]

In the real world, of course, there are substitutes for various oil products, and their prices reduce these theoretical upper limits. At one extreme, coal, natural gas, and nuclear and hydroelectric power all are relatively close substitutes for fuel oil, and thereby drastically reduce fuel oil's top price far below its value to users. At the other extreme, there is no substitute, in the short run at least, for motor gasoline in the automobile, and hence the upper limit to price tends to be closer to the value of gasoline to car owners.[2]

9

1. The Political Economy of International Oil

The theoretical *floor* to oil prices is the long-run cost of production, including some minimum rate of profit for the capital invested, for meeting different levels of demand. This long-run production cost for refined products consists of the sum of the costs of getting the oil from the ground, shipping it to the consuming center, and refining it. The most critical factor affecting these costs is that discoveries of oil made since World War II have been so enormous that for all practical purposes there should be no need for any further expensive oil exploration to meet demand for many years to come; that is, the only necessary costs in producing crude oil are the relatively minor ones of drilling development wells in proven fields and operating the wells.

The fundamental determinant of whether actual oil prices lie nearer the theoretical upper limit or lower floor is the degree of monopoly versus competition in the industry. To best understand the existing situation, it is instructive to theorize what prices would be in two different situations. Assuming a perfectly competitive situation in which there were numerous independent firms at every level of the oil industry, the price of refined products to the consumer would be built up by simply adding the separate costs of producing oil, transporting it, and refining and marketing it. This is because competition at each level would drive profits down to the minimum necessary to keep enough capital in the industry to meet demand.

If this assumption of perfect competition at all levels of the industry were modified to hypothesize one monopoly producer of crude oil (with perfect competition still assumed at all other levels of the industry), then the situation changes drastically. The monopoly producer would ask himself, "What price should I charge for my crude oil in order to maximize my profits?" In order to calculate this, the rational monopolist would compute the maximum price that could be charged to the consumer for various refined products. This, as noted, will essentially depend upon the value of the different oil products to consumers, combined with the ease of substitution by other energy fuels. Historically, since motor gasoline has been a valuable product for the consumer, with no substitute, the monopolist would multiply the amount of motor gasoline which he could profitably squeeze out of a barrel of crude oil times the maximum price which could be obtained for that motor gasoline; with the refineries designed to maximize gasoline production, the remaining products which could be derived from the crude oil would then be multiplied by the maximum price which could be charged for each of them.[3]

Once the maximum revenue derivable from refining a barrel of crude oil has been determined by the monopolist, then he can calculate the maximum price obtainable for his crude oil by "backing out" (i.e., deducting) marketing, refining, and transport costs from the final product price. For example, if the maximum revenue obtainable from consumers for a barrel of refined oil is $5, marketing and refining costs $2, transport costs $1, this leaves $2 as the maximum price which can and will be charged for crude oil by the monopolist, regardless of his cost of producing the crude oil. Note that under our previous assumption of perfect competition everywhere, if the cost of producing crude were $0.50, the price to the consumer of refined products would be only $3.50; with monopoly it is $5.00 and the monopolist gets the $1.50 difference.

Thus, the fundamental difference introduced by assuming monopoly at the crude oil production level is that the price of oil to the consumer is no longer solely dependent upon cost. Instead, it is dependent upon the value of the product to the consumer.

What are the actual facts about prices in relation to costs? First, it is clear that owing to the discovery of enormous oil fields, particularly in the Middle East, the cost of producing crude oil today is extremely low relative to its price. Various estimates have been made of the cost of producing crude oil in different areas, most of which, if anything, tend to err on the high side:

> . . . informed estimates of the producing cost of supplying different Middle East crude oils have varied, in the past, from about 8 cents a barrel to about 45 cents a barrel. . . .

One consumer government, that of West Germany, indeed had a set of independent estimates of crude costs prepared on its behalf in 1962. These suggested that costs up to the ocean terminal averaged about 33 cents a barrel in the Persian Gulf in 1959. . . . It compared this estimated average cost with similar estimates of about 90 cents a barrel for Venezuela. . . .[4]

The New York Times has reported:

The direct production costs, including transportation to a Persian Gulf port, are about 15 cents.[5]

The Organization of Petroleum Exporting Countries has expressed the following view:

Production costs estimated in Venezuela at over 50 cents per barrel, and in the Middle East at about 25 cents per barrel. . . .[6]

M. A. Adelman, a leading petroleum expert, mentions the following estimates:

. . . [It] is often said, that average or representative Venezuelan development costs are five times Middle Eastern, and that Middle East development-operating costs are 5 to 10 cents. . . .[7]

From the present author's experience a good rule of thumb for long-run production costs might be 10 cents per barrel for the Middle East[8] and 30 cents for Venezuela.

Refinery and transport costs, including both current operating costs and a "reasonable" rate of return on the investment, vary widely. Nevertheless, for analytical purposes we shall use the hypothetical but realistic figures in Table 2-1 to compare

TABLE 2-1

APPROXIMATE PRICES AND COSTS AT THREE LEVELS
OF THE INTERNATIONAL OIL INDUSTRY

	Price per bbl	Cost per bbl
Crude oil production	$1.50	$0.25
Transportation	0.25	0.25
Refining	0.50	0.50
Totals	$2.25	$1.00

with similar order-of-magnitude estimates of prices for each of these services.[9]

These figures in Table 2-1 highlight a characteristic of the economics of the international oil industry which is of great importance for the underdeveloped countries: profitability in the industry basically stems from the sale of crude oil. While this is explainable by our hypothesis of greater monopoly control in crude oil production than in refining or transport, in the real world the supply of low-cost crude is not controlled by a single monopolist but by a number of companies, which number in recent years has been increasing. This raises a danger to the companies as a group that the great profits which can be derived from selling additional amounts of crude oil will tempt individual companies to cut prices in order to get additional sales, thereby ultimately destroying the monopolistic "self-control" which is the source of the high profitability. In the words of M. A. Adelman:

> Thus, in trying to think out the price structure, we are brought to realize that monopoly price and output are what is best for an industry or group as a whole, i.e., what would be charged and supplied by a single seller, who *is* the industry as a whole. . . . A price well above incremental cost is a source of tension in the system, which, good or bad, is not to be talked away by calling it fair, reasonable, natural—or even "competitive," in view of the nearest substitute. It can be maintained only by a united front and refusal to supply at any lower price.[10]

Historically, the principal mechanism which has served to buttress the companies' monopolistic control over the world's low-cost crude oil has been their high degree of vertical integration. That is, by ownership of affiliated refining and marketing companies in various oil-importing countries, each company has secured "captive" outlets for the highly profitable crude which cannot be won away by competitors; to the extent that all of the major international companies have succeeded in doing this, the pressure on crude oil prices has been reduced.

The existence of these affiliate relationships is particularly important for underdeveloped countries, since affiliates of the major international oil companies tend to play a more prominent role in the oil industries of underdeveloped countries than they do in the developed countries. This is partly because until recent years the oil business in underdeveloped countries was largely a marketing one, with the integrated majors having the competitive advantage of relatively low-cost products from their huge refineries in the Middle East and Venezuela; it stems also from the fact that the markets were relatively small and local capital was relatively more backward.

For the underdeveloped countries, the fact that their oil industries are dominated by affiliates of the international majors is significant partly because it tends to prevent them from fully reaping the benefits which have accrued to consumers in the developed countries as a result of the competitive pressures on crude oil prices generated by the recent advent of a number of newcomers into the international oil business:

> . . . [In the 1950s, the oil industry's] relationship to the miscellaneous oil-consuming countries outside of Western Europe was still in the Dark Ages, as it were . . . there was no competition at all except for that between the few established major oil companies, and this, commerce being what it is, resulted in the maintenance of profit levels and profit margins which were considerably higher than they were elsewhere—even after allowing for the substantial cost of transport and handling incurred, coupled with the comparatively small turnover. It was said, not without some justification, that the poorer a country was, the higher were the prices it paid for its oil.[11]

That affiliates can be charged higher prices for crude oil than those paid by independent refiners was so widely recognized that it led to the expression "only fools or affiliates pay posted prices."

Not so widely recognized, or at least discussed, are the many less obvious implications stemming from the affiliate relationship. The essential point about an affiliate in any country is that its need must, by the very logic of the international corporation, be subordinated to profit maximization for the company as a whole. For the parent corporation to operate the affiliate in any different

14

manner would be to violate its trust to its stockholders, who are concerned with the total profits of the company rather than with any individual affiliate's position.

This fundamental fact of international corporate life has two major sets of negative implications for underdeveloped countries. First, affiliates may be and are charged higher prices by the parent company than would be paid by independent companies, not only for crude oil, but also for refined products, transportation, managerial services, etc. Such "overcharging" may be particularly profitable in the many underdeveloped countries where the government sets a ceiling on refined product prices on a "cost-plus" basis. The overpricing may take a variety of more subtle forms, e.g., providing the affiliate with more expensive crude oil than is economically optimal for the refinery itself.

Second, affiliates normally will not have the same economic incentive as independent companies or the goverment to set up vertically integrated oil facilities *within* an underdeveloped country. For example, a refining and marketing affiliate will generally not have as much incentive to explore for crude oil within a country as an independent might, since the parent of the affiliate already has vast quantities of low-cost crude available abroad. Again, in more backward countries where the affiliate only has marketing operations, it will have relatively less incentive to set up an indigenous refinery; the affiliate can usually import refined products from the parent company's efficient and underutilized refineries in the Middle East or Venezuela, thereby giving the parent a good profit on refining as well as on the crude oil.

The positive side of affiliate life, usually stressed in much of the oil literature, is that affiliates are the beneficiaries of the great resources commandeered by the parent company—resources of money, men, technology, and diversification. For example, an affiliate may be able to borrow money at lower interest rates than an independent, either from the parent or by using the parent's credit. Again, an affiliate may be able to export surplus products to an affiliate in another country, an option which might not be open to an independent.

We are not dealing with questions which can be answered

abstractly when we consider the extent to which these positive features are real as opposed to potential, the price which is actually paid for them, and how much they compensate for the negative features. Rather, any conclusions should be based on careful analysis of actual experiences, which we shall undertake in later sections. At this point it suffices to recognize that many of the struggles involving international oil and the underdeveloped countries have centered around the role of affiliates in these countries.

Before turning to an analysis of the major forces involved in the international oil struggle, however, it will be useful to make a brief examination of the magnitude of the stakes involved in the international oil trade. An appreciation of the enormity of the stakes, and the great impact of the division of international oil revenues upon various groups, is vital. Among other things, it will help explain the great lengths to which the major international forces operating in the oil area may be willing to go in order to achieve their goals.

Below we present an overview of the world trade in oil and its quantitative significance for the following groups: (a) the oil-exporting underdeveloped countries; (b) the oil-importing underdeveloped countries; (c) the developed countries; (d) the international oil companies. The picture presented here has been pieced together from export data for each of the major oil-exporting countries and covers a recent year for which complete data were available—1964. All value figures are F.O.B. ("free on board") and as such omit the impact of international transportation;[12] this is sizable, since oil accounts for almost half of all international sea trade in terms of tonnage. As such the figures are not directly comparable with data on imports of oil which are generally shown in C.I.F. terms ("cost, insurance and freight," i.e., exports F.O.B. plus insurance and freight charges).

Another limitation of the data[13] is that they are the officially reported "posted price" values of oil exports which are often quite different from actual market values. This divergence arises because in recent years, despite a growing glut of crude oil, posted prices for oil have been held constant due to political considerations, leading to widespread discounts off these posted prices.

Despite these limitations, the picture presented below is sufficiently accurate to indicate the general features and approximate magnitude of the international oil trade. (Even if one could precisely measure the size of the trade for 1964, the absolute levels by now would be much higher owing to the rapid growth of international oil trade.) The principal features of this trade are as follows:

(1) The value of world exports of crude and refined petroleum products totaled $10 billion: $7.8 billion of crude oil and $2.2 billion of refined petroleum products. In quantity terms, world exports of crude oil equaled 4.1 billion barrels and of refined products 0.8 billion barrels.

(2) There are fifteen significant oil-exporting countries. They are listed in Table 2-2 by regional groupings.

TABLE 2-2

OIL-EXPORTING COUNTRIES

Country	Oil exports in millions of dollars	Totals in millions of dollars
Middle East		4,800
Saudi Arabia	1,200	
Kuwait	1,200	
Iran	1,100	
Iraq	1,100	
Other (Oman, Muscat, Qatar)	200	
North Africa		1,100
Libya	700	
Algeria	400	
Venezuela		2,500
U.S.S.R.		700
U.S.A.		400
Indonesia		300
Nigeria		100
Malaysia		100
Total		10,000

(3) The Middle East thus accounts for almost one half of world exports, Venezuela 25%, and North Africa 10%. Over 85% of the Middle East oil exports are crude oil, while almost 30% of Venezuela's are refined products. The Soviet Union supplies about 7% of world exports, equally balanced between crude and refined products, and the U.S.A., the world's largest oil importer, provides 4% of world exports, totally in specialty petroleum products.

(4) The implicit average "price" for crude oil (overstated, because based on posted prices) ranged from a low of about $1.81 in most of the Middle East to $2.10 in Venezuela, $2.21 in Libya and $2.37 in Algeria. The higher F.O.B. prices in these three latter countries reflect primarily the shorter transportation haul to the major European and United States consuming markets.

Finally, from the viewpoint of the underdeveloped countries, the principal features of this $10 billion world oil flow are as follows:

(a) Of it, 90% is produced in and exported from the oil-exporting underdeveloped countries.

(b) Furthermore, 90% of this 90% goes to the developed countries and only 10% to the underdeveloped countries. However, oil imports of the underdeveloped countries amount to about 15% of their imports from other underdeveloped countries. Hence, oil is an important element of the intra-underdeveloped-country trade.

(c) For five countries, Venezuela, Saudi Arabia, Kuwait, Iran, and Iraq, payments by the oil companies to the governments aggregated $2.6 billion (in 1964), or 38% of their combined oil exports of $6.8 billion. The companies' real profits (as opposed to reported book profits) were probably near the same magnitude.[14]

(d) The oil-importing underdeveloped countries spend about $1 billion on their oil imports. On an F.O.B. basis, the oil imports of all underdeveloped areas as a group account for only about 3% of their total imports. However, for some major underdeveloped countries, oil imports are a much bigger factor. For example, on a C.I.F. basis, in various years oil imports have accounted for 17% of total imports in Brazil, 6% to 9% in India, 10% in Cuba, and 7% in the United Arab Republic and Argentina.

CHAPTER THREE

Major Forces in International Oil: An Analytical Framework

THE LAST FIFTY YEARS have seen the rise of oil as the most important world energy source and the most important commodity in international trade. At the same time, throughout its history oil has been a "political resource *par excellence*," [1] owing both to its vital significance to countries and the enormous potential profits associated with it. Thus, the basic analytical framework of this study derives from the firm conviction that the impact of oil on the underdeveloped countries cannot be understood by "pure" economic analysis alone, but must be examined within the broad context of the major international institutional forces affecting oil. The title, "The Political Economy of Oil," is an attempt to stress power relationships which dominate the international oil industry; this is an industry in which in a real sense the "economics" are simple while the "politics" are complicated.

In this section we make no attempt to "prove" or support this general point which is recognized by some students of the industry.[2] Rather, we simply present here our summary view of the structural positions and goals of what we believe are the six major international forces affecting world oil, with detailed analysis and documentation reserved for later chapters. These forces are: (1) the international oil companies; (2) Western home governments of the companies; (3) the oil-exporting underdeveloped countries; (4) the Soviet Union; (5) world organizations; (6) the oil-importing underdeveloped countries.

This bald summary statement is primarily aimed at orienting the reader to the general approach used throughout this study. Since much of this book deals with specific interactions between these major forces, only brief mention is made of some of the types of interactions. These are introduced simply as initial justification for our attempt to reorient the reader from the traditional analytical framework. The latter sees the three basic forces in international oil as the producing countries, the importing countries, and the international companies which serve as "intermediaries" between the first two. Such an approach which, for example, lumps together Great Britain and India, Italy and Brazil, for all four are oil importers, seems to us to miss the essence of international oil relationships—differential positions of political and economic power.

The focal point of any study of international oil must be the giant international companies which straddle the whole industry, both across geography and across economic levels. These companies fall into two classes:

(a) the seven major integrated international oil companies which essentially dominate the international oil industry (the "international majors")—Standard Oil of New Jersey, Royal Dutch Shell, Mobil, Texas Oil, Gulf Oil, Standard Oil of California, and British Petroleum;

(b) the 20–30 generally smaller oil companies which have ventured into the international oil industry during the 1950s and 1960s in a significant way (the "newcomers" or the "international minors")—e.g., Phillips Petroleum, Standard Oil of Indiana, ENI, etc.

The fundamental aim of all the companies is the same: profit maximization. However, there are important differences between the groups. Two attributes characterize the international majors. The first is their individual worldwide integration, which for each involves having production, refining, and marketing facilities in several countries. The second is their collective domination over the world's low-cost oil reserves.

What characterizes the newcomers to international oil as a group is that while they have had some success in finding low-cost oil in the Middle East and North Africa, they have had considerable difficulty in fitting this oil "gracefully" into world markets. This is because the newcomers lack sufficient refining and marketing facilities in the major growing markets outside the United States (whose market has quota restrictions).

The problem of accommodating the newcomers' oil is compounded by the fact that the established international majors generally have more than enough crude oil to meet the requirements of their own integrated operations (either through outright ownership or through long-term supply contracts at low prices). Because building their own refining and marketing facilities would be an expensive process, there is considerable temptation for the newcomers to take the shortcut of price-cutting. Such endeavors were an important factor leading to deteriorating crude oil prices in the last decade.

The state-owned oil company of Italy, ENI, has been a particularly "disruptive" newcomer, partly because of the aggressive policy of its guiding genius until his death in 1962, Enrico Mattei. Thus, for example, ENI was a customer for large amounts of low-priced Soviet oil. Again, in seeking overseas oil it broke the prevailing 50–50 pattern of profit sharing between the oil companies and the governments of the oil-producing countries. Finally, ENI offered technical and financial assistance to the oil-importing underdeveloped countries for exploring for oil, training personnel, and building indigenous refineries.

The second major force affecting international oil is the home governments of the international oil companies. Most of the major Western powers have one or more international oil companies either headquartered in their country or controlled by their citizens.

The United States is by far the most important Western government for the international oil industry since it is home for five of the seven majors and most of the important newcomers. Great Britain is the second most important Western oil power, having a (government-owned) majority share in one of the seven majors (British Petroleum) and a 40 percent share in the remaining one (Royal Dutch Shell). The 60 percent share of Royal Dutch Shell controlled by Dutch citizens gives Holland an important stake, albeit a relatively passive role, in international oil. France's stake resides both in the state-owned ERAP and in the partially government-owned Compagnie Française des Pétroles (CFP), sometimes considered the eighth of the international majors because, while small, it shares production facilities with the seven in the Middle East. Japan is home to one of the important newcomers, the Japanese Arabian Oil Company, as is Italy, with its ENI. Finally, West Germany, Spain, and other Western European countries have in recent years all been seeking to promote newly formed oil companies controlled by their own citizens.

The specific aims of these governments as regards petroleum resources may be summarized as follows. First, all of them seek to ensure the availability of vital energy supplies, usually on some "reasonably" low-cost basis. Second, the governments want to minimize the negative impact or maximize the positive impact of oil on their respective international balance of payments positions. Third, each government seeks to support the international oil companies controlled by its own citizens in order to foster the first two aims and also because these companies are often important domestic powers in their own right.

The general economic aims of Western governments with regard to underdeveloped countries also significantly affect the international oil industry. For example, most of the Western governments strive to promote private enterprise overseas and, in particular, investments by their own citizens; the fact that short-run balance of payments considerations may sometimes lead the governments to hinder such investments is not inconsistent with this as a fundamental long-run goal. Finally, particularly for the United States, the "Cold War" goals of preventing the spread of Soviet influence and communism and/or socialism help involve the governments even more strongly in international oil.

The third major international force affecting oil is the group of oil-exporting underdeveloped countries which are the wellspring of the enormously profitable international oil flows. The most important members of this group are Venezuela, four Middle Eastern countries (Saudi Arabia, Kuwait, Iran, and Iraq), and two North African countries (Libya and Algeria).

1. The Political Economy of International Oil

Several characteristics of their structural position within the international oil trade condition and greatly limit their potential influence. First, no single oil-exporting country dominates the international oil trade. Second, each of the major international oil companies can obtain large quantities of oil from two or more of the major oil-exporting countries. Third, in each of these oil-exporting countries the oil sector is largely isolated from the rest of the economy.

The greatest significance of the combination of these three structural factors lies as a deterrent to nationalization of the holdings of the international oil companies, with or without compensation. Thus, since no single country has a stranglehold on world oil supplies, it is physically possible for the nationalized oil to be replaced in world markets. Moreover, since the governments of the oil-exporting countries are heavily dependent on revenues from overseas marketing operations, and utilize little of their oil at home, they are highly vulnerable to overseas boycotts.

The individual countries' weakness is compounded by their proven lack of unity in the oil area. Moreover, several of the countries have dictatorial regimes whose rulers derive more than enough money from the present arrangements for their personal needs, and hence have no interest in any fundamental change.

A final deterrent to nationalization is that all the oil-exporting governments are either directly dependent upon Western governments for maintaining their power, or vulnerable to Western force. This latter danger is accentuated by the general harmony of interests and close relationships between the international oil companies and their home governments.

Short of nationalization, governments of the oil-exporting countries historically have attempted to increase their "take" from their respective oil industries by pressuring for increased production, higher prices, a greater governmental share of oil revenues as well as more indigenous "participation" in the industry. The Organization of Petroleum Exporting Countries (OPEC), formed in 1960 by the leading Middle Eastern producers and Venezuela, has become the principal vehicle for co-ordinating attempts to secure these goals. In general, it may fairly be said that the governments have had little more success in achieving these goals collectively through OPEC than individually. The companies seek to preserve the status quo and will not readily concede any points where their vital interests, particularly control over prices and production, are at stake. Since OPEC abjures nationalization and instead emphasizes negotiation and reliance on world opinion, the companies interpret this as a sign of fundamental weakness—a weakness which mirrors the basic divisions and lack of solidarity among the OPEC member countries. This reinforces the companies' unwillingness to make concessions.

The fourth major force in international oil, the Soviet Union, derives much of its influence from a resurgence in the 1950s to its historical position as an important oil exporter. With the growth of Soviet oil exports coinciding in time with increased Soviet interest in winning influence in the underdeveloped world, oil also gained important significance as a tool of foreign policy.

Thus, the Russians sought to sell oil to the developed countries, both to generate foreign exchange needed for importing Western industrial machinery and food, and to promote "peaceful coexistence" through trade. In addition, they sought to barter oil for commodities produced by the underdeveloped countries. To the extent that this oil was at times relatively surplus for the Soviets, bartering enabled them to obtain valuable commodities for improving domestic living standards, e.g., tea from India; at the same time, when payment of these commodities was deferred it provided a relatively cheap form of foreign aid to the underdeveloped countries.

The Soviets also offered and provided considerable foreign aid for governmental participation in all sectors of the oil industry, including exploration and refining. This kind of foreign aid served political goals by embarrassing the Western oil companies and "undermining" the positions of their home governments. Simultaneously, the Soviet aid may have served the general Soviet foreign policy goal of promoting socialism in the underdeveloped countries by assisting the growth of the public sector. Finally, oil served Soviet foreign policy by stirring discontent with the West, e.g., in Cuba, Ceylon, and India, while also providing means for the Soviets to support underdeveloped countries which broke completely with the West, e.g., Cuba.

The fifth major international force affecting oil is international organizations of governments. Of particular significance are the International Bank for Reconstruction and Development (World Bank), the International Monetary Fund (IMF), and the United Nations (UN).

The World Bank and the IMF, which work together closely, are both international organizations in which voting power is determined, as in any private corporation, by the amount of resources put in by the member governments. As such, overall policy is basically determined by the Western governments, which have supplied the majority of funds, and particularly by the United States.

The fundamental aim of the World Bank is to promote international investment and trade by providing loans for various "social capital" projects in underdeveloped countries. The Bank views its resources as supplemental to efforts made by private investors, and to be used particularly for projects which will help generate additional private investment. The IMF was originally set up to help stabilize the international economic system by serving as a kind of central bank to its member governments. Specifically, its basic function was to provide short-term foreign exchange loans to member governments in order to alleviate temporary balance of payments difficulties without causing internal economic activity or international trade to suffer.

Neither the World Bank nor the IMF explicitly has any role or policy as far as international oil is concerned. In fact, however, through their considerable power vis-à-vis the underdeveloped countries they frequently influence the formulation of economic policy, including oil policy. Thus, the Bank, in its roles as a major creditor to many underdeveloped countries, an important potential source of future foreign aid, and a spokesman for Western government foreign aid consortiums, has generally been opposed to the growth of public investment in the oil industry. The Bank has been unwilling to make loans to underdeveloped countries for investments in the oil sector, since it feels such investment could be provided by the international oil companies. The Bank has also conveyed the view to the underdeveloped countries that they ought not to utilize their own scarce resources for developing a public oil sector.

Similarly, the IMF, as a crucial source of funds for countries frequently on the verge of bankruptcy, has generally been opposed to public investment in the oil sector. One basis for this opposition is the IMF's penchant for deflationary fiscal policy as a means of stabilizing the value of a currency. Such a policy is inconsistent with large governmental expenditures of any kind, but particularly in the oil field which initially at least may require a relatively large foreign exchange component; moreover, expenditures by international oil companies would bring scarce foreign exchange *into* the country. Finally, governmental efforts to develop the oil sector may scare off foreign investors in all fields, thereby aggravating the monetary problems.

The basic aim of the United Nations in the economic sphere is to promote rapid economic development in the underdeveloped countries, which form the majority of its constituency. In the area of oil the UN's impact, while widespread, tends to be limited. Although numerous UN activities impinge directly or indirectly on international oil, as a supremely political organization all its roles tend to reflect and be limited by the internal power struggles. Even the "technical" economic studies of the United Nations relating to oil reflect the realities of power politics. Thus, the UN Special Commissions and groups concerned with the underdeveloped countries and energy have tended to shy away from the controversial areas of petroleum pricing, public versus private oil entities, etc. Instead they have tended to focus on esoteric energy sources such as solar or tidal energy, or have analyzed oil in a narrow, technical manner. Where the UN has touched upon the basic controversial questions, it has usually done so in a very cautious way.

The sixth and final major force affecting international oil

is the oil-importing underdeveloped countries themselves. While there are over one hundred diverse countries in this group, ranging from Communist China to Togoland, from Brazil to Tanzania, and from India to Ceylon, most of them share certain general characteristics. Perhaps most important, all are relatively poor countries, suffering from the "vicious circles" of poverty discussed at length in the economic literature. Fundamentally, each of these circles reflects the initial poverty which leads to inability to mobilize resources, which in turn tends to perpetuate poverty.

Some aspects of the vicious circles particularly relevant to oil in the underdeveloped countries are worth sketching. First, the industrialization which most of these countries desire usually requires a more rapid growth of energy and petroleum consumption than is required for the overall economy. Second, the countries generally suffer from a shortage of known indigenous energy resources. Third, the countries suffer from a shortage of foreign exchange which could be used for importing the needed energy resources. Fourth, most of the countries rely heavily on foreign aid to help offset their foreign exchange shortages. Finally, the petroleum industry of most of these countries is dominated by the international oil companies, whose home governments, particularly the United States, are the primary sources of foreign aid.

The most general goal of the oil-importing underdeveloped countries may be stated to be rapid economic development with maximum political independence. Within this broad notion on which all may agree there is a great range of specific political and economic aims. One critical question with which all must implicitly or explicitly come to grips is the role of the public sector versus that of the private. Views on this question vary widely among the underdeveloped countries and within the countries themselves. In most of the countries there is a division of public opinion, ranging from a Left which favors the public sector to a Right which favors the private sector, with the Center taking a pragmatic approach.

Finally, in developing a specific energy and oil policy, the oil-importing underdeveloped countries tend to become the focal point of conflicting pressures. These pressures emerge from the diverse relationships which the oil-importing countries have with the other major international oil forces. Some of the relationships are enumerated below.

As regards the international oil companies, the oil-importing countries may simultaneously be clashing with the established majors over questions of price or public versus private ownership, while attempting to develop relationships with newcomer oil companies. The Soviet Union may be viewed by the oil-importing countries as a lever for driving down oil prices; in addition, Soviet "barter oil" holds out the promise of reducing their foreign exchange burden while Soviet foreign aid can potentially be used for setting up a government oil sector. In turn, the oil-importing countries find themselves involved with Western governments over such questions as the country's relationship with the Soviet Union and its relative attitudes toward the public versus private sector as affecting desirability of foreign aid.

Further, the World Bank and the IMF are also strongly affected by the underdeveloped countries' attitude toward public versus private enterprise in the oil sector. The oil-exporting countries may become involved with the oil-importing countries, either directly through deals made for mutual benefit, or indirectly through their antagonistic interests on such questions as the level of world oil prices. Here the United Nations may play a role as a vehicle for expressing these conflicting interests or possibly reconciling them in the give-and-take of international diplomacy.

All these relationships and many more make up the complex fabric which is the essence of the political economy of international oil. The remainder of this study, we believe, will help show the richness and detail of this fabric.

CHAPTER FOUR

The International Oil Companies

FOR PURPOSES OF THIS STUDY, two basic groups of international oil companies should be differentiated. The first and most important group consists of the seven major international oil companies, sometimes known as the "Seven Sisters" or, more disparagingly, the "Cartel," which still dominate the international oil industry. The size ranking of these companies, based on sales, is as follows: Standard Oil of New Jersey (Jersey); Royal Dutch Shell (Shell); Mobil Oil (Mobil); Texas Oil (Texaco); Gulf Oil (Gulf); Standard Oil of California (Socal); and British Petroleum (BP).

The second group consists of the 20 to 30 generally smaller oil companies which, in recent years particularly, have entered the international oil industry in a significant way (as opposed to companies remaining overwhelmingly concerned with the domestic market of their home countries). Here this group is referred to as either the "newcomers" or the "international minors," depending on whether the stress is on their relative newness or their relatively minor position. The newcomers are primarily United States firms, such as Standard Oil of Indiana, Phillips Petroleum, Continental Oil, Atlantic Refining, Union Oil, etc. However, there are also important newcomers from other countries, such as the Japanese Arabian Oil Company and Ente Nazionali Idrocarburi (ENI) of Italy. (While in recent years the Soviet Union has also been in a sense a "newcomer" to the international oil industry, because of its very major role and complexity of goals it requires separate analysis.)

Although much has been made of the differences between the international majors and the newcomers, it should never be forgotten that the primary aim of both groups is the same: profit maximization. There should be no mistake that this is indeed the overriding goal of all oil companies, who are among the most sophisticated of modern private corporations.

Some have argued that the decline of "ownership control" and the rise of "managerial control" in the modern corporation has weakened the profit maximization motive; the modern managerial "elite" are depicted instead as balancing the claims of the various sectors of the community (including themselves) on the corporation's resources.

In fact, however, it is more likely that the development of the modern managerial elite has strengthened the drive toward profit maximization. Here it is necessary to distinguish between the "profit motive" as a subjective individual drive and "profit maximization" as an objective collective goal. It may well be that owners of a family corporation have a stronger "profit motive" than the managerial elite, since the owner derives most of his income from profits while the manager may derive most of his from salary. On the other hand, the owner can also be more "generous" or socially minded; the fact that the corporation and the individual are one will lead him to use the corporation to maximize his individual welfare, which might be at the expense of company profits. Thus, a Robert Owen (the British socialist-businessman) might pay high wages and make his factories showpieces, while a Henry Ford might undergo long strikes in order to teach his workers "a lesson," with each action being detrimental to profit maximization.

The spread of the managerially run corporation tends to eliminate the possibilities of such aberrations. It does so both by substituting collective rules for individual rule and by making it legally and morally impermissible for management to generally fail to seek maximum profits.[1]

Even more important, within the framework of collective decision making in the modern corporation, profit maximization is the only criterion which is relatively unambiguous, precise, and unarguable. Particularly in the competition among managers for top positions (both within and among firms), profit maximization for the stockholders is both the ultimate rationale for the existence of the managerial class and the indisputable yard-stick.[2] Denial of such a goal would be destructive to the morale of the managerial class and to its efforts.

In addition, insofar as modern management decision-making techniques combined with computer technology make possible greater refinements in profit forecasting and measurement, it may be increasingly difficult to justify "socially responsible" behavior. Thus, while today's management can give large gifts to education and to charity, the ultimate yardstick for justifying such behavior must be the "long run profit maximization" of the company, e.g., by giving it a "better image" in the community. As market research and measurement techniques improve, however, it will be increasingly difficult to do "good works" if they can be shown to be at the *expense* of profit maximization.

The recent development of modern business theory has had particular significance for the international oil companies, which have taken the lead in applying the new managerial techniques. A major new innovation (since the late 1950s)[3] in computing the profitability of projected investments has been use of the "discounted cash flow" (DCF) method.[4] Formally, this method defines the rate of profit as that discount rate which, when applied to the projected future stream of earnings from an investment project, makes the discounted value of this stream equal to the initial investment. This procedure is equivalent to finding that interest rate which, if paid by a bank on the original investment and compounded over the life of the project, would yield at the end the same amount of dollars as the projected investment.

In determining whether to undertake a given investment project, the problem of the relative riskiness of the projected investment has always to be taken into account. One possible approach is to make extremely conservative estimates of the likely stream of future earnings where a project is considered "risky." A second and more usual technique (sometimes used simultaneously) is to require that the expected profit rate on riskier projects exceed the company's "cost of capital" [5] by a larger margin than is required for relatively safe projects.

Thus, for example, studies of major United States corporations indicate that the typical or modal figure for the cost of capital is about 10 to 12 percent.[6] Hence, for many companies, the minimum "cutoff" rate of return for a new but relatively riskless investment project would be a DCF return of at least 10 to 12 percent per year. On the other hand, for overseas investments in underdeveloped countries, which are considered by corporations to be quite risky compared to domestic investments, the required return from a project is much higher. From our experience, even a relatively safe investment in an underdeveloped country would be expected to yield at least 15 percent per year (after taxes) in order to be undertaken.[7] At the other extreme, relatively risky projects in underdeveloped countries might need projected returns of 20 to 25 percent per year or more in order to be undertaken.[8]

The growing spread of the modern DCF profit calculation method, with its stress on the "time value of money," as opposed to previously used methods such as dividing total profits over the years by initial investment, is of considerable importance. It tends to negate what has traditionally been considered one of the great virtues of the major international companies: namely, their "long-run" interests in situations, as opposed to the presumably "short-run" drives of smaller business attempting to make a "quick kill." On the contrary, particularly for international companies like those involved in oil where investments in underdeveloped countries are associated by the companies with "high risk," the modern stress on the time value of money tends to give the companies a very short-run view of investment projects.

For example, the "present value" to a company today of one dollar received twenty years from now is 38 cents when a discount factor of 5 percent per year is applied. The present value of the same dollar received 20 years from now drops sharply to 15 cents when a 10 percent discount factor is used, and even more radically to 1 cent when a 25 percent discount factor is used. Thus, where a company considers a project sufficiently risky to require a 25 percent rate of return, it would be virtually indifferent to possible returns after 20 years.

The practical results of this are manifold in the international oil industry. In considering two ways to approach a problem, whether it be how fast to exploit a new oil field, or whether to gamble on being "caught" overpricing an affiliate, the high discount factor applied by the major companies to projected earning streams in underdeveloped countries leads to great emphasis on high, immediate profits, rather than on lower, long-run profits which may be considerably more in aggregate.

Consider, for example, two mutually exclusive ways of exploiting a newly discovered oil field in an underdeveloped country, each of which would require an initial one-shot investment of $100 million. Assume the first method would yield an annual profit of $15 million per year in perpetuity, or a 15 percent per year rate of return. Assume the second method would generate $140 million gross profit in the first year and nothing thereafter (e.g., because of the loss of gas pressure resulting from the rapid initial production); here the DCF rate of return on investment would be 40 percent per annum. An international oil company using the DCF method would reject the first method and choose the second one. This would be true despite the fact that over the long run, for example 20 years, the first project would yield a total net profit of $200 million while the second would yield only $40 million.[9] Note that under the cumulative-profits-divided-by-investment measurement technique of an earlier era the first project would be chosen, since it would show a 200 percent rate of return compared to 40 percent for the second project.

The preceding analysis has stressed that which is *common* to both the international majors and the newcomers, namely, the drive for profit maximization and its increasing rationalization. Two factors, closely connected, *distinguish* the two groups. The first is the worldwide integration of the international majors, which involves for each of them control over production, refining, and marketing in several countries. (Thus a recent article on Jersey was aptly headlined: "Australia to Zambia—The Sun Never Sets on Jersey Standard's Increasingly Profitable Empire." [10]) The second is their historical predominant control over the low-cost oil reserves of the Middle East and Venezuela. The origin and significance of these two factors have been aptly analyzed by P. H. Frankel, a leading expert on international oil:

> The origin of the predominance of the few oil companies which are big . . . can be traced back to the early days of the petroleum industry. . . .
>
> . . . The first answer . . . lies probably in the law of big numbers, or in the straight insurance element of large-scale and widespread operations. . . .
>
> The same problem of spreading risks and averaging out results applies to the subsequent phases of the oil industry, refining, transport and distribution. . . .
>
> This tendency of all the phases of the oil industry to be closely linked with each other leads to what is called integration. . . .
>
> These tendencies, inherent in the oil industry anywhere, have been enhanced by the discovery of oil fields of extraordinary magnitude in Venezuela, the Middle East, and, more recently, in North Africa. . . .
>
> . . . only a company which had an organisation strong in one or two countries and, perhaps more thinly, spread over some more, an organisation which comprised oil refineries and —most vital—direct and reasonably secure access to the ultimate consumers of petroleum products—in fact *only fully integrated oil companies* could venture to face such responsibilities and aspire to enjoy the benefits of the resources whose stupendous extent became evident in the late thirties and early forties. . . .
>
> By the simple fact that the only low cost crude oil reserves were for these reasons in the hands of a few "international" oil companies, the position of all the others was to some extent determined. . . .
>
> The new situation, which was no cartel but where the companies involved did not need a cartel to establish and enjoy a privileged and enormously profitable position, resulted from the fact that they, and they alone, had the control of the crude oil of Venezuela and the Middle East whose actual costs of pro-

35

duction, once these fields were developed, were only a fraction of those of the crude oil which had hitherto been the basis of world-wide supply and therefore the yardstick of international pricing: that of the U.S.A. . . . This meant that the prices realized at source for Venezuelan and still more so for Middle East crudes were many times their physical cost of production (including amortization of and return on capital invested locally).[11]

An isomorphism between worldwide integration and large supplies of low-cost crude oil has historically been of crucial importance in maintaining high prices and profits. This can be seen from the fact that right after World War II the seven international majors traded among themselves advantages in crude oil for advantages in extent of integration. In Frankel's words:

> The problem of extreme competition . . . was presently solved in a very workmanlike manner: those who had more oil than they had markets for made deals with those whose "downstream" positions were greater than was their control of low-cost crude oil reserves. By a number of far reaching agreements, concluded in 1947, one of them involving equity participations, the others being long-term sales contracts, the "new" producers, Texaco, Standard of California, Gulf Oil Company, and Anglo-Iranian [BP], secured large scale outlets by making use of the facilities and old-established market positions of those of the traditional international major oil companies who, at that time at least, found themselves short of oil: Esso, Socony-Mobil and Shell.
>
> By virtue of all this the need for the (comparative) newcomers to fight their way into the markets was obviated, or at least limited, and the position of those who took their oil was consolidated. . . .[12]

What differentiated the newcomers to international oil in the 1950s was the fact that, as a group, they were relatively successful in finding low-cost oil in the Middle East and North Africa, but had considerable difficulty in finding world markets for it.[13] These newcomers generally lacked refining and marketing facilities in the major consuming countries outside the United States, not only in Europe but also in the underdeveloped countries. The problem of "accommodating" the newly found oil of the newcomers in the last decade was made particularly difficult by the fact that the seven majors generally had more than enough crude oil to meet the requirements of their own integrated operations (either through outright ownership or through long-term supply contracts at low prices). Unlike 1947, when some of the established majors were short of crude oil, there was now no mutual incentive for trade between the established majors and the newcomers. Hence, the newcomers had to move their oil into international markets as best they could.

The traditional method for doing this would have been to undertake the expensive, time-consuming process of building refining and marketing facilities in the consuming countries as captive outlets for the newly found oil. Precisely because such steps are expensive both in terms of capital investment and the "opportunity cost" of deferred profits,[14] there was considerable temptation for the newcomers to take the shortcut of price-cutting. Such endeavors were an important factor contributing to deteriorating prices in the late 1950s and 1960s.

Thus, the newcomers' arrival signaled the loss of the virtually total control over low-cost crude oil which the "Seven" had until then enjoyed in the postwar period. At the same time, the newcomers as profit-maximizing organizations aspired to the same high degree of worldwide integration which characterized the established majors. A critical question for the future is the extent to which accommodation can be found between the two groups.

In an earlier era, the problem would probably have been solved by the established majors buying out some of the newcomers, or at least their surplus crude oil. Today, however, when owing at least partly to United States antitrust laws such a step is not feasible on a wide scale, a solution is not as readily apparent. True, the growing demand for oil eases the situation somewhat, since some growth for the newcomers does not necessarily require decline or even stagnation for the established majors. On the other hand, their enormous relative surplus of crude oil, combined with the high discount factors associated with

37

ownership of crude oil in "risky" underdeveloped countries, provides an enormous temptation for the newcomers to cut price; they do not have the same fear of worldwide reverberations of price-cutting as do the established majors. In any event, the continued decline in crude oil prices in the 1960s to date suggests that no means of accommodation has yet been worked out.[15]

In concluding this examination of the newcomers, a word should be said about the peculiar role played by ENI, the state-owned oil entity of Italy. ENI has been a particularly disruptive newcomer to the international oil industry, not simply because it is government owned, but largely because of the philosophy and policy of its late director, Enrico Mattei. In his words:

> This fresh outlook of E.N.I.'s accepts the realities of the new political situation created by the emergence of the under-developed countries from their state of political subjection into one of independence.
>
> From this aspect I consider E.N.I.'s formula not so much as a means for a company to penetrate into some of the most sought after oil-bearing regions, but rather as an initial step forward to more lasting relationships between oil producing and oil consuming countries. . . .
>
> As the ultimate responsibility for oil operations passes gradually over to the state, whether producer or consumer, it is the states themselves which are becoming the real protagonists and, as the chances of a compromise of interests within the framework of the agreement between the major international companies fade, favorable circumstances are being created for setting up a new system based on co-operation between producer and consumer nations.[16]

Under Mattei, ENI, whose intended function was to provide low-cost energy for Italy, played the dual role of attempting to beat down the price of imported crude oil, and at the same time gain for Italy direct ownership of overseas crude oil by providing more attractive profit-sharing deals with the governments of the oil-producing countries. Thus in the 1950s ENI became a purchaser of large amounts of low-priced "barter oil" from the Soviet Union. As a seeker after overseas oil, it shattered the standard 50–50 profit split among the oil companies and the governments of the oil-producing countries by offering a 75–25 split in favor of the governments. At the same time, both as a crusader against the power of the international oil companies and for its own profits, ENI offered better deals to the oil-importing underdeveloped countries:

> . . . whereas other oil interests built refineries to secure outlets for their crude and/or to protect their distribution networks, ENI, not having crude oil of its own, managed to get favorable consideration for its proposals mainly because it could consider refining as a business on its own . . . apart from an attractive return on capital, ENI usually had the benefit of design and engineering contracts placed with its affiliates and, having got a foot in the door, AGIP [an ENI affiliate] also managed to establish itself as an oil distributor in the country.
>
> All these activities in the underdeveloped countries, although some of them were more spectacular than substantial, were on the whole and with few exceptions sound and not unreasonable in terms of profitability.[17]

While ENI was undoubtedly a great thorn in the side of the established majors, its impact was weakened by the fact that ENI never did discover large amounts of crude oil: "Mattei was justly, if unkindly, called 'the oilman without oil.' " [18] Moreover, it is clear that even under Mattei, ENI, like the other newcomers, ultimately sought to become an integrated international oil company. As Frankel puts it:

> It was probably at that time [early 1950's] that Mattei first realized what the set-up was and I will sketch it here in a somewhat oversimplified form in which he used to describe it much later on: *there was plenty of low-cost crude oil available, which was tightly held by a few companies; it was sold at a price which was several times its real cost, the resulting large profit margins being shared equally by the producer-country governments and the oil companies.*
>
> As an Italian oil man, his reactions to what he saw might have been: 1) Since there was obviously some advantage in being in the ranks of the international oil companies, he wanted to be in on it—or, to put it in a somewhat more decorous form—he thought Italy should not be excluded from those ranks. . . .[19]

To the extent this is true, ENI basically fits into the category of a newcomer who may have to be accommodated by the international majors. And events after Mattei's death in 1962 suggest that such an accommodation is already well along. As *Petroleum Intelligence Weekly* noted in 1965:

> In the final analysis, economics don't tell the whole story behind the decision of Italy's state-owned ENI to sell out its properties in Britain to the world's biggest private oil company, Esso. . . .
>
> ENI's U.K. company, like the one in the U.S., was set up in 1961 by Enrico Mattei, at least partly as a response to an emotional drive to establish "a beach-head" in the home territories of the major international companies—called derisively by Mattei, "the seven sisters". . . .
>
> Since Mattei's death, his successor, Eugenio Cefis, has established a working rapport with the big internationals and crude oil purchase deals have been negotiated by the crude-short Italian company with both Esso and Gulf Oil. At the same time, the U.S. affiliate, "Agip, U.S.A." has been confined primarily to supervising material and equipment purchases, despite earlier plans to move aggressively into Western Hemisphere markets. And the U.K. Company has expanded much less quickly than ENI had initially expected.[20]

In the last few years the mantle of state oil maverick has increasingly appeared to shift from Italy's ENI to France's two state-controlled companies, CFP and ERAP. Superficially this might seem paradoxical given the fact that, as *Petroleum Intelligence Weekly* commented in 1966, when "drawing up a list of the major international oil companies . . . In recent years most people have added Compagnie Française des Pétroles to the list and talked about eight majors." [21] In fact, however, the paradox dissolves when one analyzes the close relationship between the international oil companies and their home governments in terms of mutuality of interests. It is to this task that we turn in the next chapter.

CHAPTER FIVE

The International Oil Companies and Their Home Governments: A Basic Symbiosis

IN CONSIDERING actual or potential relationships between under-developed countries and international oil companies, it is vital to understand the relationships of the oil companies with their home economy and government. Our primary focus will be on the United States, the home of five of the seven major international oil companies and most of the important "newcomers." To a lesser extent we will also look at Great Britain, the host country for the other two major international oil companies.

In terms of its ability to physically obtain supplies of energy products, the United States is clearly the least dependent of any developed nation on the international oil companies. Most of its crude oil is produced indigenously, with only about a third belonging to the seven majors. Moreover with Texas, the largest oil-producing state, being limited to producing at one-third of its capacity, there is additional potential oil controlled by independent United States companies. Western Europe and Japan, on the other hand, import virtually all of their crude petroleum, most of which is supplied by the seven majors, who also control the bulk of their refining and marketing facilities.[1] Finally, the relative importance of petroleum as an energy source is rising rapidly in both Western Europe and Japan, while it is unchanged in the United States.[2]

Economically, as opposed to physically, however, the international oil companies are a major factor in all their respective home countries, particularly the United States and Great Britain. The companies' importance stems from: (a) their size; (b) their role in foreign investment; (c) their impact on the country's balance of payments. In the following sections we shall analyze the significance of each of these factors in affecting the relationships between the international oil companies and their respective home governments.

By any measure of economic size, whether it be sales, profits, or assets the international oil companies as a group are by far the most important single concentration of economic power within their respective countries. Sales of the five United States majors equaled $32 billion in 1967, and among all industrial corporations in the United States, Jersey ranked number two, Mobil six, Texaco eight, Gulf nine, and Standard of California twelve.[3] In terms of profits the five companies ranked even higher: two, three, five, six, and seven. Using total assets as the measure, the United States majors rank number one, four, five, six, and eleven. The five United States majors combined had total assets of $40 billion, or almost 20 percent of the total assets owned by the hundred largest United States corporations.[4]

Net capital assets, which essentially equals total assets less long-term debt, is sometimes a better measure of corporate power. The greater are net assets in relation to total assets, the less dependent is a corporation on outside financing and hence outside influences, i.e., the relatively more autonomous it can be. Significantly, the five U.S. majors control a strikingly larger proportion of net than total assets. Within the top one hundred companies, the five internationals control 20 percent of total assets, but 25 percent of the net assets; taking the top ten industrial firms as a group, the five have 50 percent of total assets and 60 percent of the net assets.[5]

These data understate the size and power of United States international oil companies as a group since they omit the U.S. "newcomers," in themselves very large concentrations of economic power. For example, Standard Oil of Indiana, which has in recent years moved strongly into the international field, ranked twelfth among all industrial firms in total assets and net profits. Again, Phillips Petroleum ranked twentieth in total assets and eighteenth in net profits. Other important newcomers in recent years, ranked in terms of 1967 profits, are: Continental Oil, 21; Union Oil, 24; Sun Oil, 35; and Marathon Oil, 56. Furthermore, there are also long-established international minors, such as Sinclair Oil (with long-time holdings in Venezuela), which ranked 44 in profits, and Getty Oil, 33.[6] In fact, there are now few large U.S. oil companies which do not have significant interests in the international oil industry.

For Great Britain the relative size of its two international majors, British Petroleum (majority-owned by the British government) and Royal Dutch Shell (40 percent owned by British interests) is even more overwhelming. Royal Dutch Shell is by far the largest industrial company outside the United States. Its total assets of $13 billion are three times as great as its nearest corporate rival—British Petroleum. Of the top ten British industrial corporations, Royal Dutch Shell and British Petroleum together accounted for half of all assets and three-fifths of all profits.[7]

Another important role played by the international oil companies is as a major overseas investor. Of total direct United States overseas investment with a book value of $55 billion at the end of 1966, petroleum accounted for 30 percent. Even more important for the underdeveloped countries, petroleum investment comprised about 40 percent of all U.S. direct investment. The significance of this overseas petroleum investment to the U.S. economy is enhanced by the fact that it is relatively more profitable: earnings on petroleum investment accounted for 60 percent of all U.S. earnings in underdeveloped countries. In 1965 the return on United States petroleum investment in underdeveloped countries averaged about 20 percent, with a high of 55 percent in the Middle East.[8]

However, the relatively high profitability of these overseas investments, particularly crude oil production, serves to make them less of an outlet for United States capital than might otherwise exist, simply because the high profits can be plowed back directly. In fact, over the long run the international oil companies have played the role of generating capital which must find an outlet in their home country, rather than the other way around. One scholarly study, *The Economics of Middle Eastern Oil,* has pointed up that the enormous profitability of this investment for the companies has led to a huge flow of capital to the developed countries:

> When we summarize the financial results of petroleum operations in the former area [Middle East] from their establishment at the turn of the century until 1960, it is estimated that gross receipts of the oil companies from exports and local sales of crude petroleum and refined products amounted, approximately, to $32.1 billion. After deducting, from these gross receipts, an estimated amount of $5.9 billion for costs of operations, the industry's gross income, before deducting payments to Middle Eastern governments, is estimated at $26.2 billion for the corresponding period. Of this gross income, a sum of $9.9 billion was paid to these governments as royalties, rents, taxes, and share in profits; the balance, consisting of $16.3 billion, accrued to the oil companies. In turn, the oil companies reinvested about $1.7 billion of their net income in the expansion of the region's oil industry, and transferred the remaining $14.6 billion abroad. . . .[9]

Comparable figures for Venezuela indicate that of the total gross receipts of $29.5 billion (between 1913 and 1960), net investment of the companies (including liquid assets) amounted to $3.0 billion while the transfer of investments income abroad amounted to $5.7 billion.[10] Little wonder that the international oil business has been almost lyrically described by one oil executive as one where:

> . . . when through luck on an investment of $50,000 you discover oil and on this basis you acquire a concession in Arabia for $500,000 and find yourself after an expenditure of $27,000,000 owning a property worth billions.[11]

At the present time it is abundantly clear that for the underdeveloped countries as a group, the outflow from the Western countries for petroleum investment is far outweighed by the profits made on existing oil investments. This can partly be seen from the following analysis of the impact of the U.S. and British oil companies on their respective countries' balance of payments.

Increasingly, the significance of the international oil companies to their home countries and governments derives from their enormous impact on the balance of payments position. For Great Britain, which throughout the period after the Second World War has experienced currency and balance of payments problems, the effect of its two major international companies has always been decisive. For the United States, in which the balance of payments crisis has been gathering storm over the last decade, the role of the international oil companies has become ever more important.[12]

Thus, around the world, but particularly in the United States, the "oil balance of payments" has come under increasing study and scrutiny. The most definitive published study for the United States is that of the Chase Manhattan Bank (for 1964).[13]

The most striking fact emerging from the study is the huge *positive* contribution made by the U.S. petroleum industry as a whole to the United States balance of payments position. The petroleum industry affects the U.S. balance of payments in three separate areas: trade (exports of petroleum less imports of petroleum); services (inflows of dollars for royalties, licensing, and managing fees, less outflows of dollars for transportation and shipping); and capital investment (outflow of dollars for investment versus inflow of profits from these investments).

In 1964 the U.S. petroleum trade account showed a deficit of $0.6 billion, due primarily to large crude oil imports. The service sector was virtually in balance. The capital account, however, showed an enormous positive balance, with foreign investment income of $1.9 billion, more than offsetting a $0.9 billion increase in capital investment abroad, so as to provide a positive flow of $1 billion. As a result, the net positive balance for the petroleum industry as a whole was $0.5 billion. Since the total U.S. balance of payments in 1964 was minus $2.8 billion, without the contribution of the petroleum industry the U.S. balance of payments deficit in 1964 would have been one-sixth greater.

The above figures, which are the usual ones presented in "oil balance of payments" calculations, understate the full impact of the operations of the U.S. international majors on the country's balance of payments. They do so because they include transactions which would likely take place irrespective of

45

whether the U.S. oil industry had international affiliates and investments, e.g., imports of crude oil from nonaffiliates. As the Chase study notes, one must distinguish between affiliate and nonaffiliate transactions;[14] fortunately, the detailed data provided by this Chase survey enable it to make just such a distinction. What emerges from this is a true picture of the even greater impact of the investment in foreign affiliates of the integrated U.S. international oil companies.

Thus, while the total contribution of the oil industry to the balance of payments as a whole in 1964 was $467 million, this was made up of a positive contribution on the balance with overseas affiliates of $743 million, which more than offset a $277 million deficit for the petroleum balance with other nonaffiliate foreigners. That is to say, without the overseas affiliates of the international oil companies, the United States balance of payments deficit in 1964 would have been one-quarter greater than it was.

An analysis of the factors which make for the large positive contribution of overseas affiliates to the oil balance of payments also sheds some additional light on the "facts of affiliate life" in the international oil industry. In the trade account, as would be expected, imports and exports of crude and refined petroleum products are largely within each company's own affiliates. In the crucial capital account, affiliated companies provided a surplus of $1.2 billion, while nonaffiliate transactions involved a deficit of $0.1 billion. In the service sector, affiliates produced a positive balance of $150 million, while nonaffiliates generated a deficit of $100 million.[15]

Further, the positive impact of overseas oil investments on the U.S. balance of payments has been increasing in recent years. The Chase survey calculates that in 1960 the total net balance of payments of the United States petroleum industry as a whole was only $207 million, made up of an overseas affiliate's surplus of $384 million and a nonaffiliate's deficit of $178 million. Thus in 1960, when the United States total deficit was $3.9 billion, the "oil-affiliate" offset was only 10 percent of the total.

This trend toward an increasing significance for the "oil-affiliate" balance of payments appears likely to continue in the remainder of the 1960s and early 1970s. The Chase survey has projected that the oil-affiliate positive balance will increase from $743 million in 1964 to $1,181 million in 1975, or almost 60 percent.

The fundamental factor underlying the growth of the oil-affiliate positive contribution to the balance of payments is, of course, the fact that the enormous discoveries of crude oil in the Middle East and Africa yield huge profit flows, relatively little of which has to be plowed back for further crude oil development. Meanwhile, with the demand for oil outside of the United States rising at a rapid pace, there is a strong tendency for profits of affiliates, primarily the crude-producing affiliates, to generate a huge and growing profit flow back to the United States. The Chase survey states:

> In the post-war period the earnings remitted to U.S. parent petroleum companies have been larger, and have tended to grow more rapidly, than outflows for new overseas direct investments. Thus the yearly average net inflow of $65 million in 1948–50 grew to about $1.2 billion in 1964. Our projections indicate further growth of this surplus to roughly $2.3 billion by 1975.[16]

The only things which could prevent the rapid growth of this oil-affiliate positive balance would be loss to the U.S. companies of the overseas oil fields, and to a lesser degree the loss of markets in the developed and underdeveloped countries. Because oil plays such a vital role in the U.S. balance of payments, it makes the United States government particularly sensitive to the need for assisting its home companies in preventing any such occurrences.

The importance of the British overseas oil investments has generally been openly acknowledged by the British government:

> "We are, of course, aware of the immensely important contribution which the British oil companies make to our economy and the balance of payments," said a government spokesman in the Finance Bill debate this week.[17]

In fact, sorely beset by balance of payments difficulties since the end of World War II, the British government is so concerned with the oil sector that "the details of its 'oil balance of payments' remain a closely guarded secret." [18]

However, some figures were released in 1966 which help to place Britain's oil balance of payments in quantitative perspective.[19] These figures indicate that over the 1955–64 period oil imports of the two British internationals (Shell and British Petroleum) amounted to $5.3 billion, which was more than offset by their earnings of $6.0 billion, leaving a total positive inflow of $0.7 billion. Superficially, since the annual average was only about $70 million, the net effect of their overseas oil affiliates would seem relatively small. However, as was pointed out previously, in assessing the overall impact of overseas oil affiliates, it is necessary to eliminate those flows which would have taken place whether or not there were overseas investments. Clearly most if not all of the crude oil imported from the two companies' overseas affiliates would have had to be imported in any event, simply because Great Britain has no indigenous oil supplies. Hence, the real contribution of the two companies to Great Britain's balance of payments is the $6 billion which they earned for the country, primarily from their overseas crude oil production.[20]

From this perspective it is clear that Great Britain's overseas oil investments have been the difference between a shaky solvency and bankruptcy. For example, Great Britain's total reserves of gold and foreign exchange have in recent years ranged between $2 billion and $4 billion, so that the $6 billion contributed by the overseas oil investments over the past ten years is far greater than her total reserve position. Any doubts about oil's crucial impact on Britain's precarious monetary position must have been resolved by the close temporal proximity between the closing of the Suez-canal in June, 1967, and the devaluation of the British pound in November, 1967; the latter was widely recognized as being partly triggered by the big jump in oil import costs stemming from the former.

Given the economic facts of life described above,[21] it is hardly surprising that Western governments have always been vitally concerned with oil and their international oil companies. The British government has a long history of relatively open involvement in the international oil industry. This goes back to the period before World War I when Winston Churchill, as First Lord of the Admiralty, brought about the purchase by the British government of a majority interest in the Anglo-Persian Oil Company (now British Petroleum). In the words of one writer, J. E. Hartshorn:

> In the long run, the consequences of this decision were momentous and complex. It certainly assured oil supplies for the Navy; it also assured governmental backing for the development of Anglo-Persian (which became first Anglo-Iranian and later British Petroleum) throughout the Middle East, notably in securing a half share in the concession in Kuwait. . . .
>
> What is germane in this context is that the decision to buy control of this company represented the first open manifestation of the vital interest to the government of a country lacking indigenous oil supplies of securing a source of supply that it recognized as strategically vital. But it was the prototype. . . . Nor was BP its only lien upon international oil: the 40 per cent British in Royal Dutch/Shell, in spite of hard words over that "combine" when Anglo-Persian was being bought, has linked this group's interests with Britain's almost as firmly as those of BP.[22]

The close relationship between the British government and its oil companies has been described by a U.S. official as:

> . . . so intimate that it is difficult to discuss where the oil companies end and where the Government begins. . . . The Government, wherever it has a shadow of influence, uses its power by fair means or underhand to secure markets or concessions for its British owned companies. Every conceivable subterfuge, use of assumed authority over mandates, threat, intimidation, distortion of the laws, or bribery, is used to aid their companies in securing oil concessions in weak countries and holding their market against all comers, even British companies that are foreign owned.[23]

France, too, has a long history of involvement with its oil

companies, dating as far back as 1924 when CFP was formed to hold France's share of former German oil holdings in the Ottoman Empire; the French government owns 35 percent of CFP's shares and has 40 percent of its votes. In recent years, however, with the growth of de Gaulle's nationalist policies, the government has moved from a relatively passive stockholder into an aggressive prime mover in the oil area. Thus, it has set up a wholly state-owned oil company, Entreprise de Recherches et d'Activités Pétrolières (ERAP), both to operate at home and abroad and to coordinate France's oil interests. (As *World Petroleum* put it in January, 1966: "the state will be able to reinforce its power on those companies which had long followed their own commercial policy for the reason of 'protecting' the private interests of its shareholders.") At home the French government has worked unceasingly to increase the French companies' share of the oil market. Abroad, the government has striven to coordinate political and economic policies so as to obtain additional French-controlled oil supplies, even at the cost of coming into major conflict with the U.S. and Great Britain. (See Chapter 6.)

The connection between the government and the international oil companies in the United States has generally never been as open nor as close as in Great Britain or France. This is partly because the existence of a large indigenous oil sector has historically made the role of international oil less crucial and also has generated conflicting interests between independent domestic oil companies and the internationals. (In addition there is an important oil equipment sector whose interests may sometimes conflict with the internationals, e.g., in desiring foreign aid even for state-owned refineries abroad.) Nevertheless, over the years the United States government has also acted vigorously to promote the interests of her international oil companies. The following extracts from a study on *The Politics of Oil* by Robert Engler indicate the wide range of early United States government efforts in the sphere of diplomacy:

In an earlier period, companies regularly received the fullest protection of the flag in their extensive and profitable marketing operations in China. . . . Summing up its role, the State Department has said, "Many such cases required long and difficult work . . . A study of the cases in which requests for aid were made and of the action taken in such cases by representatives of this government would reveal that the latter consistently did their best to render effective assistance to oil companies; that measured in terms of dollars and cents, the cumulative value of such assistance would reach a very impressive figure. . . ."

After World War I the State Department had ardently advocated the full implementation of the "open door" policy. When it appeared that the Netherlands Government was planning to turn over to Shell the German concessions in the Netherlands Indies, ignoring Standard Oil (New Jersey), which was also in the area, American diplomats argued for equal opportunity. . . . A State Department memorandum tells what happened next: "With a view to exerting further pressures on the Netherlands Government the American Government took steps to block the issuance of further concessions to the Royal Dutch Shell in public lands in the United States. The final result was that the New Jersey company was given additional producing concessions in the Indies which turned out to be some of the richest in the islands. . . ."

When there has been pressure for greater control of the companies and when new legislation has been planned, as happened in Colombia, Venezuela, and Mexico during the twenties, the American embassies have engaged in what the State Department has described as "the closest collaboration" with the corporations and producing countries representing and safeguarding private investments and profits.[24]

In the words of John Loftus of the U.S. State Department, writing in 1945:

. . . a review of diplomatic history of the past 35 years will show that petroleum has historically played a larger part in the external relations of the United States than any other commodity. . . .

There is no reason to think that the importance of petroleum in international relations and the probability of international problems arising over matters related to production and sale of oil will be less in the future than they have been in the past. Rather it seems probable that at least for the first half dozen years after the war the problems will be greater and more difficult than at any time in the past. . . .

Distribution problems from the longer-run point of view involve a very large number of difficult issues, such as the terms and conditions upon which American companies are allowed to participate in the refining and distribution of petroleum in France and other countries which have an internal system of refining and distribution quotas and in which a government-owned company participates in the commercial operations of the oil industry; the nature of the settlement whereby the expropriated properties and rights of American nationals in Italy will be either returned or compensated (and in the event of return the conditions under which American companies will participate in the petroleum business in Italy); and the various proposals under discussion which contemplate the establishment of state monopolies for the distribution of petroleum in various countries, and the effect of such proposals upon legitimate property rights of American nationals.

The importance of these longer-run distribution problems is particularly great in view of the widespread trend toward complete or partial nationalization of the petroleum industry or any of its branches in various countries. While recognizing the sovereign right of any country to assume ownership (upon payment of prompt and adequate compensation) of the petroleum industry or any of its branches, this Government must nevertheless recognize and proclaim that international commerce, predicated upon free trade and *private enterprise (which is the conceptual core of United States economic foreign policy), is, in the long run, incompatible with an extensive spread of state ownership and operation of commercial properties.*

Another major category of problems concerns the support given by the Department on behalf of the United States Government to American nationals seeking to obtain or to retain rights to engage in petroleum development, transportation, and processing abroad. This is the traditional function of the Department with respect to petroleum. It has continued to be significant, though of temporarily diminished importance, during the war period. As normal economic conditions return this function will come to be of very great importance. Recently significant exploration concession rights have been obtained by American companies, with the assistance of the Department, in Ethiopia and Paraguay. In Iran the negotiations which were apparently near to culmination last fall have been temporarily suspended for political reasons. *In China there are great possibilities for the post-war period.* Large, potentially productive areas in Colombia are yet to be concessioned out to private enterprise; and in Brazil, where there may be very great potentialities of petroleum production, no concessions at all have yet been granted. In both Colombia and Brazil there is a fair probability of basic legislation being enacted which would permit the obtaining of concessions by private companies on a mutually satisfactory basis. The foregoing cases involve areas where concession rights are being sought. There are other critical situations where concession rights are in jeopardy and where the Department's vigilant attention is required. Furthermore, there are other areas where after the war there is a genuine possibility of securing an amelioration of the unfavorable discriminatory conditions under which American nationals were able to obtain rights before the war.[25]

Finally, knowledgeable Washington reporter Jack Anderson states in 1967:

> . . . *the State Department has often taken its policies right out of the executive suites of the oil companies.* When Big Oil can't get what it wants in foreign countries, the State Department tries to get it for them. In many countries, the American Embassies function virtually as branch offices for the oil combine. . . .
>
> *The State Department can be found almost always on the side of the "seven sister," as the oil giants are known inside the industry* . . .[26]

Continuing diplomatic efforts by all the Western home governments on behalf of their international oil companies will be shown in detail in later chapters.

In addition to the diplomatic efforts, the Western governments have, in crisis situations, provided more tangible support, including their force, to back up international oil companies. Examination of these efforts, however, are deferred to the analysis in Chapter 24 of "oil boycotts."

53

Engler has argued:

> It would be simple to conclude that in the international sphere the oil industry is always the Machiavellian manipulator and the American Government the innocent bystander or agent. Yet ample evidence has been offered thus far to suggest that the alliance between the private government of oil and the public government of the United States derives in part from an assumption of mutual needs.[27]

This view is supported by our preceding analysis of the vital contributions made to the economy of their home countries by the international oil companies. In our opinion it correctly indicates the essentially symbiotic relationship which exists between the governments and the companies. Fundamentally, the international oil companies, operating as they do in many underdeveloped countries with potentially revolutionary conditions, and in rivalry with other companies frequently backed by their home governments, often feel the need for governmental assistance. They also feel that the government has an obligation to provide such assistance:

> Oilmen assume that the American Government accepts as a first article of faith the proposition advanced by the National Petroleum Council that "the participation of United States nationals in the development of world oil resources is in the interest of all nations and essential to our national security." The primary task of the American Government is to create a "political and financial climate both here and abroad . . . conducive to overseas investment," the chairman of Texaco has explained.[28]

Both in deeds and in words the United States government has implicitly and explicitly acknowledged its interdependence with U.S. international oil companies. In the words of Andrew Ensor of the State Department, "a healthy oil industry overseas is as vital to United States security as a sound domestic industry." [29]

The acceptance of the symbiotic relationship between the international oil companies and their home governments is undoubtedly strengthened by two factors. First, the governments and the oil companies both operate within conceptual frameworks which take for granted free enterprise and private property. Second, this is made even more significant by the fact that there is a continuing two-way flow of personnel between the government and the international oil companies. (Andrew Ensor of the State Department, quoted above, for example, worked for years for Standard of California.) The importance of this flow was enunciated openly by Standard Oil of New Jersey's treasurer after World War II:

> As the largest producer, the largest source of capital, and the biggest contributor to the global mechanism, we [the United States] must set the pace and assume the responsibility of the majority stockholder in this corporation known as the world . . . American private enterprise . . . may strike out and save its own position all over the world, or sit by and witness its own funeral. . . . As our country has begun to evolve its overall postwar foreign policy, private enterprise must begin to evolve its foreign and domestic policy, starting with the most important contribution it can make—"men in government." [30]

The two-way flow between the oil companies and the government in recent years has been described by Engler as follows:

> The permeability of oil has extended throughout the peacetime machinery of the federal bureaucracy. . . . The number of public officials with oil backgrounds or relations could begin with former Secretary of State John Foster Dulles. Until 1949 he was the senior member of Sullivan and Cromwell, the major law firm for the Jersey Standard empire. . . . Herbert Hoover, Jr., a petroleum engineer and director of Union Oil whose major associations have been with oil, was the State Department's representative in the secret Iranian negotiations. He later became Undersecretary of State and was involved in questions of Middle Eastern policy and represented his Department in many of the top-level Suez arrangements. Winthrop W. Aldrich, head of the Chase Bank, which has long been tied to the Rockefeller and related oil interests, was sent to London in 1953 as United States Ambassador. In addition to the general sympathy within the State Department for oil positions, there

has also been an interchange of personnel with the industry. For example, William A. Eddy, former educator, OSS chief in North Africa and the first full-time United States resident Minister to Saudi Arabia, became a consultant to the Arabian American Oil Company, handling governmental and public relations. Harold B. Minor, once ambassador to Lebanon, became assistant to James Terry Duce, Aramco's vice president for government relations. Brigadier General Patrick J. Hurley has been an attorney for Sinclair and a special envoy to the Middle East. Henry F. Holland, former Assistant Secretary of State for Inter-American Affairs, has represented oil groups involved in Latin America.

Walter J. Levy, the first chief of the European Cooperation Administration's Petroleum Branch, was a petroleum consultant whose clients included Esso, Caltex, and Shell. He had also been on Averill Harriman's staff when the latter was sent by President Truman to seek a solution to the Iranian dispute. . . .

Robert B. Anderson, formerly Secretary of the Navy, Deputy Secretary of Defense, and appointed in 1957 Secretary of the Treasury to succeed George M. Humphrey, was a Texan who had been active in oil production. . . . Anderson had been manager of the $300 million W. T. Waggoner estate with its extensive oil operations, a member of the National Petroleum Council and a director of the American Petroleum Institute. . . .

Wherever there are consultants called in on national policy, the position and power of oil within the business community are sure to be recognized.[31]

One reason, of course, why people close to the oil industry are frequently relied upon by government is that they tend to have a monopoly of knowledge about the many technical problems associated with oil.[32] Similarly, it may well be that government employees, particularly from the State Department, are hired by international oil companies partly for their expertise in dealing with public relations and international relations. Whatever the reasons, it is hard to dispute that there has been and continues to be a steady two-way flow of personnel between the international oil companies and government. However, it should be made clear that the problem is *not* one of "conspiracy." As Engler has so well stated:

> For the real problem is one of frame of reference rather than of conscious motivation. Where do their choices lie and how are they made when the stake of oil in profits and the going system of private enterprise come into conflict with public considerations? . . . Advancement and the ability to take one's opportunities are still honored as motivating factors for American life. The business of business is still business. The governmental process offers one more hurdle or channel toward fulfillment of this purpose.[33]

Finally, it should be pointed out that, though there may be a symbiosis between government and the international oil companies in general, this is certainly not always the case in specifics. In many situations it may well be the case that the government and the companies will have conflicting views, with the government pursuing general "overriding interests" as opposed to the more narrowly conceived ones of the companies. Moreover, differences will exist among governments as to which or whose needs are "overriding"; a Kennedy administration, for example, may take a different view of a particular problem facing oil than either an Eisenhower or a Johnson administration.

Nevertheless, we would not want to conclude this chapter without emphasizing that the *effective* differences among Presidents on vital oil questions are far smaller than the similarities. This can perhaps best be shown by quoting verbatim comments made by *World Petroleum* about Presidents John Kennedy and Lyndon Johnson, respectively:

> Administration not Tough on Oil—The USA oil industry finds little to complain about in the Kennedy Administration. The president did ask changes in the income tax depletion treatment for oil production at home and abroad, to add to oil's tax bill, but he could have asked far more. And he did not fight too hard when the House Ways and Means committee rejected most of what he asked in the area. . . .
>
> The president even went out of his way to write a nice letter to the National Petroleum Council, an industry advisory

group of 95 top oil and gas executives [see Chapter 7], a letter almost without precedent in the 18-year history of the council. . . .

In addition, the president has named few persons to high government positions that the industry objects to. . . .

All in all, the Kennedy administration feels it is in a good position to go to the oil industry in coming months and ask its endorsement and its support in the national elections coming up in 1964. Many, and perhaps most oil executives are Republicans; some others make contributions to candidates of both national parties. But a sizeable number of oil men, many of them from the traditionally Democratic states of the Southwest, may stick with the Democratic party [September 1963].

* * *

Lyndon Baines Johnson, age 55, probably knows more about oil and gas than any man ever to serve as chief executive of the United States.

He has the interests of the industry at heart, as shown many times during his career of 31 years in political life in Washington. Time and again, he came to the aid of the industry in battles over the income tax percentage depletion provision, the drive for a natural bill about producer prices, and other matters. As much as anyone, he is the father of the current program to limit oil imports to the United States. The oil and gas industry is regarded as in a sounder position, both economically and politically, because of the efforts of Johnson [January 1964].

CHAPTER SIX

The Oil-Exporting Underdeveloped Countries

THE OIL-EXPORTING UNDERDEVELOPED COUNTRIES cannot help but have some impact on the international oil industry, since as a group they are the primary source of the world's oil. The most important suppliers are Venezuela, followed by four countries in the Middle East, Saudi Arabia, Kuwait, Iran, and Iraq, each of which supplies less than one-half of the dollar value of Venezuela's oil exports. Recently, the North African countries of Libya, now in the same league as the big four in the Middle East, and Algeria have become sizable exporters. Other underdeveloped countries with significant oil exports are Indonesia, Malaysia, and Nigeria, although only the latter appears to have significant possibilities for major expansion. To understand the role and goals of the oil-exporting underdeveloped countries, it is vital to bear in mind several characteristics of their structural position within the international oil trade. These characteristics, we believe, help to limit greatly the real impact of these countries on the international oil industry.

First, no single oil-exporting country has a dominant position in the international oil trade, in the sense that without its oil, world energy supplies would be drastically affected. That is to say, while there is no question but that stoppage of the oil flow from any single major oil-exporting country would cause additional expense in rearranging supply sources, the basic point is that the supply sources *can* be physically, and relatively economically, rearranged.

Second, and related to this, each of the seven major international oil companies has available large supplies of oil from at least two of the major oil-exporting countries, either through outright ownership or through long-term supply contracts. Thus, all of the international majors have an ownership share in Iran. In addition, four of the American majors (excluding Gulf) own Saudi Arabia's oil. While Jersey leads in diversification with oil owned in every major producer except Kuwait, Mobil has oil in Iraq, Libya, and Venezuela; Texaco is an important producer in Venezuela, as is Gulf Oil, which also has half of the production from the fabulous Kuwait reserves; Royal Dutch Shell has major production from Iraq, Venezuela, and Nigeria, as well as long-term supply contracts for oil from Kuwait (as does Jersey). Finally, British Petroleum has major production in Iraq plus one-half of Kuwait's production.

Third, in all of the major oil-exporting underdeveloped countries, the oil sector is essentially isolated from the rest of the economy. For one thing, most of the oil-exporting countries have small populations, with only Nigeria's and Iran's over ten million, so that internal demand for petroleum would tend to be relatively low in any case. Moreover, petroleum has never been integrated into the indigenous economy in the sense of becoming a leading sector for economic development, so that internal demand has not been built up over the years.[1]

The significance of these three factors lies in their relationship to a common goal of oil-exporting underdeveloped countries: the maximization of the country's revenues from its oil sector (whether it be to increase the opulence of the rulers, or to promote economic development, or both). Opinions vary, however, among regimes and within countries as to the best method of maximizing the country's return from its oil resources. The extreme positions are complete commercial freedom for the companies, with the government simply collecting royalties and taxes, versus nationalization. Among the masses of the oil-exporting countries, it appears that nationalization would often be the preferred policy. For example, David Hirst in his study, *Oil and Public Opinion in the Middle East*, reports with reference to Iraq under the Kassem regime:

Kassem himself can have had no doubts about the short-term success that nationalization would have brought. It is significant in this connection that in November the government evidently received a flood of letters congratulating it on a step which it had not taken—the nationalization of the Khanaqin Oil Company. The government had in fact terminated the company's (a BP subsidiary) concession in agreement with the company itself. The mistake arose as a result of the way in which the Minister of National Economy made the announcement. But the popular approval which it evoked cannot have been lost on the Iraqi leaders.[2]

Several factors have deterred nationalization of the oil industry in oil-exporting countries, despite its popular appeal. First, some of the major oil-exporting countries have personalistic, dictatorial regimes which are primarily interested in oil moneys for their own luxury living.[3] Moreover, the rulers of Saudi Arabia, Kuwait, and the other sheikdoms of the Middle East usually derive more than enough money from the present division of oil revenues between the companies and the governments for their personal needs. They thus have no interest in any attempts at "rocking the boat" which might jeopardize these revenues. Further, these rulers tend to be greatly dependent on Western governments for maintaining their power, and would be in grave danger of overthrow if they were to undertake such a policy.

Second, the structural factors discussed previously make nationalization by any single oil-exporting country, without assistance from the other oil exporters, a perilous economic prospect, at least in the short run. Since no country has a stranglehold on world oil supplies, it is physically possible for the nationalized oil to be replaced in world markets. Since each of the major companies also has oil elsewhere, no one of them would be put out of business by nationalization in one country; moreover, if one company would be severely hurt by such nationalization, it is likely that compensatory arrangements would be worked out by the other majors in order to ensure that the damaged company would have no strong incentive to cede to the nationalization. Third, since no oil-exporting country either controls significant overseas marketing operations or utilizes much oil at home, each is quite vulnerable to oil boycotts.[4]

Finally, if these factors were not sufficient, there is always the underlying threat, owing to the close relationships between the international oil companies and their home governments, of the use of outside force in the event of nationalization. As Hirst notes:

> Apprehension about the security of oil supplies is one of the basic instincts of Western governments and a cardinal factor determining their policies in the Middle East. . . . The Aden base and others like it are a permanent reminder that, if the worst came to the worst, the West would use physical force to insure that Arab oil continues to flow to Western markets in conditions of reasonable security (reasonable security, of course, being a fluid concept which varies according to particular circumstances and the outlook of Western leaders).[5]

Short of nationalization, each government in the oil-exporting countries has attempted to increase its "take" from its own oil industry by pressuring for one or more of the following goals: (a) increased crude oil production; (b) higher crude oil prices; (c) a larger share of oil revenues, or profits, for the government; (d) "participation" in the oil industry, both at home and abroad.

In general it may fairly be said that the governments have had little success in attaining any of these goals, except for getting a bigger share of the profits derived from their oil production. The companies have zealously resisted pressures by individual countries for higher production or prices, or greater participation, because yielding in these areas would impinge upon each international company's freedom of maneuver.

One may ask why the oil companies as profit-maximizing organizations are completely unwilling to allow any interference with production, price, and participation, but will give in on sharing the ultimate payoff, oil profits? One answer is that the tax laws of their Western home governments, the United States and Great Britain, allow the companies to reduce their tax payments at home by the amount of taxes paid to overseas governments; hence, in many cases, an increased share of the profits paid to the governments of oil-exporting countries in the form of higher taxes will cost the companies little or nothing.[6]

Where, as a result of these tax law "distortions" taxes in the oil-exporting countries lose their profit-penalizing significance, the focus of the international companies may shift toward maximizing pre-tax rather than after-tax profits in these countries. Since unit costs of production are generally small and constant, this is tantamount to maximizing total revenues from oil exports. And, since total revenues from oil exports are dependent upon the level of production and prices in oil-exporting countries as a group, in order to maximize total oil revenues it is vital for the companies to maintain complete control over prices and production in each country.

For example, if demand is growing rapidly in a country like Japan, it is important to the companies to be able to supply the increasing demand from the Middle East rather than from Venezuela, which has much greater transport cost. Or, again, if demand is growing rapidly in Western Europe for fuel oil as a substitute for coal, it is vital for revenue maximization to use the most competitive crude oil from nearby Libya, rather than oil from the Middle East. Neither of these company goals could be optimally achieved if companies were forced into agreements to maintain certain price levels, or certain annual production increases, in any particular country. It may thus be cheaper for the companies to provide increased payments to any particular oil-exporting government by giving it a greater share of the country's total oil revenues than by making that particular country's revenues greater at the expense of those in other countries.

It should also be noted that this control over an individual country's production and prices is important for minimizing potential frictions among the major international oil companies. Some of these potential frictions arise from the differing arrangements among companies sharing joint production facilities in the Middle East. In fact, since production costs in all the Middle Eastern countries are low relative to the price of oil, these inter-company sharing arrangements[7] are probably more important than costs in determining growth rates for oil exports from various Middle Eastern countries.

Thus, in Kuwait, both British Petroleum and Gulf Oil,

which jointly own this oil, want to export as much of it as possible. Here either company can "lift" unlimited amounts from the ground at cost and hence capture all the profits from its sales. Similarly, Saudi Arabia, where the other four major U.S. companies jointly own all the oil, is normally the most favorable supply source for each of them. This is because the ownership agreement essentially provides that each company will get more than half of the profits derived from its own liftings of oil; moreover, there is no penalty for "underlifting," i.e., taking less than one's quota.

Iraq and Iran, on the other hand, are generally less attractive in the short run, at least, as sources for expanding supply. In Iraq, where the oil is jointly owned by four of the majors (Jersey, Mobil, Shell, and BP), while one's quota can be lifted at cost, additional amounts usually have to be paid for at more than one-half the market price. Moreover, there are major penalties for underlifting one's quota, and, since the quotas are set five years before they come into operation (e.g., quotas for 1962–1967 had to be set in 1957), there is a strong tendency to propose conservative quotas.

Iran is also relatively unappealing as a source for meeting growing demand. Here for any amounts in excess of its quota the company has to pay the full posted price (i.e., more than it can sell the oil for), while there are penalties for underlifting one's quota. Additionally, the quota for each company is set by a nominating procedure, with votes weighted by ownership share, which reportedly leads to the following results:

> Over the years, there is often a fairly regular division between those companies that nominate high totals for the next year's production schedule for Iran production—mainly those who are "short of crude"—and those that nominate low, because they have more crude available elsewhere, either at lower costs or in countries where they are more concerned to increase total offtake.[8]

At the same time, all the companies recognize that for political reasons they cannot let production stagnate either in

Iran or Iraq. Therefore, there will be implicit pressure on the companies whose short-run economic interests lie in low quotas for these two countries to nominate higher quotas in order to maintain political stability and the group's long-run position. Clearly, any control which the governments exercised in terms of forcing increased production would make it that much more expensive (and hence less likely) for individual companies to sacrifice their short-run profits in favor of the group's long-run interests. At the extreme, for example, if BP and Gulf were forced by the government to sharply increase production in Kuwait next year, their short-term interests might lie in drastically reducing the total production quota for Iran, which would automatically reduce their own required liftings of Iranian oil. Since BP and Gulf together have 47 percent of the votes for the Iranian quota, they could probably do that.[9] Such a move could lead to political chaos and disaster for all the companies in Iran.

That the companies as a group are fully aware of the need to keep production rising in all of the Middle Eastern countries, despite immediate short-run interests, is clearly indicated by production figures over the 1959–65 period. During this period, of the four major Middle Eastern exporting countries, the rate of increase of exports was highest in Iran (103 percent). The fact that Iran was the one country in the Middle East which ever nationalized its oil industry (1951–54) and still has a significant radical movement and a relatively shaky regime has undoubtedly contributed to the willingness of the companies to raise their quotas at a rapid pace.

The second most rapid growth in exports has been in Saudi Arabia (86 percent), which is consistent with the direct short-term economic interests of the four U.S. owning companies. Similarly, it is not surprising that Kuwait's oil exports have increased by only 57 percent during this period, since BP and Gulf are both handicapped by their lack of marketing outlets. Iraq's production, which had the lowest relative increase, 54 percent, no doubt suffers from the fact that her quotas tend to be quite conservative because they have to be set five years in advance. In addition, however, there are indications that, just as Iranian production may have been sharply increased for "political" reasons, Iraqi production may have been held back for different political reasons.

This leads into the broader long-range reasons why the companies consider it vital to maintain complete freedom of operation in the areas of price and production. This control is potentially an important weapon for the companies in their recurrent struggles with the oil-exporting countries. Control of production in particular can be used as a deterrent or a punishment, either by slowing the country's growth rate or actually reducing production. The companies have been accused, even by relatively conservative groups, of using this weapon. As Hirst has reported:

> Between the break-off of the IPC–Iraqi negotiations in October 1961 and the downfall of General Kassem in February 1963, for example, IPC production remained more or less static, a fact which the Baghdad daily *al-Bayan* (whose comments, under the editorship of oil expert Muhammad Hadid, were the most constructive in the Baghdad press) stigmatized as a typical instance of the companies' time-worn monopolistic tactics. . . . To take a more specific instance, IPC was once accused of lowering production in direct retaliation for the imposition of higher port duties at Basrah.[10]

Price-cutting too is a potential weapon for the oil companies. Normally it cannot be used to single out individual countries within an oil-producing region like the Middle East, but it could be used to strike at widely separated oil-exporting countries, as Arab oil economists have charged:

> When the companies reduced the price of Venezuelan crude in February 1959 after Venezuela had increased its income tax two months earlier [December 1958], the impression which this action left on the producing countries was that the companies intended to "punish" Venezuela for increasing its income tax. The timing is, indeed, revealing. The 1959 price reductions in the Middle East preceded the first Arab Petroleum Congress by two months. The "preventive punishment" was perhaps intended to serve as a warning to the participants in the con-

gress, who were sure to raise the question of increased income from the industry. In this atmosphere of pressure, oil prices have been the pivot of manoeuvre.[11]

Without firsthand knowledge it is difficult to know whether the motivation for such price cuts is coercion, or short-run profit maximization, or both. In 1959, for example, if the cut in Venezuela's posted price simply reflected the fact that her actual prices were already lower, then the short-run cost to the companies of such action might be nil. In the Middle East, where profit sharing with the governments is based on posted prices, the companies would increase their after-tax profits by lowering posted prices to actual market prices. Hence, it may well be the case that in 1959 the companies were simply reaping an additional psychological benefit from profit-maximizing actions, owing to the tendency of oil-exporting countries to see all oil company actions as aimed against them.

Another possible reason for the companies' desire to maintain complete control over prices and production is essentially negative. The companies would be reluctant to have formal agreements with the individual countries on production or prices which the companies, under crisis conditions, might want to abrogate. Since the notion of "sanctity of contract" is the legal cornerstone of the companies' existence in the oil-exporting countries, it would set a dangerous precedent for them to fail to fulfill any existing contracts.

As an illustration of the possible dangers to the companies of contractual agreements, assume that such agreements specifying increasing rates of production had existed between the companies and Iraq in 1961–63. Then, in the course of their dispute with the government during that time, the companies would have found themselves faced with the choice of either continuing to reward Iraq by increasing production or penalizing it, as they might have liked, by slowing production. The latter course, which would by hypothesis have breached a formal contract, might well have provided a legal basis for the Iraq government in turn to nationalize the companies; even if it did not provide a totally sound legal basis, the companies' breach of contract would certainly improve the Iraqi position vis-à-vis public opinion, both at home and abroad.

Too great weight probably should not be placed on this motive of preserving "sanctity of contract." It is doubtful whether legal considerations have ever blocked either the companies or the oil-exporting governments from taking actions which they considered vital to their respective interests. This may be particularly true now that the Middle Eastern governments have developed a legal philosophy which challenges the notion of the sanctity of oil concession contracts.[12]

A final reason why the companies struggle to maintain complete control over prices and production is their widely recognized secretiveness about most aspects of the workings of the international oil industry. Hirst stresses the psychological impact upon Middle Eastern countries of the oil companies' "unwillingness to publicize their accounts":

> Quite apart from the purely economic considerations involved, this secretiveness is quite irritating to the Middle East producers for its human implications. . . . This "policy of silence," as it has been called, amounts to more than canniness with regard to certain trade secrets. It is but one aspect, possibly the most striking one, of a behavioural pattern which, taken all in all, is wounding to the self-respect of the producing countries. . . .
>
> With memorable explicitness he [Shaikh Tariki at the Second Arab Petroleum Congress, 1960] goes on to make it perfectly clear how this attitude is resented: "They treat us like children," was the remark, said one Western correspondent, that exploded like a firework in front of this sophisticated audience. The same correspondent went on to say that Shaikh 'Abd Allah al-Tariki was no extremist for all his fervent nationalism.[13]

In reality there are numerous bases for the oil companies' "policy of silence"; an analysis of these bases will be useful because some of them also operate toward a similar policy vis-à-vis the governments of oil-importing countries.

Discussion of oil companies' "policy of silence" too often tends to emphasize "irrational" causes. Thus, one view is that "the companies' aloofness is a reaction to a hostile environment—an argument which they themselves are inclined to use. . . ." [14] Another interpretation, e.g., as expressed by President Ben Bella of Algeria, sees the oil companies' policy as a manifestation of racism: "their arrogance is reminiscent of that of the colons. . . . Oilmen believe that the Arab mind cannot grasp the mysteries of oil technology and the subtleties of economic science." [15]

While not denying these irrational factors, there should be little doubt that in the modern corporation the rational bases for such a strongly resented policy are decisive; as such, the irrational simply tends to reinforce the rational. The motivation of profit maximization alone is quite sufficient for explaining each company's desire to maintain as much secrecy as possible.

In the short run, knowledge about specific prices, production costs, and techniques, etc., are valuable "trade secrets," not just for competitors but even more for governments. For example, an oil-exporting government's information about the extent of price discounting may make the companies operating in that country subject to pressure to give greater discounts in order to increase the country's oil exports. Similarly, market price data in the hands of governments of the oil-importing countries may place the companies under greater pressure to provide more favorable terms in its oil imports.

In the long run a policy of secrecy also helps to promote a "mystique" surrounding the international oil industry. This mystique, in turn, helps justify the presence of the international oil companies in the oil-exporting countries, and especially their enormous profits. Such justification may become increasingly important since most of the companies' concessions expire before the end of the twentieth century, at which time they will need some rationale for their continued participation in the country's oil industry.

This "mystique" of the international oil companies has two basic components. First, it is argued that the oil business is so technologically complicated and economically sophisticated that many highly skilled personnel and great storehouses of knowledge are required. Hence, of course, it is a business which is not well suited for the governments of underdeveloped countries to undertake. Second, there is a stress on the vast size, and widespread ramifications, of an *international* industry such as oil. This not only makes the industry appear more complex, but emphasizes the vital role of the international oil companies as the controller of refining and marketing facilities in the oil-importing countries. Both of these points are related to a fundamental aim of the oil companies, which is to deter governments of the underdeveloped countries from participating in the oil business.

Finally, the relative lack of success of the oil-exporting underdeveloped countries in achieving their various goals can be seen by a brief analysis of the Organization of Petroleum Exporting Countries (OPEC). OPEC was formed in 1960 in direct response to cuts in the posted prices for crude oil which reduced the revenues received by the governments of the oil-exporting countries; the founders were the four leading Middle Eastern producers and Venezuela, with Indonesia, Libya, and Qatar later becoming members. The general approach of OPEC has been described by Hirst as follows:

> Above all, OPEC has always sought to justify its existence through its own competence and expertise. . . .
>
> On the other hand, the existence of a responsible public opinion versed in oil affairs would constitute an invaluable basis of support for the technocrats themselves. Their appeal to public opinion was from the beginning essentially apolitical, in that they did not identify themselves with any particular political group or interest in their home countries. Their sole object, were it in revolutionary Iraq or absolutist Saudi Arabia, was always to improve their country's "take" from the oil industry. This involved, above all, a thorough understanding of the workings of this industry and an ability to challenge the companies on specific issues.[16]

In pursuing its general goal of "improving the take" OPEC has sought a variety of specific measures. These have included prevention of further cuts in posted prices for crude oil, restoration of older higher levels of posted crude oil prices, and increased profit shares for the governments. In addition, OPEC has sought greater national participation in the oil industry through increased employment for nationals, the building of new refineries within the producing countries, and government equity in the industry.[17]

In general, it may be said that OPEC's greatest success has come in areas where the companies traditionally have been most willing to make concessions to the individual governments; conversely, where OPEC's demands have most threatened the vital interests of the companies, it has had virtually no success. Specifically, since the formation of OPEC there have been no further cuts in posted prices, although market prices have continued to decline. While this may perhaps be attributed to the formation of OPEC, it is also true that the companies have not lost their freedom of actual market pricing. The success of OPEC's stand here has effectively been to increase the governments' share of actual profits, without formal recognition of such a change, while still leaving the individual governments as essentially passive tax collectors. Similarly, OPEC's other main success, negotiating changes in the tax treatment of royalties, has also served to increase the government's revenues as a tax collector.

On the other hand, OPEC has failed to get posted crude oil prices restored to their former high levels. Moreover, in perhaps the most critical question, national participation in the oil industry, OPEC has had relatively little success with the established majors: "except for the Shell group, all the international major companies have set their face against any deals involving sizeable government participation." [18]

Again, OPEC has failed in the field of obtaining specific price information from the companies:

> . . . the major companies agreed to consider a further reduction after 1966 of the discount for tax purposes, taking into account such evidence of the state of the market as the host governments could put before them. The companies, therefore, received letters from the governments asking for details of the prices paid on their actual sales of oil during 1964 and for con-

tinuing data. The majors answered negatively since, as they say, they did not promise to provide price details.[19]

Furthermore, despite high hopes of obtaining governmental participation in all levels of the international oil industry, little success has been achieved in this area. Plans for governmental marketing of oil abroad, ownership of tanker fleets, and control of overseas refining and marketing facilities have either fallen by the wayside or been carried out on a token basis.

OPEC's relative lack of success in achieving major gains has called into question its own existence:

> After OPEC had been in existence for three years, and—apart from the very real but not very inspiring achievement of preventing a further decline in prices—had little concrete to show for it, it was natural that voices should begin to be heard questioning its methods and outlook. . . .
>
> . . . Ironically, it was OPEC's chief architect, Shaikh 'Abd Allah al-Tariki—who, in a theatrical outburst, called OPEC the "daughter of the [Arab Petroleum] Congress" and went on to invite his audience to commit infanticide. OPEC, he said, had been wasting too much time on studies and negotiations while the oil companies had been collecting profits to which the Arab people were entitled. . . . Tariki, the former advocate of moderate and scientific methods, now outdid Egypt in the popular vein: "Don't depend on the agreements," he warned the companies, "the governments are under pressure; if the people don't like the agreements they'll tear them apart with no compensation." [20]

The basic inability of OPEC to obtain fundamental changes in the position of the oil-exporting governments may be summarized as due to the following factors. First, the companies, true to their obligations to their stockholders, will not readily concede any points where their vital interests, particularly control over prices and production, are at stake. The companies will only yield where there is a real danger of being forced into a worse situation if they do not. Second, OPEC's great emphasis on negotiation and world opinion is interpreted by the companies as a sign of weakness. The companies have amply demonstrated that they would not hesitate to use force (oil boycotts) or appeal to their home governments' force to protect their vital interests vis-à-vis the oil-exporting countries; OPEC's virtual abjuration of these ultimate weapons can logically be seen as stemming from a position of weakness. When combined with a high time value of money to the companies this situation makes stalling on their part a highly profitable business. Third, OPEC's weakness seems to reflect basic divisions and lack of solidarity among the OPEC member countries.[21] The divisions between the OPEC members, and particularly the conservative nature of most of the member regimes, in turn virtually forces OPEC itself to take a moderate approach.[22]

Finally, as a result of all these factors, OPEC increasingly looks like an organization which is of more long-run value to the oil companies than to the oil-exporting countries. While this may appear paradoxical, it is analogous to the theory that trade unions, properly moderate, play a sufficiently useful role in providing labor force "stability" to outweigh the losses to employers from paying out increased benefits. This view may be beginning to prevail among the oil companies, as indicated by *Petroleum Intelligence Weekly*'s report on the Sixth Arab Oil Congress held in Baghdad in 1967:

> Amid Baghdad's changing facets, OPEC seemed an element of stability to many oilmen, much as they'd huffed and puffed about its existence only a few years ago.[23]

OPEC, as presently operated, has thus far utilized moderate methods and obtained only marginal gains. At the same time, however, it has provided a rallying point and an outlet for voicing some of the aspirations of the oil-exporting countries. The danger for it, and for the international oil companies, is that OPEC will appear ultimately impotent and lose all influence:

> In appealing to world opinion, OPEC runs essentially the same risk as it does in appealing to a moderate body of opinion in the Arab world itself. That is to say, it inevitably strikes a note of patient exhortation, one might almost say entreaty, which tends to jar on those, more nationalist in outlook, who believe that

73

the Arabs and all in their situation should adopt forceful methods as a matter of sovereign right.[24]

This danger was implied by OPEC itself in a 1965 paper analyzing the organization's experience with the oil companies in the drawn-out negotiations over "royalty expensing," an issue which involved a few cents a barrel in the division of profits:

> . . . a brief examination of the development of the OPEC royalty negotiations was attempted with particular emphasis on the practices and procedures employed by the oil companies. Generally, the tactics used include: (a) delaying tactics; (b) tactics to undermine OPEC collective bargaining; (c) the making of obviously unacceptable offers; (d) attempts to divide Member Countries; and (e) last-minute modifications of offers before OPEC Conferences.
>
> In view of OPEC's own experience of negotiating with the major oil companies, the following conclusions may be drawn: . . . the major oil companies, by their own behavior, have made the principle of negotiations, which OPEC generally endorses, increasingly hard to apply. . . . Therefore, this experience automatically opened an unavoidable debate within OPEC regarding the very principle of negotiation itself in future dealings with the major oil companies on issues that might be much more important than the royalty issue.[25]

Finally, no analysis of the oil-exporting countries would be complete without mention of a relatively recent trend which has potentially profound implications: the accelerating rivalries among the major Western countries, and in particular between the Common Market countries and the Anglo-American "alliance." The most obvious manifestation of this rivalry has been the financial crisis which has rocked the capitalist world, and at this writing has already toppled the British pound and appears likely to dethrone the dollar, perhaps at the price of international economic collapse. Underlying this financial crisis, in our view, is the attempt of the Western European powers, led by France but covertly supported by West Germany and the others, to restore the balance between the economic and political power of Europe and the United States which was lost as the result of

World War II; seen in this perspective, the "gold crisis" represents the results of efforts by continental Europe, strongly resisted by the United States, to turn its accumulated financial power into *real* economic power. Another important manifestation of this power struggle has been the rivalry for control of Middle Eastern oil.

The earliest signs of this rivalry were the successful efforts of Italy's ENI in 1956–57 to obtain oil concessions in Iran by offering the Iranian government better terms than the "traditional" 50–50 profit split. The most recent and perhaps even more historic series of developments has been the 1967–68 struggle of Italy and France to obtain a bigger position in Iraq's oil, at the expense of the United States and Great Britain. The stage for this was set back in 1961 when the Iraqi government took back from IPC, the consortium of international oil companies which controls Iraq's oil, the concession for oil production in areas not then being exploited. This revoked concession area covered over 99 percent of the country, including the North Rumalia field which was estimated to have reserves of 5 billion barrels. A dispute between the companies and the Iraqi government over this action simmered for years while oil exploration in Iraq was at a standstill. Then, in early 1967, ENI entered into negotiations with the Iraqi government to exploit the disputed area, which in turn led the governments of the United States, Great Britain, the Netherlands, and France to make diplomatic "representations" on behalf of their oil companies (CFP being a member of IPC) to the Italian government.[26]

In late 1967 France mounted a two-pronged attack to enlarge its position in Iraq. First, the wholly state-owned ERAP signed an agreement to explore all unproven areas for Iraq's state oil company. Soon after, CFP itself began separate negotiations with the Iraqi government with the aim of exploiting the North Rumalia plum. This latter step particularly set the international oil pot to boiling, as indicated by a report in *Petroleum Intelligence Weekly*:

> Most of CFP's partners in Iraq—British Petroleum, Shell, Esso and Mobil—are indignant, to say the least. They consider the

75

CFP move as an "unauthorized breach" of the Iraq Petroleum Co. convention and both the U.S. and British Governments have protested strongly to the French Government.

In Paris, diplomatic sources confirm CFP's claim that its action has been taken at the urging of the French Government (which owns 35% of CFP and 40% of its voting rights). The government maintains IPC has no chance of getting back North Rumalia, and it doesn't want France to lose out to the Italians, Japanese and other potential bidders. And the move, moreover, fits with France's overall policy of trying to augment its interests in Middle East oil.[27]

A columnist for *The Oil and Gas Journal* commented with skepticism about CFP's justification of its move in terms of preemptive necessity, noting that "This was the same alibi used last spring by Italy's ENI when it first came to light that ENI was trying to make a similar deal with Iraq." [28] More important, the writer stated, was the indication "that de Gaulle has now decided that France can go it alone in oil affairs, and no longer needs to be cautious about stepping on Anglo-American toes." [29]

Later reports prior to the mid-1968 coup d'état in Iraq indicated that the Iraqi government itself might exploit the North Rumalia field, rather than CFP; however, at the time of this writing the new government's oil policy was not yet clear. Nevertheless, regardless of the ultimate resolution of this issue, France's activities in Iraq may already have proven to be a historic and irreversible milestone in international oil history. The reasons for this have been clearly stated by John Buckley, executive editor of *Petroleum Intelligence Weekly*, in a 1967 talk to the New York Society of Security Analysts, which was summarized in *The New York Times* as follows:

. . . [Mr. Buckley] believes France may have shattered the foundations of the long-standing arrangements under which the international oil companies pump black gold throughout the world. . . .

He said that when France's state-owned Entreprise de Recherches et d'Activités Pétrolières took over confiscated concessions in Iraq, it broke faith with other consuming nations which had previously refused to support similar seizures by producing Governments.

E.R.A.P. has agreed to finance a search for oil in concession lands once owned by Iraq Petroleum Company. Mr. Buckley said that if France does this, other Arab nations are likely to break concession agreements that permit oil companies to reap big rewards. . . .

Mr. Buckley said that the United States is now in danger of losing its control of Arab oil, which has put the nation in a very strong international position.[30]

In order to fully appreciate the historic significance of Mr. Buckley's charge that France "broke faith with other consuming nations," it is necessary to analyze the role of the oil boycott in the international oil industry. This task is undertaken in Chapter 24. At this point, however, it might suffice to note the irony of recent events. Thus, one of the earliest and strongest triggers to ENI's shattering of the 50–50 profit split was the refusal of the international companies to allow ENI into Iran in 1953, following the overthrow of the Mossadegh government which had nationalized the Iranian oil industry. This, despite the fact that ENI had steadfastly joined with the other major consuming countries in boycotting the "nationalized oil," thereby helping to contribute to the ultimate overthrow.[31] Hence the actions of ENI and France in Iraq can be seen to have once again brought back Mattei's spirit, via de Gaulle; whether it triumphs yet, at the expense of the international majors, remains of course to be seen.

CHAPTER SEVEN

The Soviet Union

INTERMITTENTLY during the last seventy years Russia has been an important factor in the international oil industry. In fact, at the turn of the century she was the world's largest producer. Each World War, fought as it was on Russian soil, temporarily removed her as a factor in world oil markets. Thus, unlike the arrival of other "newcomers" to the international oil industry in the 1950s, that of the Soviet Union was essentially a resurrection rather than a birth.

By 1950, while Soviet oil production had returned again to the same levels as those before World War II, the country was still a net importer of oil. Since then two periods need to be distinguished. The first is the era of extremely rapid growth of Soviet oil exports, particularly from 1955 through 1961 when Soviet oil exports increased dramatically from 8 million tons to 40 million tons, or over 30 percent per year. The second period covers 1962 to date, during which time the annual growth rate of Soviet oil exports has dropped sharply, averaging about 10 percent. The sharp drop in the growth rate undoubtedly partly reflects the fact that by 1961 Soviet oil exports had risen to a very large absolute volume, making it difficult to match the high growth rates which were associated with the low starting point in the early 1950s. Nevertheless, the fact that the change in the growth rates is so closely correlated with major changes in the Cold War and in Soviet economic and foreign policy suggests that these too may have played a role in the slowdown of Soviet oil exports. In the following analysis we will deal separately with each of these two periods.

During the 1955–61 period the Soviet Union succeeded in

finding oil markets in a large variety of countries, including not only Eastern Europe but also Italy, West Germany, Sweden, Japan, France, Austria, Greece, Egypt, Cuba, and Brazil. In order to break into and expand in these markets, the Russians used two basic tools: (1) they offered lower prices; (2) they offered to barter oil for other commodities produced by the importing country.

Price-cutting was naturally attractive to all petroleum importers. Barter deals, however, had particular appeal for the underdeveloped countries, since one of their great problems was a scarcity of hard currency foreign exchange for importing oil. Even Italy, hardly an underdeveloped country but a major importer of oil and in 1960 suffering from severe foreign exchange problems, was happy to barter pipeline equipment and synthetic rubber for Soviet crude oil.

The economic attractions to the underdeveloped countries of Soviet barter oil deals were considerable. Frequently the underdeveloped countries would be trading commodities which had, in technical economic terms, an inelastic demand. This means that the normal market was so constricted that offering this additional output for sale would drive down the price sharply. The final result might be a decrease rather than an increase in the country's total export revenues. Moreover, it was felt frequently by the underdeveloped countries that their bartered exports were a net addition to total exports. The reasoning was that if the barter deal were not made, the Soviet Union would not buy these consumer commodities on the world market, because the Soviet Union itself had relatively little foreign exchange, most of which was allocated for importing machinery and capital goods. Thus, the Russian willingness to accept Brazilian coffee, Cuban sugar, Indian tea, etc., gave it a powerful competitive tool in selling its oil. This is particularly true because the international oil companies were unwilling to engage in such barter, viewing themselves as sellers of oil and not commodity traders or general merchants.[1]

The aims of the Soviet oil export drive during this earlier period seem to have been both economic and political. With respect to sales to the developed countries, particularly Western Europe, which accounted for the great bulk of Soviet oil exports, the fundamental aim was probably narrowly economic. In the words of a U.S. expert, Harold Lubell:

> The primary aim of the U.S.S.R. in expanding exports to the economically advanced countries of the world is probably to obtain the gains from trade that result from exchanging a commodity that is relatively cheap to produce at home for others that are relatively dear to produce at home, either because the raw materials are not available or because the techniques of production are better elsewhere or because the investment cost of establishing production capacity at home is higher than desirable. In trading with Western Europe and Japan, the U.S.S.R. is mainly after industrial goods that it cannot produce in sufficient quantity at home or as cheaply as they can be bought abroad.[2]

With regard to the underdeveloped countries, it would appear that the Soviet oil export drive had a political component. It is true that the Soviets could frequently use the bartered goods to be exchanged for their oil for raising standards of living at home. Moreover, to the extent that Soviet oil offered to the underdeveloped countries on a barter basis was relatively "surplus," i.e., would not reduce needed home consumption or exports to developed countries for hard currency, the value to the Soviets of the barter oil might be lower than would otherwise appear. At the same time, there are good reasons for believing that even if the barter oil offers were somewhat costly to the Russians on a straight economic basis, they would still have made them.

For one thing, the resurrection of the Soviet Union as a major oil exporter in the 1950s coincided with the post-Stalin era when under Khrushchev the Soviet Union undertook a policy of active wooing of the underdeveloped countries, particularly following the 1955 Bandung Conference. In addition, the Soviets had demonstrated a willingness to meet underdeveloped countries' needs for oil by providing both men and equipment for indigenous refineries and oil exploration:

In addition to direct sales of oil, its ability to provide the equipment and the techniques for oil exploration and oil refining has enabled the U.S.S.R. to open up new areas of influence. The availability of oil contractor services from a source other than the West, particularly if financed by the extension of a Soviet credit, provides underdeveloped countries with a partial alternative to dependence on foreign investment by the Western oil companies for the creation or expansion of a domestic oil industry.[3]

Assistance in building up both oil exploration and refining in the underdeveloped countries was clearly costly to the Soviet Union, since it meant providing men and equipment which could be used at home for expanding the Soviet's own oil industry. While it might be argued that one of the aims of the Soviets in building up an indigenous public refining sector was to ultimately provide an outlet for Soviet crude oil, no such "diabolical" case could be made for assisting in indigenous oil exploration; the latter effort, if successful, would offer direct competition to potential imports of Soviet crude oil.

Another reason for believing that the Soviet oil export drive to the underdeveloped countries had a major political component was the fact that Soviet political theoreticians placed heavy emphasis on the key importance of the oil industry as a source of Western power:

> It should be borne in mind that oil concessions represent, as it were, the foundation of the entire edifice of Western political influence in the (less developed) world, of all military bases and aggressive Blocs. If this foundation cracks, the entire edifice may begin to totter and then come tumbling down.[4]

Finally, during this earlier period the Soviets had shown a willingness to ignore narrow economic interests and use oil as a weapon for furthering political aims. For example, they unilaterally canceled a contract to deliver oil to Israel in November 1956, following the Israeli invasion of Egypt. Again, in late 1958 the Russians were disturbed by formation of a government in Finland which they believed to be hostile to the Soviet Union and reportedly embargoed oil exports to that country.[5]

81

All these factors together suggest that the Soviets had both the means and a demonstrated willingness to use their oil resources for other than immediate economic gains. To the extent that there were direct short-run economic benefits to the Russians from their barter oil deals, this does not rule out long-run and/or political motives; instead, the Soviet barter oil may be viewed as somewhat analogous to the United States food aid program, which served the immediate U.S. goal of reducing internal farm surpluses.

While a negative reaction to Soviet oil on the part of the West would be anticipated, it may be somewhat difficult for those without a long memory to appreciate the furor which accompanied the "Soviet oil drive" during this period. The position of the U.S. oil industry as regards Soviet oil in this earlier period was most clearly articulated in a report prepared by the National Petroleum Council, most of whose members are top executives of the major oil companies.[6] The general tone of the Council's approach to the question is well captured in the preface to the report:

> In considering the problem of Soviet oil, it should be emphasized at the outset that, as the result of an absolute state monopoly over its foreign trade, the Soviet Bloc is in a unique position to use trade for political purposes. Politics and trade cannot be considered apart when dealing with the communists.
>
> The ultimate goal of the Soviet Bloc is to expand its political control, destroy freedom, and communize the world, and it uses its monopoly of foreign trade to further these objectives. This, in short, is the problem the free world faces when trading with the Soviet Bloc.[7]

The conclusions which the Council drew from its study are worth quoting at length:

> In view of the above factors, the Committee feels that continued expansion of communist oil exports is a serious problem to the Free World. The impact of Soviet oil on Free World countries is far greater than the volume figures would indicate and goes beyond its immediate implication for the oil companies involved.

Without a doubt, Soviet oil is the most important element in the Soviet politico-economic offensive in the Free World. The communists are using it to procure vital equipment and technology, to create political unrest and spread communism. It is a weapon with which they hope to destroy the private oil industry.

The seriousness of the Soviet economic offensive requires a concerted effort by the leading countries of the Free World to restrict further imports of communist oil and the export of strategic materials to the Soviets. Individual action is insufficient. . . .

The political alliances which Free World countries have formed to combat Soviet aggressions must now be extended fully to the equally important economic field. It is unrealistic to leave the Free World's economic flanks unprotected, particularly as the Soviets have clearly indicated that their trade is conducted "most for political purposes." [8]

The attitude of Western governments toward having Soviet oil go to any underdeveloped country can be understood only within the framework of their attitude and position in the Cold War which was raging at that time. The United States in particular viewed the Soviet "oil drive" as part of an overall drive for worldwide domination. One of the earlier public formulations of this notion was in a 1958 speech by Donald K. David, then Chairman of the Committee for Economic Development, and former Dean of the Harvard Graduate School of Business Administration. It was appropriately entitled "A Plan for Waging Economic War":

A few weeks ago . . . I was bold enough to outline an idea in the field of foreign economic policy that I had been discussing for some time with a few friends. . . . My intent was merely to send up a trial balloon. . . .

Some of the things I said at Harvard I wish to repeat, and then to amplify. And I think the place to begin is with Mr. Khrushchev and the following statement of his:

"We declare war. We will win over the United States. The threat to the United States is not the ICBM, but in the field of peaceful production. We are relentless in this and will prove the superiority of our system."

What I have to suggest is based on the belief that Mr. Khrushchev meant *every word* of that statement. War *has* been declared. The Soviet government is *deadly earnest* in its intent to defeat the United States and all it stands for, relating to man's purpose in life and the institutions we have built to serve the goals of *our* civilization.

. . . The weapons of economic warfare, such as dumping and preclusive buying, are kept at the pushbutton readiness that dictatorships can maintain. And now a program of loans, grants and technical assistance has been added to the Soviet arsenal of economic weapons. Each shot from this arsenal is aimed at a political target.[9]

Similarly, in November 1958 the International Management Division of the American Management Association held a three-day briefing session on Soviet economic competition. Over a hundred top executives attended the meeting and listened to a roster of illustrious speakers including Under Secretary of State C. Douglas Dillon, a Congressman, the Assistant Secretary of Commerce for International Affairs, the *New York Times'* Soviet specialist, the director-general of the Federation of British Industries, the president of the Federation of German Industries, and an executive of the National Committee of French Business, in addition to executives of U.S. companies with major overseas interests. The same theme of a Soviet economic war was developed at this meeting; in addition, there was expressed the need for cooperation among Western governments and between government and business to meet this Soviet challenge. Petroleum was particularly singled out as a target of the Soviets; Dillon cited this as an area in which there was a need for cooperation between the United States government and U.S. oil companies. The same general call had also been sounded by Mr. David in the previously cited speech:

All these conditions make me conclude that the central issues of foreign economic policy cannot be dealt with effectively, as in the last decade, by the mere export of capital. The answers will be found, I am convinced, in the *massive* export of the managerial and entrepreneurial talents as well as the productive skills that reside in the American business community.

Achieving this purpose requires the building of a *new partnership* of business and government dedicated to promoting economic development abroad. This partnership *could be effected by having responsible government agencies contract with private companies and private management* to plan, build, organize, operate, and train local people for operating, business enterprises abroad.[10]

The companies felt that the problem of Soviet oil was one which they could not deal with without taking measures, such as drastically cutting prices or agreeing to handle Soviet oil, which would sharply reduce their worldwide profits. Being unwilling to take these radical steps, the companies sought to get their governments to use state power to help keep Soviet oil out of the underdeveloped countries:[11] thus, *World Petroleum* reported in early 1962 that "Jersey Standard, for instance, in a paper submitted to State and other departments, suggests a free world boycott against Russian oil." [12] One outcome of the companies' efforts was that in late 1962 NATO embargoed exports of oil pipeline to the Soviet Union in an attempt to reduce Soviet oil shipments to central and western Europe. (Ultimately the Russians built their own pipe factories, and the embargo was lifted in 1966.)

At the same time that the oil industry mounted a campaign for Western governmental backing, it also sought the support of the oil-exporting countries, whose sales and revenues would presumably be reduced as a result of the Soviet oil exports.[13] It would appear, however, that this particular effort was unsuccessful. For one thing, as Lubell points out, the Russians knew that they had little to fear from the governments of the oil-exporting countries:

As Russian pressure to sell oil mounts, it is likely to generate negative reactions in the oil-producing countries of the underdeveloped world. . . . This may not bother the Russians at the present time since the governments of oil-producing countries (such as Iran, Venezuela, Saudi Arabia, Kuwait, and even Iraq) are wholeheartedly or halfheartedly anti-Soviet anyhow.[14]

Probably more important, Arab public opinion was far more anti-West than anti-Soviet. As Hirst has noted:

> It is in the same spirit of mistrust that most Arab personalities, with one notable exception, have consistently refused to share the companies' alarm at an alleged Soviet oil threat. This they consider to have been grossly exaggerated so as to provide a convenient argument with which to resist the demands of the producing countries. The exception is the late Lebanese deputy and businessman, Emile Bustani, who was well known for his resounding denunciations of Russia's Machiavellian oil politics. Coming as they did from so conspicuous a friend of the West, whose highly successful business career was not unconnected with the prosperity of the Middle East oil industry, these onslaughts did not have much effect on the rest of Arab opinion. If anything, to judge by Bustani's habit of crossing swords with the much more influential Shaikh 'Abd Allah al-Tariki, they probably served to strengthen the general belief that the Russian oil threat is a scare manufactured by the companies.[15]

Of far more significance to the oil companies was the support of international organizations and particularly the World Bank. Discussion of this is deferred however to later chapters.

It would seem that during this period a considerable amount of political pressure had to have been generated to prevent widespread use of Soviet oil in the underdeveloped world. After all, the substantive arguments put forth by the strongest opponents of Soviet oil could not have weighed very heavily in the oil-importing countries. The basic charge was that sales of Soviet oil were "politically motivated" rather than based on economics; specifically, it was claimed that Soviet oil prices were "subsidized." On the question of Soviet pricing, an analysis by Adelman points out that even according to the National Petroleum Council Soviet production costs were $0.68 per barrel, compared to 1960 F.O.B. Soviet prices ranging from $1.34 to $1.83 per barrel. Adelman states:

Any respectable discussion of the Soviet oil export drive starts with the basic fact that, at present prices, oil finding, development, production, and export are a rational, profitable activity for the Russians to undertake. . . .

As for Soviet prices, it ought to be easier today to find acceptance for the proposition that they are set low enough to make the sale, and no lower. . . .

Summing up: exports are profitable business for the Russians, even at lower than present price levels.[16]

The broader question of political motivation might have been serious, however, in that no independent country wants to be dependent on any other country for such a crucial commodity as oil. Moreover, as we have seen, there was evidence that the Soviets had used oil as a political (coercion) weapon with various countries. Nevertheless, the underdeveloped countries knew that reliance on private foreign oil companies for their oil will not necessarily insulate them from these political pressures.

After all, the policy of Western Europe countries has been to seek a diversity of oil supply sources, including the Soviet Union. Moreover, the oil companies themselves seek a diversity of supply sources, in order not to be dependent upon the pressures from governments in particular oil-exporting countries, and this is considered by the companies to be a major selling point in their favor. In this light, importing at least some of their oil requirements from the Soviet Union might well have been viewed by the underdeveloped countries as providing political as well as economic benefits. In terms of the goals of maximizing economic development and political independence, importing Soviet oil might have been seen as a transition step on the road to developing an indigenous integrated petroleum industry.

In fact, however, despite its apparent attractions, the Western fears that Soviet oil would flood the world in the 1960s never materialized. In particular the growth of Soviet oil exports to the underdeveloped world slowed markedly during this period.[17] At the present time Soviet oil undoubtedly accounts for well under ten percent of the oil imports of the noncommunist underdeveloped countries.

The slowdown in the growth of Soviet oil exports to the underdeveloped world, despite the great attractiveness of this oil to many of these countries, is undoubtedly partly related to developments in the Cold War. For one thing, the U.S.-Cuban struggle over the introduction of Soviet oil into Cuba in mid-1960 clearly revealed the great lengths to which the United States would go in backing its oil companies. This must have served as some deterrent to other countries, e.g., India, which might have been considering attempts to force the oil companies to handle Soviet oil.

Moreover, since the 1962 Cuban missile crisis, Soviet foreign policy has been far more cautious. In retrospect it would seem that from the Soviet viewpoint their oil offer to Cuba, as a political move to win influence in the underdeveloped world, was almost too successful. It was an important link in a chain of events which ultimately brought the Soviet Union into direct confrontation with the United States, threatening a nuclear war which Soviet foreign policy now apparently seeks to avoid at any cost.

From all evidence Soviet barter oil is no longer available to underdeveloped countries on the extremely attractive terms of the earlier period. Thus, Adelman stated in 1963:

> It should have occasioned no surprise that the Soviets were repeatedly underbid in 1962, and in 1963 were complaining that the international companies were practicing "artificial competition" and meanly underselling; that when an offer was made to take ships for oil, the deal was not to be sweetened by "cut prices" for the oil . . .[18]

In assessing the reasons for the slowdown of Soviet oil exports to the underdeveloped countries it is difficult to estimate the role of a more cautious Russian foreign policy. For one thing, as oil demand within the Soviet Union and Eastern Europe grew, it may have become increasingly costly for the Russians to offer low-price barter oil. Another factor which also operated was the series of right-wing coups d'état in the 1960s in the oil-importing underdeveloped countries themselves. (See Chapter

25.) Whatever the reasons, to mix a metaphor, it would seem that the specter of Soviet oil which haunted the Western World in the late 1950s and early 1960s went out not with a bang but a whimper. Nevertheless, it should be borne in mind that in the historical periods under analysis, at a minimum the "threat" of Soviet oil was always an operative factor.

CHAPTER EIGHT

World Organizations

No REALISTIC STUDY of the political economy of international oil would be complete without an analysis of the respective roles played by three international organizations of governments: the International Bank for Reconstruction and Development (World Bank), the International Monetary Fund (IMF) and the United Nations (UN). While the first two are formally affiliated with the United Nations family, in practice they have virtually total autonomy; despite the lack of publicity given their influence on oil policy, the World Bank and the IMF actually are considerably more influential than the various other specialized agencies of the UN.

The World Bank is traditionally portrayed as an organization primarily devoted to promoting the economic development of the underdeveloped countries by borrowing money in the developed countries and lending this money for projects in the underdeveloped countries. In fact, however, according to the Articles of Agreement drawn up at the 1944 Bretton Woods Conference which organized the Bank, its primary mission was described as follows (Article I):

> (i) to assist in the reconstruction and development of territories of members by facilitating the investment of capital for productive purposes . . .
> (ii) to promote private foreign investment . . . and when private capital is not available on reasonable terms, to supplement private investment by providing, on suitable conditions, finance . . .
> (iii) to promote the long-range balanced growth of international trade and the maintenance of equilibrium in balances of

payments by encouraging international investment for the development of the productive resources of members . . .[1]

In the words of one study of the Bank, "Article I thus makes clear that the functions of the IBRD were to be those of assisting, promoting, facilitating, and encouraging international investment."[2]

Several characteristics of the Bank ensure that it will ultimately remain faithful to its primary goal of promoting private international investment. First, voting power of the member governments in the Bank is basically determined by the amount of money which each has subscribed to the Bank's capital. The developed countries which have provided the bulk of the capital thus have the dominant voice in the Bank's operations; the United States has 29 percent of the votes, Great Britain 12 percent, the Common Market countries 16 percent. Second, the bulk of the money actually loaned by the Bank is raised through issuance of Bank bonds, most of which are purchased by private investors in the United States and Western Europe. Third, in recent years United States foreign aid loans have turned out to be an important source of debt servicing for the underdeveloped countries; in the words of *Fortune*, "Indeed, without the side flow of federal dollars loaned on soft terms, the World Bank would have faced shattering defaults."[3] Finally, each president of the Bank since its inception has been a U.S. citizen with a business background. Even a "friendly" study of the Bank by the knowledgeable Andrew Shonfield of *The London Economist*, noted that "the World Bank remains a very American institution."[4]

As an organization dedicated to the general goal of promoting international investment, the World Bank has no specific policy relating to oil or any other commodity in the underdeveloped countries. In practice, however, the general policies of the Bank strongly tend to favor minimizing the public sector's role and maximizing the private sector's role. The Bank sees its role in the world economy as providing loans to governments to provide the infrastructure necessary to attract private investment (e.g., roads, electric power, etc.), not loans for assisting a government's own enterprise. This doctrinal position of the Bank is noted by Shonfield in discussing its former president, Eugene Black:[5]

> His limitations appear most obviously in his attitude towards socialism and state enterprise. Black, at any rate in certain moods, is inclined to talk as if there were some kind of exclusive relationship between political freedom and private enterprise, and as if he believed that his main task is therefore to defend the only true economic faith throughout the world. It is no accident that the right-wing opposition movement in India, led by Mr. Masani, has adopted a slogan taken from Mr. Black: "People must come to accept private enterprise not as a necessary evil but as an affirmative good." . . .
>
> The Banks [sic] fails to make any significant contribution to the advance of manufacturing industry in the undeveloped countries. It is inhibited in its attitude towards industrial investment banks in Latin America, which could, in some countries at any rate, be a potent engine of useful enterprise—and private enterprise at that. It appears that the Bank has not grasped the fact that in undeveloped countries the introduction of government enterprise in unaccustomed fields can often be the means of stimulating private business initiative which would not otherwise have emerged.[6]

While the Bank's policies have a generally anti-industrial bias,[7] particularly in relation to government participation in manufacturing, this bias would be even stronger against assisting a relatively capital-intensive industry like oil in underdeveloped countries. In the words of the Bank itself:

> The establishment or expansion of appropriate manufacturing and processing industries is an essential aspect of sound development in almost every case. It will normally be advisable, however, to lay initial stress on light consumer goods and processing industries which employ small amounts of capital equipment per worker and can often build upon traditional skills in the introduction of mechanized techniques.[8]

Thus, while the Bank has been a major lender to underdeveloped countries in the energy field, with more than one-third of its loans going to that sector, it has never lent any money for petroleum refining or exploration in underdeveloped countries. The Bank's unwillingness to grant loans for exploration purposes may be attributed to its strict criteria of "safety" in choosing among possible investment projects. In addition, since the Bank views its capital as a supplement to rather than a substitute for private capital, and since private capital is ostensibly available for exploration and refining in the underdeveloped countries, this too may be a factor in the Bank's decisions. Nevertheless, it seems clear that the Bank's strong private enterprise philosophy also plays a role. As Shonfield reported:

> . . . the Mexicans claim that their request for a development loan for Pemex, the state-owned petroleum industry, was turned down by the Bank for ideological reasons, and there has been further trouble about the division of investment in electric power between the private foreign company, Mexlight, and the nationalized power corporation.[9]

The corollary of this bias against the public sector is the Bank's apparent willingness, if not eagerness, to promote the interests of international companies operating in the resource field. Thus, we have the following portrayal of the Bank's role as given in Shonfield's book (which, it should be noted, was based on extensive interviewing with various professionals in the development field, "including especially the World Bank"):

> At the same time it is able to act as a kind of shield to foreign companies which want to develop some mineral or other natural resource in a distant spot, but are anxious to be protected against the political risks that are run by a rich and isolated alien concern operating in the territory of a poor nation. Thus for example the American steel companies, who wanted to develop the manganese deposits of Gabon on the west coast of Africa, were able to avoid a direct connection with the business, because the Bank took the initiative, once the American interest in the manganese had been firmly established, and made a loan—knowing that it would be able, almost immediately afterwards to sell rather more than half of the loan obligations to American insurance companies, who were standing ready to buy. These com-

bined operations of the finance house, the large industrial corporation and the Bank are an interesting extension of modern
investment practice. It is clear that some of the big American
aluminium companies, who are taking an increasing interest in
Africa as a source of bauxite and cheap power, would now like
to move in under Mr. Black's umbrella. In time this may, indeed,
become the standard formula for large-scale company investment in the natural resources of an undeveloped country—a
formula which seems much preferable to the old type of arrangement, with its overtones of anxiety and arrogance and its under-
tones of political intrigue.[10]

Aside from its negative role of refusing to loan money for
petroleum development, the Bank has also played an active role
in attempting to affect petroleum policy in various underdeveloped countries. This is consistent with the Bank's own view of
its wide-ranging responsibilities, as expressed by Black:

> The Bank, however, could not logically confine—and has
> never confined—its attention to the accomplishment of particu-
> lar projects. For the Bank's financing of economic development
> could never be more than marginal to investment from other
> sources. The Bank could not hope that its loans would stimulate
> private investment in a member country whose policies discour-
> aged that kind of investment. . . .
> The Bank has been at least equally concerned with the eco-
> nomic environment in which its loans are to be put to work.[11]

Most of this Bank activity takes place "behind the scenes." One
advantage of operating this way is that officials in underdeveloped
countries are thereby protected from adverse popular pressures
which might be generated by openly following Bank policies.

However, much of the "backstage" activity of the Bank in
the petroleum area is known to knowledgeable people within the
industry. This is hardly surprising considering the outspokenness
of Black, who publicly referred to "the nationalistic oil policy
of Brazil" as an example of "obstacles in the way of growth." [12]
Thus for example it was widely known within the industry that
in 1960 the Bank commissioned and distributed among the gov-

ernments of the underdeveloped countries a study prepared by an outside consultant which generally took a negative approach toward active governmental participation in oil exploration in underdeveloped countries. (See Chapter 10.) Again it was well known that Bank pressure helped deter the Indian government from undertaking overseas oil exploration in the early 1960s. As another example, in 1966 the petroleum trade press reported that the Bank had recommended to the Venezuelan government that it not increase taxes on the oil companies: "The Bank's report said Venezuela already is overly dependent on oil and should develop other sources of tax revenue." [13]

In this latter case, it should be noted, the Bank's recommendations were ignored by the Venezuelan government. The different responses of the Venezuelan and Indian governments to the Bank's "suggestions" partly reflect of course different degrees of power and/or influence which the Bank has with various underdeveloped countries. This, in turn, is partly a function of how indebted, both at present and in the likely future, a country is to the Bank. Thus, India has been the largest single borrower from the Bank, accounting for over ten percent of its loans, while Venezuela, which gets considerable foreign exchange from its oil exports, has been a far smaller borrower.

In addition, the Bank's influence is partly a function of a country's dependency upon the Western aid-giving nations and the International Monetary Fund. The degree of a country's dependence upon the Fund is particularly significant because of, in the words of a joint Bank-IMF publication, "the close relationship between the Fund and the World Bank in the process of setting internationally accepted standards of legitimacy for the claims of member countries for financial support." [14] Hence, at this point we turn to an analysis of the impact of the IMF on international oil.

In the IMF, founded at the same time as the Bank, voting power is similarly determined by the amount of money subscribed by each government. Here too, the developed countries have a dominant voice, but in this case Europe has had a relatively larger role. Shonfield writes that:

Right from the beginning the convention was established that of the two financial institutions set up at the Bretton Woods Conference in 1944, the International Monetary Fund should be predominantly controlled by Europeans with a European president, and the World Bank by the Americans.[15]

Nevertheless, a comprehensive academic study of international organizations states flatly:

[In] the International Monetary Fund, the price of life has been continuous deference to the views of the United States. Any other nation making a comparable subscription to the capital funds of these agencies, however, could command comparable deference to its views.[16]

The theory underlying the IMF was that it would help stabilize the international economic system by serving as a kind of central bank to its member governments. The hope was that, by being able to lend various "hard" currencies to nations with temporary balance of payments problems, the Fund would help prevent the only alternatives historically open to a beleaguered country, unilateral devaluation or trade restrictions. In this way the Fund could contribute to international monetary stability and promote international trade and investment.[17]

While the Fund has even less of a direct relationship to petroleum than the Bank, the fundamental thrust of its general policies has always run in the same direction, reinforcing those of the Bank. The essence of the Fund's approach to the underdeveloped countries has consistently been to prevent inflation and if necessary to devalue the currency in order to ultimately promote a stable currency. This can be seen from Shonfield's report of an interview with Per Jacobsson, the Fund's late managing director:

"What would you suggest, Mr. Jacobsson, to help the undeveloped countries to move forward faster?"

"Well, the main thing is to stabilize their currencies." [18] Shonfield also comments:

. . . Jacobsson's economics were largely shaped by the pre-1914 world. This world, in which there was no escape from the rigours of the gold standard, in which sinners (those who ran a deficit on their balance of payments) were automatically punished, and the virtuous (thc ones with a surplus) were able to indulge in expansion on a full stomach, is really the Fund's ideal. Jacobsson's watchword is "discipline." . . . One of his main fears seems to be that economic aid too generously given to the undeveloped countries will simply entrench them in a number of nasty habits.[19]

The Fund's general policies and prescriptions for the underdeveloped countries impinge on the petroleum sector of underdeveloped countries in the direction of opposing governmental participation. For one thing, the Fund is strongly opposed to barter trade between countries, and uses its power to stop it. Shonfield gives tangible evidence of this:

Indeed, one would have thought it elementary that these nations should be encouraged to engage in barter trade, wherever this helps them to put their actual or potential surpluses of production to some useful purpose. It came as a shock, therefore, when I was in India at the end of 1959, to run into a mission from the International Monetary Fund engaged in a vigorous effort to break up an Indo-Burmese barter agreement. The IMF started by hurling anathemas. What the Indians were doing was a sin against the principle of untrammeled multilateral trade, with equal freedom for all exporters to enter a market. . . . A barefaced piece of bilateralism, certainly: the IMF no doubt congratulated itself on having stamped on it in time.[20]

Thus, the IMF must clearly line up against the underdeveloped countries' acceptance of Soviet offers to barter oil for their various commodities.

More generally, the Fund's preoccupation with monetary stability also tends to align it against governmental participation and in favor of private foreign development of an underdeveloped country's oil sector. This is true if for no other reason than that the anti-inflation bias of the Fund (and the Bank as a creditor institution) tends to make it favor minimal government expenditure. This would be especially so as regards the oil sector since private foreign capital is presumably available for this investment, and this foreign capital would in the short run at least contribute to a favorable balance of payments for the country, thereby helping to stabilize its currency. Conversely, governmental participation in the "normally private" oil sector might scare off foreign investment from other sectors of the indigenous economy.

Finally, it might be noted that the power of the IMF over an underdeveloped country's policy may frequently be even greater than that of the Bank or of major creditor nations. This is because the Fund usually provides only short-term loans, and a country which has borrowed heavily from it and must soon make repayment generally has little alternative but to accept the Fund's prescriptions. Moreover, since the Fund is dealing directly with the whole economy of the country rather than individual projects, its suggestions or demands frequently strike at the fundamental policy decisions which are normally the prerogatives of sovereign governments. *The New York Times'* reports on the Fund's negotiations in 1967 with two countries, Colombia and the United Arab Republic, serve to indicate both the Fund's power and its wide ranging demands:

> An ambitious economic development program in Colombia is awaiting the solution of an awkward dispute with the International Monetary Fund.
> The immediate issue is exchange policy, but the underlying question is whether the international lending community, including the United States, will accept President Carlos Lleras Restrepo's strategy for putting this country of 18 million people on the road to rapid growth. . . .
> The International Monetary Fund, which has a mission here now, has recommended a strong exchange devaluation as the remedy. . . .
> . . . Mr. Lleras Restrepo expressed his disagreement with the fund's recommendations by enacting strict exchange controls . . .
> Colombia's problem is . . . [that] the exchange restrictions have dried up new private investment. They have also disturbed such public lenders as United States Government agencies and the International Bank for Reconstruction and Development.[21]

*　　*　　*

The United Arab Republic has agreed in principle to a program of stringent economic reforms that will permit a resumption of international monetary support for its hard-pressed economy.

The agreement, between the Cairo Government and the International Monetary Fund, is expected to be concluded formally within a few weeks. It follows months of difficult negotiations.

Cairo already owes the fund $105-million. Now the fund, a 103-nation organization, will extend about $50-million in credit to help the United Arab Republic overcome its shortage of foreign exchange.

But in return, according to Western officials and Egyptian diplomats, the fund has exacted pledges of economic reforms that, according to Western diplomats, must have been difficult for President Gamal Abdel Nasser to accept.

One reform, probably the most difficult, amounts to a selective devaluation of the Egyptian pound, now officially worth $2.30. . . .

Many of the reforms run counter to Mr. Nasser's Socialist policies. Only three months ago, in an angry speech, he was attributing his country's economic difficulties to foreign pressures and proclaiming that Egypt would not "pay our debts to whoever exerts economic pressure on us"—a threat that was believed to be aimed primarily at the fund.

In the opinion of Western diplomats, Mr. Nasser was brought around to accept the fund's proposals by the pressure of foreign debts and by a shortage of foreign reserves with which to pay them. . . .

Cairo apparently hopes that the agreement with the fund will pave the way for foreign development loans, possibly from the International Bank for Reconstruction and Development, to improve exports of agricultural food products. . . .

The United States has been holding off on a rescheduling of the debt pending the outcome of Cairo's negotiations with the fund.[22]

Only a few weeks later the *Times* further reported that:

The crisis in the Middle East has probably blocked an impending loan to the United Arab Republic by the International Monetary Fund. . . .

The managing director of the Monetary Fund, Pierre-Paul Schweitzer, may conclude that the military situation could force expenditures sharply upward, thus starting renewed inflationary pressure. In these circumstances, the financial conditions behind the original agreement would not be met.[23]

The United Nations potentially represents one of the most important international organizations affecting oil in the underdeveloped countries. Part of this potential lies in the fact that voting power in the United Nations, unlike in the Bank or the Fund, rests with the underdeveloped countries. And, most of the underdeveloped countries have had hopes that the United Nations would play a significant role in helping them to solve many of their problems associated with oil. In the words of the *Far Eastern Economic Review:*

> Petroleum the commodity is not on the agenda of this General Assembly. But petroleum the political weapon and development asset has been used and debated in the Assembly plenary and four of the Assembly's committees of the whole—Economic, Special Political, Trusteeship and Legal. . . .
>
> New and not making headlines is a proposal to give a new role to the UN, that of monitor of the nature and adequacy of capital flowing to the developing countries. . . . What emerges from these moves is a sense of need by many countries for crutches to lean on in their dealings with big international companies. Time and time again the call is heard in UN meetings for a scrutiny of profits and for technical assistance in measuring how much of the money flowing in and out of their countries sticks to their development ribs.[24]

In fact, however, as has increasingly been recognized and articulated by the underdeveloped countries, the UN's potential has been far from realized. Below we present an overview of the UN's actual role vis-à-vis oil and an analysis of the causal factors which have determined it.

UN activities in the oil area could potentially take several forms: actual operations, personnel training, presentation of information or assistance in policy formation. In general, it may fairly be said that the extent of effective United Nations action in any of these areas has been inversely correlated with its real significance for the underdeveloped countries, as measured by the degree of controversy surrounding a proposed course of action.

Thus, UN activities to assist member governments in actual oil exploration or petroleum refining, transportation or distribution, have been virtually nil. Aside from a small amount of technical assistance, such as providing an advisor for a refinery in Tanzania or a drilling engineer for Taiwan, the bulk of the meager UN efforts in the petroleum field consisted largely of surveys and compilations of existing technical knowledge and data. A typical, albeit larger than usual, proposed UN project was a five-year survey of energy (and other nonagricultural) resources, at a cost of about $10 million. However, as a knowledgeable trade journal published at the UN reported, "Although surveys in these fields are much needed, developing nations tend to think that UN help in less peripheral areas, such as oil and its exploitation, would be more beneficial." [25] The paucity of the UN effort in the oil field may partly be indicated by the following *World Petroleum* roundup of UN activities relating to oil in 1965:

> *UN Secretariat, New York:* Petroleum activities were again intensified last year. . . .
> Thus the staff of the Energy and Electricity Section [of the Natural Resources and Transport Division of the Department of Economic and Social Affairs] . . . is now represented by three economists . . . and four technical advisors . . . [26]

Even within the relatively small UN effort in the energy field, comparatively little attention is paid to oil as opposed to more esoteric sources of energy such as geothermal (natural steam), oil shale, wind power, tidal energy, and solar energy. Although today some of these energy sources may in special cases be significant, and in the far future may be significant for all countries, there is little doubt but that oil is likely to continue to be the most important energy source for the underdeveloped countries over the next ten or twenty years at least. This fact is widely recognized and the propensity of the United Nations to be pre-occupied with the peripheral energy sources seems only explain-able as an attempt to steer clear of the controversial oil area.

The defensive approach of UN officials toward the UN's role in the petroleum area is pointed up in the conclusion of an article in *The Oil and Gas Journal*, which sees this role as "grad-ually" expanding:

> What's ahead? How far can the U.N. go in its work before it collides with private oil and petrochemical interests?
>
> No one in U.N. headquarters will hazard a guess on this possibility. They stress that their activities, so far, have been largely administrative, advisory, and educational.[27]

The same sort of caution can be seen in the occasional UN study on economic or financial aspects of the petroleum industry. (See Chapter 10.) There is a strong tendency for these docu-ments to present large amounts of potentially useful technical and economic data, but without organizing them within a theoretical framework which could be utilized for making policy decisions.

The fundamental reason for the UN's lack of effectiveness in assisting the underdeveloped countries is of course that UN decision making is primarily determined by the power of the countries involved rather than a mere head count. Thus, for example, even the previously discussed proposed UN survey of natural resources is ultimately dependent upon the agreement of the major financial powers:

> Mr. Thant asked the governments for their comments. He also warned them that the feasibility of making the surveys would depend on "the concrete support of those governments which are in a position to provide financial contributions, ex-perts, consultants, and other resources." [28]

Moreover, the Western home governments of the inter-national oil companies, which still have the largest say in UN activity, generally work hand-in-hand behind the scenes with the companies to keep the UN from encroaching on the com-panies' activities in the underdeveloped countries. Sometimes the activity is so vigorous, however, that bubbles rise to the surface, as in the following case cited by Engler:

> Late in 1958 the United Nations General Assembly received Soviet-backed resolutions, first proposing that the UN provide aid for nations wishing to develop their own petroleum resources and then one more simply suggesting a study of international cooperation in such development. . . . "In no time at all the oil lobbyists were swarming around the United Nations," the *St. Louis Post-Dispatch* reported. "There were so many conferences between the oil men and members of the United States delegation that one American diplomat said he told the oil people to 'let us alone so we can protect your interests.' " [29]

With the dominant financial, political, and administrative power in the UN still residing with the major Western powers, it is natural for UN officials to stay clear of the controversial oil area by focusing their energies on other energies.

The activities of the special regional economic commissions of the United Nations, particularly the Economic Commission for Asia and the Far East (ECAFE) and the Economic Commission for Latin America (ECLA), have always been of special concern to the international oil companies and the Western home governments. Part of this fear stems from the fact that these commissions are relatively more independent of the UN headquarters; moreover, while the Western powers are not members of these commissions the Soviet Union is a member of ECAFE who can and has used it for attacking Western petroleum interests. In general, in the words of one scholarly study:

> Many decisions in ECAFE reveal the extensive corroding effect of Cold War considerations. . . .
> . . . the West has generally taken a conservative, go-slow position on matters which the ECAFE Secretariat and the Executive-Secretary have urged. . . . [30]

A brief analysis of the various ECAFE symposia on petroleum may help indicate the character of the Western effort and its impact on the UN role in the petroleum area.

Generally speaking, the international oil companies and the

Western governments, particularly the United States and Great Britain, have viewed these ECAFE symposia as dangerous hurdles to be overcome. Thus, they have favored postponements and lengthening the periods between the symposia as long as possible. They have sought to keep controversial economic questions such as pricing, governmental participation in the industry, etc., off the agenda, hoping to limit the symposia to forums for technical papers. Where this will not work, the companies have attempted to take the offensive by presenting papers portraying the positive role played by the oil companies in the underdeveloped countries in terms of capital investment, transfer of technological knowledge, training of nationals, effects on balance of payments, etc.

The tactics used to reach these goals have been varied. In general, however, a great deal of effort is put into insuring coordination of position, among the oil companies both in the United States and Great Britain, and between the companies and the official Western governmental delegations to these meetings. Typically the U.S. oil companies have tended to operate behind the scenes, with their employees either not appearing at all or as advisors to the official United States governmental delegation; in Great Britain, on the other hand, the company-government relationship is considerably more open, with company officials appearing as prominent members of the official delegation. The companies will wherever possible seek to have trusted indigenous employees rather than Westerners appear as spokesmen and will also use "sympathetic" consultants to prepare reports for these meetings. In short, a complete "diplomatic" effort is utilized for influencing these regional symposia on petroleum.

The efforts that went into the first symposium on petroleum, which took place in New Delhi in December 1958, are indicative of the importance that the oil companies attach to these meetings. This meeting had developed out of an earlier ECAFE meeting in 1957, where one of the topics under discussion was petroleum legislation; it was suggested that since the ECAFE member countries were relatively ignorant about the oil industry they ought to first hold a factfinding meeting before considering legislation. For this first symposium, assistance of Western governments and the oil companies in providing materials was solicited by the ECAFE Secretariat. For many months before the actual meeting the State Department and the oil companies were in constant consultation about the personnel to be in the United States delegation and the material to be submitted. The companies with important interests in the ECAFE area (all of the United States international majors except Gulf) took the leading role in preparing for these meetings. The companies were able to play the dominant role partly because they had greater knowledge about the industry and partly because they had more financial resources than the State Department had allocated for this meeting.

All in all, from the companies' viewpoint, their efforts were successful, judging by the relatively innocuous results of the symposium. The principal recommendations were to study possibilities of establishing a regional petroleum institute for the area, standardize and publish petroleum industry statistics, prepare an oil and natural gas map of the area, study whether it was feasible to manufacture indigenous exploration equipment, and compile case histories of oil fields in the area, for presentation to a second symposium proposed for 1962.

Again, a considerable amount of oil company and Western governmental effort went into shaping the agenda for the second ECAFE symposium which was held in Teheran, Iran, in 1962. Through the efforts of Western delegations the majority of controversial items which had been proposed for the symposium were dropped from the agenda, and all the rest were lumped into one item dealing with "economics of petroleum exploration, production and distribution." The actual papers presented on this key item, and the ECAFE report, shed little light on the crucial questions of pricing, public policy on government versus private, etc. The same may also generally be said about the results of the third symposium, held in Tokyo in 1965.

Finally, the ineffectiveness of ECAFE in the petroleum area ultimately stems from the basic political and economic realities governing the United Nations proper. While ECAFE has a good deal of independence from UN headquarters, ultimate control over ECAFE's substantial activities lies with the headquarters Secretariat through its power to approve budgets and orders of priority for various projects. At the headquarters, the United States, which supplies a large part of the UN's budget, is generally opposed to expanding ECAFE's activities, particularly in the petroleum arena. Furthermore, within the regional secretariats there is a widespread fear of stirring up the same kind of attack which was launched against the Economic Commission for Europe when its 1955 study of oil prices in Western Europe implied overpricing by the international oil companies.

The opponents of expanded activities of the regional economic commissions have been flexible enough to yield on certain occasions, particularly in meeting the demands of highly influential underdeveloped countries. For example, almost ten years after the original suggestion for setting up a petroleum exploration training institute, one was approved and located in New Delhi, India. Ultimately, the Western approach is that budgetary control is the surest method of limiting encroachment on the companies in the petroleum area, while the symposia may serve some useful purpose, if properly controlled, as an outlet for the grievances and frustrations of the underdeveloped countries in this area. Barring a fundamental reshaping of the structure and control of the United Nations and/or greatly increased pressure by the oil-importing underdeveloped countries, there is little likelihood of any significant change in this situation in the near future.

CHAPTER NINE

The Oil-Importing Underdeveloped Countries

THE OIL-IMPORTING UNDERDEVELOPED COUNTRIES, the final major force affecting international oil, are the focal point of this study. While there is great diversity among these countries, most of them share certain common energy and petroleum problems. This commonality centers around several interrelated vicious circles of underdevelopment in the energy and petroleum resource area which impede economic development.

First, industrialization and modernization generally require even more rapid growth in energy consumption than in the total output of the economy. Second, many of the countries have a shortage of indigenous energy resources, due either to nature, lack of knowledge as to the resource base, or failure to exploit proven resources. Third, most of the countries are short on foreign exchange which could be used for importing additional needed energy, particularly petroleum. Fourth, many of the countries rely to a lesser or greater degree on foreign aid to help fill their foreign exchange gap. Fifth and finally, their internal petroleum sector, whether well or poorly developed, is generally dominated by the international oil companies whose home governments, particularly the United States, are presently the primary sources of this foreign aid.

In addition to these common energy problems, the oil-importing underdeveloped countries as a group share the twin general goals of rapid economic development with maximum political independence. However, while all may agree on these general goals, or at least pay lip service to them, there are a wide variety of views as to the best specific economic and political policies for attaining them. Petroleum policy is both an important component of general policy, and also a major influence on these other elements of policy. Hence in analyzing or arriving at petroleum decisions, a number of factors should be considered, especially the following: (1) expected or planned changes in the output structure of the economy as a whole in the course of development; (2) costs of various energy sources; (3) flexibility of different energy sources; (4) locus of ownership and control of energy resources; (5) economic, political, and social repercussions of different energy growth patterns. These five sets of considerations are crucial for any underdeveloped country attempting to establish an energy and petroleum policy consistent with its other goals in the area of overall economic development. Below we sketch out some aspects of these considerations.

The pattern of economic development emerging in many underdeveloped countries is one of relatively rapid industrial growth and relatively stagnant agricultural production. Perhaps the most crucial cause thus far has been the apparent inability of any underdeveloped country, whatever its social system, to satisfactorily deal with the vast problem of agriculture. Whatever the reasons, the pattern in the underdeveloped countries of an industrial sector growing more rapidly than the agricultural sector has profound implications for energy policy. Because energy input per unit of output is generally far higher in industry than in agriculture, this pattern means the energy sector will almost always have to grow far more rapidly than the overall economy.

In addition, the specific composition of the overall economic development plan will have a crucial relation to energy and oil policy, both as cause and effect. For example, the problem of transport development—whether to stress rail, road, or river —is intimately bound up with energy resource availability.

The cost of different energy sources is of course a key factor in the development of energy policy. Several dimensions of "cost" should be taken into account. First, what are the "social costs" (i.e., the ultimate costs to the country as a whole) as compared with the private costs (i.e., the immediate money costs to the individual or firm). Second, what are the short-run and long-run costs, i.e., the current operating costs and the capital costs? Third, what are the foreign exchange costs as opposed to the indigenous currency costs?

The latter question is particularly significant since most underdeveloped countries face severe foreign exchange shortages. It might be noted that the concern with saving foreign exchange reflects in effect an implicit recognition that, for the purpose of importing foreign goods at least, the value of the indigenous currency is overstated in official exchange valuation with other currencies. Hence, in considering an energy resource development pattern, government planners must often attempt to relate foreign exchange costs to indigenous costs in terms of a "true" exchange rate between indigenous currency and "hard" foreign currencies (e.g., the dollar, the mark, and the franc).[1]

Another important consideration in energy policy is the "flexibility" of different energy sources. Generally speaking, oil is the most flexible fuel resource available to underdeveloped countries. It can be used directly for heat energy (in industrial boilers, home cooking, etc.), for all forms of transportation, for lighting, and for generating that most important secondary energy source, electricity. Oil can fill specific requirements that cannot be met by other energy sources, at the present stage of technology (e.g., petroleum fuels for road or air transport). Specific petroleum products can also be used for various purposes, e.g., kerosene for home lamps or for aviation fuel, diesel oil for locomotives or stationary engines. And the proportion of various products derived from a barrel of crude oil can be varied somewhat within a refinery.

Further, oil can be more easily and cheaply transported than coal, although usually less cheaply than hydroelectric or nuclear power (except for extremely long distances). Additionally, the latent energy in oil is more readily captured than that in coal. In India it has been officially estimated that the proportion of latent energy actually utilized is only 6 percent for coal burned in a steam locomotive compared to 22 percent for diesel oil in a diesel locomotive.[2] Coal does have one advantage in that as a mining operation with relatively little further processing required, small scale production can take place at numerous sites on a relatively efficient and cheap basis.

Hydroelectric and nuclear power are also inflexible energy sources because they are a specific form of energy. Hydroelectric power is further limited since it can only be generated at waterfall sites and because efficient hydroelectric dams generally have to be relatively large. In addition, droughts may occasionally make hydro power unavailable. Nuclear power has no locational limitations which is one of its great potential advantages for underdeveloped countries. Because the amount of nuclear fuel input is extremely low relative to the energy output it can be used economically in areas where it would be difficult and/or expensive to transport other energy sources.

On the other hand, nuclear power has the disadvantage that to be economic it must be produced on a very large scale. Moreover, nuclear power plants cannot readily be turned on and off with the same ease as thermal power plants (or hydro power plants by using unneeded energy to pump up water which can be used to make energy again later). Thus, the efficient use of nuclear energy requires a high ratio of demand to potential capacity, e.g., the load factor—the ratio of utilized energy to potential energy—must be 80% or greater, compared to 60% or less for conventional or hydroelectric energy sources. (Hence, nuclear power is best for providing the base load of demand, with other sources meeting fluctuations in demand.)

The locus of ownership and control of energy concerns energy planners and governments the world over. There are three basic dimensions to the problem: (1) physical location —foreign or indigenous; (2) ownership location—foreign or indigenous; (3) social control—private or government.

In the developed countries it is widely agreed that indigenous energy resources should be favored at the expense of foreign energy sources. In the United States, for example, one rationale for restricting oil imports (albeit not the basic reason) is to encourage the domestic oil industry so that in event of an emergency indigenous oil supplies will be physically available. The Common Market countries, lacking indigenous energy resources, have been moving toward requiring large-scale stockpiles of oil within the countries as a cushion against emergency cutoffs.

One view of the physical location problem is that of Pierre Maillet, a Common Market energy official:

> Energy is such an important factor in the economy of a country that it is imperative that regular supplies be ensured in the best possible conditions. Such security must be achieved as regards both quantities and prices.
>
> From the point of view of quantities, the risk may be a shortage of supplies. In fact the risk is only significant when it is a case of large economic units using considerable quantities of energy as compared with world production. Such is the case of the United States, for instance, or the European Economic Community, or Europe as a whole. On the other hand, in the case of an individual country, particularly a small country, or a country with limited economic potentialities, and which does not consume large quantities of energy, the risk of short supplies is small and often negligible, except in certain conditions due to strictly political considerations.[3]

It might be noted that the exception of "certain conditions due to strictly political considerations" may be at some point in time an extremely important problem for any given country, and cannot really be ignored in developing an overall energy policy. At various times such diverse countries as Cuba, Israel, Finland, Italy, Rhodesia, and the Union of South Africa have been faced with real or potential attempts to force changes in their basic economic or political policies by cutting off energy imports. Hence, Maillet's solution to the problem of "security of supplies," quoted below, deals really only with narrow economic considerations:

> In these [small] countries the problem of security of supplies is essentially a question of stability of prices. Such a country should therefore not be dependent on a single supplier for the whole of its requirements. This would amount to a sort of monopoly. Such a monopoly could in fact apply not only to the supplier but to the carrier and to avoid the possibility of one or

the other raising prices to a notable extent and over a long period, the importing country may have to obtain supplies from various sources, either from different countries or from different undertakings in the one country.[4]

In most underdeveloped countries the decision on foreign versus indigenous ownership generally follows from decisions on the other dimensions of the location problem. For example, a policy decision to rely primarily on indigenous energy resources may effectively rule out foreign ownership because of lack of foreign interest in hydro or coal development. Again, a decision to have government ownership of energy resources automatically rules out foreign ownership. Nevertheless, one should bear in mind the following realistic choices as to predominant type of ownership and control: (a) private foreign ownership of external energy resources; (b) private foreign ownership of indigenous energy resources; (c) private domestic ownership of indigenous energy resources; (d) government ownership of indigenous energy resources. These alternatives may be viewed as running the spectrum from complete reliance on an international trade system in private hands to total government control of national resources. (Of course there are a number of possible variations, e.g., crude oil may be controlled by foreign private sources with refining capacity owned by the government and the marketing network controlled by either.)

The overall policy considerations which help determine how any given country answers the question about the control or ownership of energy may be illustrated by some extreme examples. A communist country such as China would by definition insist on government rather than private ownership of the crucial energy sector. At the same time, it would likely opt for development of domestic energy resources rather than relying on obtaining external energy through international trade, if for no other reason than fear of boycotts. A country devoted to free enterprise, such as Liberia, might rely on private foreign energy sources because they were cheaper than domestic ones. At the same time, it would probably rely on private capital to develop intermediate energy sectors, such as oil refineries or electric power stations, because of the greater supply of private capital and/or because of the belief in the efficiency of private capital.

Intelligent policy on energy must also take into account the important economic repercussions which may result as a "byproduct." For example, development of hydroelectric resources frequently serves nonenergy goals, such as irrigation and flood control. Probably the most universal economic byproduct consideration in formulating petroleum policy is its impact upon the potential for an indigenous petrochemical industry. Because of their vital significance it is worth noting the interrelationships between petroleum policy, petrochemical industry, and economic development.

Petrochemicals consist of all those chemicals which are produced from raw materials of petroleum origin, including both oil and natural gas. Petrochemicals are playing an increasingly important role in the developed countries and hold out enormous potential for the underdeveloped countries. In the words of the United Nations:

> Among the various industries that may come in for consideration in an accelerated programme of industrialization, some sectors, because of their particular technical and economic characteristics, the type of resources on which they are based, and the nature of their products, are of particularly dynamic character. The establishment of industries in these sectors creates, in addition to the direct economic effects, an impact which exerts an over-all stimulating effect upon the rest of the economy. The petrochemical industry is an example of such a dynamic industry.[5]

More specifically, the UN notes:

> This industry is considered to be of strategic importance in inducing further industrial development because most of its output goes to other producing sectors. It shares this characteristic of intermediate manufacture with other industries such as iron and steel, paper and its products, and petroleum products.[6]

Petrochemicals are used extensively for making rubber products, textiles, and a wide range of other products. Two petrochemical-based products are worth separate discussion because of their enormous potential significance for the underdeveloped countries: fertilizers and plastics.

It is widely agreed that fertilizer is probably the single most important physical factor in determining the level of agricultural productivity. A study made by the UN's Food and Agriculture Organization found that three quarters of the variation in per acre grain yields among a group of 40 countries was associated with variations in per acre fertilizer consumption. In the words of the foreword to the study, ". . . the investigation makes it clear that any country which aims at increasing the production of food and economic crops must plan to increase the consumption of fertilizers." [7]

The potential significance of the plastics industry to underdeveloped countries has not been as widely heralded; this is probably partly because in the developed countries plastics tend to be associated with a wide variety of consumer products which would be relative luxuries for underdeveloped countries. In fact, however, the technology has already been developed to allow plastics a major role as a construction material in the underdeveloped countries, particularly for buildings and for piping.[8] In the words of one expert study prepared for the UN:

> Although the promise of plastics has long been known, the strides now being made in the application of the various plastic materials in replacing more traditional materials, such as wood, paper, glass, aluminum, and steel are truly astounding. We may now be at the point where countries will start to consider a plastics industry as one of the basic foundations to build on, and may thus defer the installation of some of the other industries until a later time. Since hardwood is not always available and the capital investment for glass, steel, paper, or aluminum industry is one or two orders of magnitude greater than for plastics, it is not surprising that developing nations are giving most serious attention to the latter. . . .
>
> This is a good example of the use of new technology in a "leap-frogging" role. Japan has partly followed this path, and has created an extraordinary plastics and fibre industry, fully competitive with the world industries in these fields; and in some respects more advanced.[9]

The direct relationship between petroleum policy and the development of the petrochemical industry derives partly from "the trend towards utilization of naphtha as the 'universal' petrochemical raw material since practically all petrochemicals can be produced from it." [10] In addition, in many underdeveloped countries where the demand for motor gasoline—highly refined naphtha—is generally low, development of an indigenous refining industry frequently leads to a surplus of naphtha; when as is frequently the case there would be no alternative use for the surplus, this makes the real cost to the economy of using naphtha for petrochemicals virtually nil. Given this situation it follows that "close co-ordination between petroleum refining planning and nitrogen fertilizer facility planning is essential in order that the project may approach the optimum utilization of raw material as closely as possible." [11]

Whether the original petrochemical raw material is naphtha or refinery gases or natural gas is not as crucial as the fact that raw materials are a major cost component in the petrochemical industry. A United Nations study of the chemical industry in Latin America estimated that, even for small size plants where capital and labor costs tend to be relatively high, the raw materials' share of total production cost for various petrochemicals was as follows: alcohol 38 percent; PVC 80 percent; butadiene (for rubber) 46 percent; carbon black 44 percent; polyethylene 58 percent; ammonia (natural gas) 35 percent; urea (fertilizer) 68 percent.[12]

As noted above, the petrochemical industry is closely related to petroleum refining, both for technical and economic cost reasons. Historically this has led to close institutional ties as well, with the private oil companies owning much of the world's petrochemical industry.[13] This situation further strengthens the view that governmental petroleum policy should not and cannot be developed in isolation from petrochemical policy. Moreover, as we will argue in a later chapter, formulating an intelligent energy policy requires that petroleum refining and petroleum exploration policies should also be considered as interrelated. Since natural gas is a frequent byproduct of oil exploration efforts, and is also a basic raw material for petrochemicals, the importance of a petrochemical industry further strengthens our belief that all petroleum and petrochemical policy must and should be considered as a whole in relation to economic development.

Finally, aside from directly economic repercussions from any given petroleum and energy policy, there are often important political and socioeconomic ramifications to be considered. For example, does the policy leave the country vulnerable to an energy boycott, and how important is that to the country? What would be the impact of choosing government ownership of energy resources on foreign relations? Does the country want a nuclear power industry as a base (or cover) for developing nuclear weapons? How will resolution of the public versus private sector question affect the balance of political forces within the country?

The social repercussions of different energy patterns may also be very significant in the long run. For example, undertaking a major hydroelectric project complete with irrigation and flood control may ultimately revolutionize conditions in the local area affected. It may greatly increase productivity, raise the standard of living of local inhabitants, and radically change the social relations among various groups within the area. Again, since most rural areas are not electrified and oil lamps are the only source of light, a decision to allow and promote growth in kerosene consumption may be a necessary adjunct of any campaign to promote literacy. Conversely, blocking growth in kerosene demand may have a negative impact on literacy promotion, and through that may slow down increases in productivity and positive changes in social relations.

The potential ramifications of energy and petroleum policy can thus be seen as widespread, affecting both internal and external relations.[14] At the same time for just this reason development of petroleum policy itself is ultimately shaped by the internal and external power relationships. The rest of this study, we believe, provides ample evidence of this.

CHAPTER TEN

Major Issues: Oil Exploration in Underdeveloped Countries—Public versus Private

GOVERNMENT POLICY toward indigenous oil exploration is one of the most important and controversial decisions facing any underdeveloped country. The most basic policy decision which government officials must make is whether oil exploration should be undertaken by the state, or instead left to private enterprise. In this chapter we present and critically analyze the widely accepted case against government oil explor. cion.

The case for private and against government oil exploration in oil-importing underdeveloped countries rests on three major premises: (1) Oil exploration is a very "risky" business. (2) Because of this "riskiness," considerable diversification of efforts is needed to have a chance of a successful oil exploration program. This necessarily entails great expenses, both in men and money. (3) The international oil companies have the money and technical skills which are required and are sufficiently diversified so that they can afford risk; conversely, the governments of the underdeveloped countries have neither large amounts of capital nor trained manpower and hence cannot afford the "risk" of oil exploration. It logically follows from the above that governments of underdeveloped countries should not embark alone on an oil exploration program, but rather should leave it to the resources of the international oil companies; at most a government might participate with the companies on some joint-venture basis, where the companies would provide the scarce manpower and foreign exchange.

While the international oil companies are naturally the strongest proponents of this case, they have a wide range of supporters. For the governments of major oil-exporting countries, any exploration for oil in the importing countries represents a threat. No matter who undertakes the exploration, governments or companies, discovery will mean a reduction in world oil exports, and possibly in their own revenues.[1]

The United States and Great Britain have also been opposed to government oil exploration. They are unwilling to provide foreign aid for oil exploration in the underdeveloped countries for two basic reasons. First, and probably more important to the United States, assistance to overseas governments exploring for oil would run counter to the general philosophy of fostering private enterprise. Second, successful governmental oil exploration would reduce the profits of the international oil companies, and also increase their home countries' balance of payments problems. Given this it is not surprising that these governments have never provided any foreign aid for oil exploration.[2]

The World Bank too has refused to lend money for any government oil operations in underdeveloped countries. In addition, the Bank has also played an active, albeit subsurface, role in trying to dissuade underdeveloped countries from using their own capital for oil exploration. One indication of this is the study commissioned by the Bank, entitled, *The Search for Oil in Developing Countries: A Problem of Scarce Resources and Its Implications for State and Private Enterprise.*[3] A brief analysis of the history and content of this report sheds some light on the Bank's role and influence.

The report was prepared for the Bank by Walter J. Levy, Inc., a prominent oil consulting firm headquartered in New York. The avowed purpose of the report was to provide data for the use of underdeveloped countries (which were members of the Bank) in working out a petroleum exploration policy.

Despite the usual disclaimer that the report was for "informational" purposes only, and not a statement of the Bank's official position on the question, it is hard to believe that the governments of the underdeveloped countries to which the Bank circulated the report would not, in fact, view it as just that. If not its particulars, at least the general tone was likely to be seen as some expression of the Bank's general position. After all, on such a highly controversial question as governmental versus private exploration for oil in underdeveloped countries, why would the Bank circulate that with which it was not largely in agreement.[4] Moreover, the timing of the report (1960–61), the period when Soviet offers of assistance to governmental oil exploration in underdeveloped countries was spreading, suggests that this was no mere academic study.

Finally, the choice of consultants to prepare the report for the Bank is in itself significant. While Walter J. Levy, Inc., is a highly renowned oil consulting firm, it is well known that a great part, if not the bulk, of its work has always been for the international oil companies.[5] This is not to call in question the integrity of this consulting firm, but rather to recognize that its long-standing relationships with the international oil companies suggest that its philosophy was basically consistent with theirs (and the Bank must have known this).[6]

Not surprisingly, the reader of this report comes away with the impression that the worst approach to "the search for oil" is for the government to undertake oil exploration on its own.[7] Further, there is a strong implication that if the government insists upon an oil exploratory program, and is unwilling to provide the necessary incentives for private enterprise to undertake it alone, at least it should operate in conjunction with private enterprise.

The basis of the report's judgments is presumably its case studies of oil exploration in three different groups of countries: those in which private enterprise alone participated, those in which the government alone participated, and those in which there was some type of joint participation.[8] For the first group of countries, where private enterprise operated alone, the thrust of the analysis is to show that the international oil companies are willing to "bear the risk" of exploring for oil in underdeveloped countries. The analysis of this group also emphasizes the risk and uncertainty; the companies spent $32 million in Guatemala in 1957–59, without any significant oil finds and some $70 million on exploration in Turkey in 1955–60, with only partial success. Libya, at the other extreme, clearly represents a major success story for the companies.

On the other hand, the report finds the achievement of government "oil monopolies," specifically Petrobas in Brazil and Pemex in Mexico, to have been much less satisfactory. It might be noted that the criterion for evaluating the governmental petroléum effort in these two countries is different in each case. For Brazil, the government effort is found to be "disappointing" because sufficient oil was not found between 1953 and 1959 to replace all imports of crude oil. In Mexico, where the governmental effort has made the country essentially self-sufficient in crude oil, the weakness is indicated to be a failure to earn an adequate return on capital invested in the oil industry. (See Chapter 22.)

The third group consists of three countries which had originally reserved oil operations to the government and then "found it advantageous" to allow private enterprise to operate. However, the experience of these countries (Argentina, Bolivia, and India) fails to fully support the report's major thesis. While Gulf did discover oil in Bolivia, in India, as we shall see, the main impetus to production came from the government which has made the important discoveries. In Argentina, it is true, crude production more than doubled between 1959 and 1962, the period in which private oil companies were first allowed in. However, it must also be pointed out that the companies largely developed and produced oil from already proven fields, rather than making major new discoveries. (See Chapter 25.)

The subsequent history of this report is worth noting. An abridged version of it was presented to the United Nations "Interregional Seminar on Techniques of Petroleum Development," held in early 1962.[9] Interestingly, while the abridged report does devote some space to specific countries in the first and third groups there is no specific mention of the government monopoly cases of Brazil and Mexico. What was left instead from the larger report was simply the general critique of government oil monopolies:

A government oil entity often runs into very real difficulties in staffing its operations. . . .

Further, the government organization often finds itself outside the main stream of technological progress. . . .

This leads to a basic question, whether the centralization of judgement that is inherent in a government oil monopoly is the best means to successful oil operations. It is not just the possibility of political intrusion into operations that should be rigorously oriented toward maximum effectiveness. The problem goes much deeper. It would still exist, even if the government oil entity were set up as a quasi-independent body with safeguards against outside interference. It would exist as well under private operations if one company alone exercised a monopoly in operations. . . .

Thus, the very substantial capital requirements for major exploration effort would entail expenditures on a far larger scale than most national budgets could support. The alternative is apt to be a more modest effort in which resources are allocated to exploration by dint of sacrifice elsewhere. But where prospective areas are vast, chances of success are increased by thorough survey and extensive testing. A very limited programme, constrained moreover by the single judgement of one government organization, can hardly be expected to yield the best results.[10]

After these remarks, the paper continues:

In reviewing these problems, it is not my intention to discourage a national effort but only to help you assess realistically the choices that lie before you. In that respect, I would now like to turn to the role of private enterprise in oil development, and particularly about the many different ways in which private undertakings have been adapted to the interests and orientation of various countries.[11]

The same general approach can be seen in a 1962 report prepared by the UN on the Financing of Petroleum Exploration, which appears to draw heavily on the original Levy report for the World Bank.[12] Besides citing it in several places, the UN report contains the following statements:

While the centrally planned economies, as well as Mexico and Brazil, have retained a complete monopoly, the development of exploration has led the other countries studied to seek out ways of associating private capital and public funds. The systems thus being established in these countries closely resemble those developing in countries which have started from a diametrically opposed position, i.e., where exploration and development of petroleum resources was originally entirely in private hands and where government participation in the development of these resources has tended to increase. . . .

If the "petroleum adventure" demands the allocation of a large part of a country's economic potential, the uncertain profitability of this investment will have to be weighed against the safer yield of other investments in different sectors of the national economy. Such an analysis becomes easier, if only by reason of the alternatives which may be available, if it is undertaken in an international context. The problem of exploration, production and supply in a given country can then be studied in broader perspective. Except in special and essentially temporary conditions, it may be stated that isolationism makes for insecurity and that autarchy is inconsistent with the optimum use of a country's economic potential. This is particularly true of petroleum development. For both technical and economic reasons, petroleum development in every country, requires to be undertaken in an international framework.[13]

The basic weakness of all these analyses is that they fail to provide a theoretical framework for analyzing the effects of various oil exploration policies on underdeveloped countries, or data necessary for making policy judgments within such a framework. Thus, the critical questions which a government energy planner would like to have answered before deciding whether or not to embark on an oil exploration are barely raised, let alone evaluated.

The logical way to evaluate the economic merit of a government oil exploration program is by subjecting it to the same yardsticks used for evaluating other possible government investments. The best criterion for evaluating investment projects in underdeveloped countries is presently an area of controversy among economists. A number of criteria have been suggested as the best measurement yardstick, including: (1) the rate of return on investment; (2) the "capital-output" ratio; (3) the effect on the balance of payments; (4) the effect on savings rates in the country.

To our mind the most useful concept for evaluating government investment programs is the rate of return on investment, appropriately defined for the economy as a whole, i.e., the social rate of return on investment. For one thing, we believe that the rate of return on investment is the broadest, most general concept; some of the other criteria can be incorporated within it by appropriate modification.[14]

The main attraction of the rate of return on investment as a criterion for choosing among investment projects is that it correctly measures the contribution of the investment to the rate of growth of the economy, which is generally what government planners are attempting to maximize.[15] Further, it can take into account the effects of different time patterns for the flow of returns from various projects, allowing greater weight to more immediate rather than later returns. This is particularly important in the case of oil exploration, for once a field is discovered the oil is normally extracted at a declining rate over the life of the field; since a capital-output ratio assumes a constant stream of output from a given investment, it would be misleading for evaluating oil exploration.

Finally, another advantage of the rate of return on investment criteria is that it is in principle the same as the DCF criteria used by private oil companies for evaluating investment profitability in their own oil exploration projects. This can help the planner estimate the likelihood of private oil exploration in various situations, assuming that the government abstains from the oil exploration field. Hence, use of the rate of return yardstick for analyzing government investment simultaneously assists in determining the feasibility or desirability of private as opposed to government exploration.

In later chapters we will examine in detail some different oil exploration policies, and evaluate their results within such a rate-of-return-on-investment framework. First, however, it would be useful to examine and evaluate the specific arguments in the case against governmental oil exploration in the light of a wide variety of data.

The "Great Risk" in Oil Exploration

There is no more trumpeted notion in the litany of the international oil industry than the "great risk" associated with oil exploration efforts. In the words of Walter J. Levy:

> Oil exploration in new areas is an undertaking that requires costly effort, and promises uncertain rewards. . . .
>   The record is marked with evidence of the uncertainty that accompanies the search for oil. . . .
>   . . . for every Burghan, Ghwar, or Agha Jari, there are literally thousands of determined efforts that result only in dry holes.[16]

In the words of the United Nations:

> Even when all the theoretical conditions seem to be present, exploration is often unsuccessful and sometimes results only in the discovery of reservoirs of water, occasionally accompanied by a little gas. Many examples illustrating the speculative nature of petroleum exploration will be given later in the report. At this point, attention needs to be drawn to the risks involved and the impossibility of predicting results until a deposit has been reached by drilling, and its size has been estimated.[17]

What are the facts about the "risk" involved in oil exploration? The data presented below are utilized, not to establish precise risks, but to indicate orders of magnitude of risks involved in oil exploration programs. These figures should be viewed against the background of the following warning on their usefulness:

> Various of the oil industry's journals periodically publish data and statistics on exploration results. These may be very useful if properly understood but have, owing to the lack of statistical background of some users of such statistics, led to remarkable nonsense being talked and even published. . . .

Wildcat (exploration wells) Success Ratios for the U.S.A. are regularly published and are even given for different sizes of accumulations discovered. These are the numbers of successful exploration wells divided by the total number of wells drilled; they refer to a region and to wells, not ventures as a whole. For an exploitable discovery the ratio has recently varied between about 1/8 to 1/9 but any one successful venture may have had success with the first well or only after drilling say thirty wells; thus 1/9 is not a venture success ratio and furthermore there is no reason to believe that success in one specific region or total of geologically vastly different areas is in any way related to a success ratio in another area with its own peculiar geological attributes.[18]

The one thing all authorities agree upon is that the probability of success can be greatly increased, relatively cheaply, by scientific survey methods. According to *The Petroleum Hand Book,* prepared by Royal Dutch Shell, the situation is as follows:

. . . the enormous contribution made by geology and geophysics to the success of modern oil exploration may be judged from what is called the success ratio (that is, the ratio of successful wells to the total number of exploration wells drilled), taking into account the various exploration methods, singly or in combination, used to determine the drilling site of the exploration well. This is shown in the following figures:

No scientific exploration   success ratio 1:30
Geological exploration   success ratio 1:10
Geophysical exploration   success ratio 1:6
Geological and geophysical exploration   success ratio 1:5

These ratios are based on figures from the United States. If those from the world outside, particularly from the Middle East, were available they would put modern exploration methods in an even more favourable light.[19]

The relative cheapness of the scientific survey stage of oil exploration is well known also:

. . . the preliminary reconnaissance of an area is relatively inexpensive compared with the investments needed for exploration and production. The cost of preliminary reconnaissance was one-twenty-fifth of the cost of exploration and appraisal drilling in the first case, and one-eighth in the second.[20]

The significance of the preceding is twofold. First, it indicates that risk in oil exploration is not "immeasurable" but is estimable, at least within a wide range. This should not be surprising after all, since the international oil companies must themselves use preliminary (if perhaps more refined) estimates of risk as screening devices for considering oil exploration.

Second, and more important, "risk" in oil exploration is not fixed once and for all time by nature, but rather is susceptible to marked reduction through relatively inexpensive preliminary investment. Oil exploration in underdeveloped countries can thus be approached as a two (or more) stage proposition, in which variable amounts of money can be invested, and depending on the information feedback from this investment, the program can either be aborted after a relatively low capital outlay, or continued with greater investment but a higher probability of success. As an independent Australian oil man, E. W. Avery, has pointed out:

> A group engaged in exploration in any area will be continually reviewing its actual results against its ideas of what constitutes sufficient encouragement to justify risking further funds.
>
> A magnetometer survey over an unknown area could yield evidence that the sedimentary sequence was shallow, and, therefore, commitment of further funds would be unjustified. A seismic survey can encourage or discourage an operator to continue.[21]

The figures presented above suggest that an expenditure of anywhere from four cents to thirteen cents on oil surveying for each dollar spent on oil exploration increases the chances of finding an oil well by a factor ranging from 3 to 6. It seems indisputable therefore that a major surveying program should be a preliminary to any oil drilling program. In the words of Avery:

> The most satisfactory approach to oil exploration is that oil may be expected in every unknown geological basin and it is the

commercial factors, after the initial risk resolution appraisals have been carried out, which determine whether a viable industry will be established.[22]

The "Great Expense" of Oil Exploration

A second concept enjoying widespread popularity in the international oil industry is the notion of the "enormous investment" required for oil exploration. Presumably the need for this enormous investment stems from the riskiness of oil exploration, which requires considerable diversification of efforts. Unfortunately, all too frequently the literature prepared for underdeveloped countries simply presents staggering cost figures or vague generalities instead of a careful analysis of the range of possible oil exploration costs, particularly in relation to possible returns. For example, the Levy paper for the UN states:

> In terms of over-all capital outlays, it has been estimated by the Chase Manhattan Bank that some $90,000 million were spent in the past decade on oil facilities outside the centrally-planned economies. Over the next decade, expenditures could run to $140,000 million. These numbers are striking. . . .[23]

In the words of the United Nations report on financing petroleum exploration:

> There has been very substantial progress in drilling techniques during the last few years. . . . All of these factors may well lead to substantial improvement in the conditions of exploration. Nevertheless, exploration will continue to be a risky and generally costly operation.
>
> The production of petroleum is the culmination of a process which is invariably lengthy and which necessitates the employment of considerable technical and financial resources.[24]

Another paper prepared for the UN in 1962 by P. Leicester, a British petroleum consultant, had the following to say:

> In his report to the International Bank, Walter Levy states "recent experience (outside the United States and the Soviet bloc)

suggests that expenditures to establish production have averaged almost $2000 per barrel per day."

From statistics of a number of ventures kindly supplied by a very large operator, the writer and one of his erstwhile colleagues discovered that the average coincided remarkably closely with Levy's figure.

It will be seen that, taking this ratio, to establish a production of 100,000 barrels per day (5 million tons per year approximately), the average order of investments is some $200,000,000.[25]

Now, far from being discouraged by these figures, governments in underdeveloped countries considering undertaking oil exploration should find them encouraging. Placing a conservative import parity value on crude oil of about $1.50 per barrel,[26] for each capital investment of $2,000 to establish a productive capacity of one barrel per day, a country would realize a gross return of $550 per year (365 times $1.50), or a social rate of return on investment of over 25% per year.[27] Moreover, imported crude oil is a complete foreign exchange drain, while at least some of the capital cost of oil exploration can be met from indigenous resources; therefore, if foreign exchange were valued above the official exchange rate, the rate of return would be even higher.

Finally, it is interesting to note some cost figures presented in *Petroleum Outlook*, an oil trade journal.[28] The data are presented here in terms of the average cost experience (based on all wells—productive and dry) for finding and developing (but not actually producing) one barrel of crude oil for various regions. The highest per barrel costs, $1.45, were in the United States which, largely because of tax incentives, is the most over-drilled country in the world. More important, in the underdeveloped countries of the world, the finding and developing costs per barrel were far lower. The lowest, of course, was the Middle East, with a cost of only $0.04 per barrel. However, in the Western hemisphere, excluding the United States, Canada, and Venezuela, the costs were still only $0.42 per barrel, while in Africa they were $0.15 and in the Far East, $0.12.

The "Great Resources" of the Oil Companies for Oil Exploration

It is another cliché of the oil industry that only the international companies have the money, skills, and required amount of diversification to be able to afford the risks of oil exploration in underdeveloped countries. In the words of the Levy paper for the UN:

> Technical competence reflects the training and experience that is built up only in the course of repeated operations in many areas and under all possible conditions. . . .
> On the financial side, I would stress two vital contributions of private enterprise—in providing essential capital and subsuming foreign exchange costs. As noted earlier, the flow of capital required to finance oil operations is very large indeed. . . . In effect, world-wide operations of the industry enable companies to draw on their earnings in established areas to finance exploration and development elsewhere.[29]

In the words of the UN report on financing petroleum exploration:

> Although individual exploration programmes inevitably involve an element of risk, it might perhaps be possible, on the basis of a series of operations spread over a long period in different areas, to achieve a rate of return that would offset the risks of single operations. This method is, of course, of mainly academic interest so far as the developing countries are concerned. It can only be put into practice by companies or institutions operating internationally. As will be seen, it is the practice of such companies to rely on their own resources to finance exploration. . . .
> . . . The companies thus possess relatively complete experience on the technical and commercial level, and consequently on the economic level too.[30]

Both of these papers correctly point out that the international oil companies have vast resources in money and men, which *could* enable them to efficiently carry out oil exploration programs. The critical point which both reports fail to analyze is

129

to what extent *would* the international oil companies want to explore in the underdeveloped countries.

The most crucial factor which would deter the international oil companies from exploration in the oil-importing countries is the existence of vast reserves of low-cost crude oil particularly in the oil-exporting countries. The conservatively estimated addition to Middle Eastern reserves alone between 1947 and 1962 was 187 billion barrels, or over thirty-five times the current annual world trade in crude oil. The seven majors, who control three quarters of the noncommunist world's 300 billion barrel oil reserve, have the greatest imbalance between reserves and production: their reserves equal 45 years of their current production, compared to 24 years' worth of reserves for all other producers.[31]

A priori, then, the international majors, who control most of the refining and marketing facilities in oil-importing underdeveloped countries, should be particularly opposed to indigenous oil exploration programs: any crude oil found will normally "back out" the companies' imports of their low-cost external crude oil. If the government undertakes the oil exploration program, any oil discovered will involve a complete loss to the companies. If the companies undertake it, *and even if they are ultimately successful*, they would still suffer a net loss except in the unusual case that: (a) the profits from sale of any indigenous crude oil, after subtracting exploration and development costs, are greater than the profits from importing an equal amount of their external crude oil; or (b) large enough quantities of crude oil are discovered by a company to allow sufficient increase in its market share within the country to offset any loss from backing out its present crude imports.

However, while the established majors would generally like to see little or no exploration in the oil-importing underdeveloped countries,[32] they are realistic enough to recognize that in some countries, particularly where foreign exchange shortages are severe, the government may insist on some oil exploration. Since any oil found by the government is a total loss for the companies, rather than seeing large-scale government exploration, the companies would sometimes prefer to undertake some exploration of their own, to forestall large scale government entry (which might well carry over to refining and marketing). Exactly how much exploration they are willing to undertake in such a situation depends of course upon the specifics of each country.[33]

In sum, the logic of the situation suggests that the international oil companies would have relatively little incentive for oil exploration in the oil-importing underdeveloped countries. In the tactful words of Mr. Avery:

> If a company is required to carry out the broad appraisal of an area then it will usually require very large tracts of land for a relatively long period of time. Again, while not wishing to be in any way disparaging, it must be kept in mind that a company's objective must be to maximise its overall world profit, and that this may require holding an area with the minimum expenditure.
>
> . . . there are instances, not necessarily in Australia, of large tracts of sedimentary area being in the hands of a group of explorers, where the Authority had not compelled a diligent investigation of the possibilities.[34]

In addition, it might be predicted that newcomers with no established position in a given country might be somewhat more willing than the established majors to explore for oil, since the newcomers would not be backing out any of their external crude.[35] Available data on capital expenditures for oil exploration seem to support these theoretical deductions.

For the 1947–62 period, total capital expenditures in the non-communist world for oil exploration and development equaled $73.5 billion. Of this, 73% was devoted to oil exploration efforts in the United States and another 6% to those in Canada, or almost four-fifths in North America. Venezuela received 7.2% of the total oil exploration effort and the Middle East 3.7%. In the oil-importing underdeveloped world, Latin America received 4.2% of the effort, Africa 2.9%, and the Far East 1.8%, for a total of one-eleventh of the world effort.[36]

Physical data on oil drilling indicate more clearly the relative lack of interest of the international oil companies in exploring for oil in the oil-importing underdeveloped countries.[37] In 1963 some 51 thousand wells were drilled throughout the noncommunist world. Of this number, 90 percent were drilled in the U.S. and Canada; 37,000 of these were development wells for bringing up oil from proven fields, while 9,000 were new exploratory wells.[38] Of the remaining 5,000 wells drilled, only 164 were drilled in Asia (excluding Indonesia, Japan, British Borneo, and Australia), 91 in Africa (excluding Libya, Nigeria, and Algeria), and about 2,000 in Latin America (excluding Venezuela, The Netherlands Antilles, and Trinidad).

Moreover, it is clear that a good proportion of the drilling taking place in the new oil-exporting underdeveloped countries is being done by governmental oil companies. Thus, of the 2,231 wells drilled in 1963 in the underdeveloped world (as defined above), 554 were in Mexico, 202 in Brazil, and 100 in India, all of which were undertaken by government-controlled oil entities. Another third of the oil drilling in the underdeveloped countries took place in Argentina, much of which was development drilling by private companies operating under contract from the government company which had originally discovered the fields. Thus, less than one-third of the oil wells drilled in the underdeveloped countries were accounted for by private companies operating on their own efforts; in effect, less than 2% of the world's oil drilling was done by the international companies in the oil-importing underdeveloped world.

Finally, fragmentary data indicate that of the relatively small amount of oil exploration done by the private oil companies in the underdeveloped countries, a disproportionately high proportion is accounted for by the newcomers as compared to the established majors. For example, Standard Oil of California in its 1965 Annual Report states that, of 1,152 wells drilled in 1965, 97% were drilled in the Western hemisphere (overwhelmingly the United States). This was true despite the fact that 34% of its refined products were sold in the Eastern hemisphere, and 39% of its refinery capacity was located in the Eastern hemisphere. Obviously this relative lack of activity is related to the fact that Standard Oil of California has proven reserves of at least 22 billion barrels of oil in the Eastern hemisphere or 74 times its Eastern hemisphere production and over 150 times its Eastern hemisphere refining capacity.

On the other hand a newcomer, Phillips Petroleum, drilled 315 wells in 1965 of which 23% were outside the United States. Again, ENI has been propelled into a very active oil exploration program by its lack of crude in relation to refinery capacity. This imbalance has led ENI into active exploration in Italy, Iran, Libya, Morocco, Nigeria, Sudan, Tunisia, and the United Arab Republic.[39]

Summary and Conclusion

The preceding analysis has suggested that the "conventional wisdom" case against government oil exploration is, for a variety of reasons, questionable. While it is true that there is "risk" in oil exploration, the risk can be reduced considerably by relatively inexpensive preliminary surveying efforts. Moreover, risk only has economic meaning in relation to the amount to be expended and the potential or likely returns from such expenditure. The fact that diversification of drilling may be required does not mean that it is beyond the means of the government to undertake. Again, the "vast expense" involved in oil explorations is a relative question, depending upon the extent of the effort and also upon the possible or likely return. Finally, the fact that the international oil companies have the resources and diversification to "afford the risk" is no guarantee that they will undertake the effort. Rather, both the structural situation of the international oil industry and available data suggest that they will generally avoid oil exploration in oil-importing underdeveloped countries, largely because of their control of vast low-cost oil reserves in oil-exporting underdeveloped countries.

The question, then, of whether a government should "go it alone" cannot be solved *a priori* but only in the light of the specific possibilities and resources facing the underdeveloped country. There should be no presumption that oil exploration is beyond the means of a country nor that if the government does not undertake it private companies will. To the contrary, the presumption should be that an underdeveloped country cannot afford not to undertake an oil exploration program. This is because, viewed in the light of a social rate of return on investment analysis, oil exploration has uniquely attractive features for economic development.

The uniqueness stems from the fact that, once discovered, the real cost of producing oil is relatively very low, equaling perhaps 10 percent of the total value of the output. Hence, obtaining some 90 percent of the value of the output requires no current resources which might be utilized elsewhere in the economy. For other potential investments, e.g., a steel mill, a much higher proportion of the total output will embody resources which potentially could be used otherwise (the iron ore might be exported or the labor might be needed for construction and/or operating other factories).[40]

To restate the principle in formal economic language, oil has a low *social cost* of production relative to its *private* (monetary) *cost*. (The crucial distinction between social cost and private cost rests on the fact that when human or material resources are unemployed, use of them might be costless to society as a whole; for private entrepreneurs such use would not be free since they have to pay the going market rates to obtain these resources.)

A further unique aspect of investment in oil explorations is that it has a comparative advantage over other investment projects in terms of the potential ability to harness the increased output for further reinvestment. Because of crude oil's low current production cost relative to its value, potentially 90 percent of the value of crude oil output can be utilized for reinvestment in the economy. In a steel mill, on the other hand, even if all of the resources utilized would have been unemployed otherwise, thereby making the real social cost of operating the mill as small a proportion of total output as in oil exploration, the problem still remains that the bulk of the monetary value of the output will have to be paid out to the labor and other businesses which provide the various resources. Hence, there will normally be a good deal of "leakage" from the investment in a steel mill, through a large part of the value of the additional output ultimately being used for consumption.

The critical factor which makes oil exploration such an appealing investment for an oil-importing underdeveloped country is the fact that, in the present world economy, oil in the ground represents a major source of "locked-up capital" which can potentially speed up the rate of economic development. This potential arises from the fact that oil is unlike most commodities which move in international trade, where the price is relatively close to the average cost of production.

An underdeveloped economy cannot normally earn a high rate of return from an import-substitution investment project unless it heavily utilizes unemployed resources. Thus, for example, if it costs $100 to import a ton of steel, then an underdeveloped country which can produce it indigenously for $120 in monetary cost but $70 in real cost to the economy, can perhaps save 30 percent of the real value ($100) of the steel by investing in a steel mill. For oil, on the other hand, which costs $2 per barrel to import and 20 cents per barrel to produce indigenously, the real potential for saving by investing in petroleum exploration is 90 percent of the value of the output derived from the investment.

Thus, an investment in petroleum exploration potentially allows an underdeveloped country to take advantage of the enormous gap between the price and the average cost of finding and developing oil—a gap which is caused by the monopolistic control of the international oil companies over the supply of crude oil. Put another way, the "locked-up capital" in crude oil is equal to the rent and monopoly profits which the international oil companies and the oil-exporting underdeveloped countries share and which is maintained by artificial scarcity. As long as the gap between the average cost of finding and developing oil and the price of oil remains anywhere near as great as at present, there is an enormous incentive for an underdeveloped country to invest in exploring for crude oil.[41]

CHAPTER ELEVEN

Major Issues: Oil Refineries—Location and Ownership

IN RECENT YEARS oil planners in underdeveloped countries have faced two basic policy questions with regard to oil refineries: (1) Should the country build indigenous refineries? (2) Who should own any such indigenous refineries? Different approaches to these questions have led in many underdeveloped countries to continuing struggles among the various international forces affecting oil. In this connection, the basic aim of this chapter is to provide data and an analytical framework for underdeveloped countries in evaluating the range of possible answers to these questions.

To put the problem in its historical context, it will be recalled that up to World War II, when many of the underdeveloped countries were still colonies, the typical pattern was for petroleum products to be imported by the international majors from their large refineries in the oil-exporting countries. The era following World War II has been marked by a strong trend in all but the smallest oil-importing countries toward the building of indigenous refineries in order to replace product imports. This can be seen from the fact that while in 1950 the oil-importing underdeveloped countries were consuming about 44 million tons of refined oil fuels annually, only 20 million tons, or less than half, were produced indigenously.[1] In 1965, on the other hand, even with a vastly increased consumption level of 145 million tons in these countries, over five-sixths of the products were refined indigenously.

The basic cause of this locational shift has been the pressure brought by the governments of the underdeveloped countries on the international oil companies to build indigenous refineries. This government pressure, in turn, has commonly been motivated by a desire to save foreign exchange through substitution of imported crude oil plus local labor for imported refined products.

The established international oil companies, on the other hand, generally resisted this pressure as much as was feasible because they normally would make more profits with less risk under the old system. For a company already marketing in an underdeveloped country, building an indigenous refinery would not provide any additional crude oil profits. At the same time, an integrated company's overall refining profits would generally be reduced, because operating costs in the new, smaller indigenous refineries were usually greater than that of the existing large refineries located in the oil-exporting countries. Finally, the new investments in refineries in underdeveloped countries were always potentially susceptible to nationalization, which would tend to make a company more vulnerable to other governmental pressures, particularly in the area of pricing.

While the international majors were thus reluctant to build these indigenous refineries, at all times their main goal was to insure maintenance and expansion of their outlets for crude oil, the key to profitability. Moreover, they recognized that the period following World War II had brought with it fundamental structural changes. First, the newcomers to the international oil industry, eagerly seeking outlets for their crude oil, had a strong incentive to build refineries in the underdeveloped countries. Second, the governments themselves often were eager to build refineries in their own countries, and Soviet aid was increasingly available to them for this purpose. Either one of these alternatives now open to a country could lead to the established majors being shut out from a crude oil outlet, a fate far worse than the relatively minor diminution in overall profits resulting from a shift in refinery location. Hence, the established majors tended to respond to the changed situation flexibly and pragmatically. Their stance ranged from one of eagerly seeking the right to build refineries in countries where the danger of government and/or newcomer refineries was great to dragging their feet in those countries where the danger appeared slight.

The extent of the companies' willingness to undertake indigenous refineries, and of government pressures for such refineries, is partly intertwined with the question of the economies of size in oil refineries. Ordinarily, the smaller the proposed indigenous refinery the higher are the per barrel costs (both capital and operating), and hence the greater the reluctance of the companies to build the refinery and the less the pressure from the country to do so. In order to attempt to sort out the role of size in relation to these location decisions, and as one guide to future decision making both in terms of location and ownership, it will be helpful now to briefly examine the extent of economies of scale in oil refineries.

No attempt is made here to develop precise numbers as to the costs of refineries in underdeveloped countries. Various studies of refinery economics have noted that:

> . . . it is difficult to present meaningful capital cost figures for modern refineries. Such factors as geographic location, prevailing material and labor costs in the area, owners' mechanical specifications, owners' space equipment philosophy, type of crude run and products required, availability of purchased electricity, availability of cooling water, tankage requirements, etc., have such a profound effect that any rough estimates may be misleading.[2]

While strictly speaking this may be true, the petroleum planner still needs some very rough order-of-magnitude figures along with an analytical framework to serve as an initial "screening device"; i.e., as a basis for determining whether or not it is worthwhile to undertake more detailed and costly studies of the precise refinery economics for the given country.

Table 11-1 presents the estimated investment cost (in millions of dollars) and the crude oil capacity (in thousands of barrels per stream day) for various sized oil refineries now under construction in ten underdeveloped countries; we have further translated these figures into investment cost per annual barrel

TABLE II-I

INVESTMENT IN MEDIUM TO SMALL REFINERIES
UNDER CONSTRUCTION IN UNDERDEVELOPED COUNTRIES[a]

Country[b]	Crude oil capacity thousands of barrels per stream day	Total investment cost in thousands of dollars	Unit investment cost/bbl[c]
Argentina (s)	75	$30,000	$1.22
India, Madras (j)	50	32,000	1.95
Features:[d] vacuum unit, vis-breaker, catalytic reformer, hydrotreaters, lube oil plant, asphalt unit.			
Brazil (s)	45	38,000	2.57
Features: vacuum unit, fluid catalytic cracker.			
Ceylon (s)	34	29,000	2.60
Colombia (s)	25	25,000	3.05
Guyana (p)	20	20,000	3.05
Gabon (p) Features: catalytic reformer.	12.5	16,400	4.00
Saudi Arabia (s)[e]	12	6,700	1.70
Peru (p) Features: catalytic cracker.	10	10,000	3.05
Israel (p) Features: simple products: gas oil, marine diesel oil, fuel oil.	4.5	1,500	1.02

[a] Based on all data available on such refineries in survey compiled by *Oil and Gas Journal*, December 26, 1966, pp. 140–144.

[b] Owner: State (s), Private (p), or Joint Venture (j).

[c] Unit investment cost expressed in dollars per barrel of refined product produced in a year. Figure is computed by dividing total investment cost by estimated annual production of refined products; latter is estimated by multiplying daily crude capacity per stream day by 328, which allows 5 percent of the year for refinery "downtime" and 5 percent of the crude for "refinery fuel and loss."

[d] Features described usually add product flexibility and considerable investment cost above and beyond cost of simple refinery; Israeli refinery is extremely simple.

[e] Saudi Refinery owned 75 percent by state and 25 percent by Saudi investors.

of refined product (assuming five percent refinery fuel and loss and five percent "downtime"). Two things stand out from this table. First, the range of capital costs for even the same size refinery can be considerable; for example, a 50,000-barrel-per-day refinery to be built in Madras, India, will cost $32 million, or $1.95 per annual barrel, while the same sized refinery to be built in Brazil will cost $38 million or $2.57 per barrel.[3] Second, there is a clear tendency for larger refineries to have lower per barrel capital costs than smaller refineries.

For order-of-magnitude purposes, we have assumed that a relatively large but simple refinery in an underdeveloped country, one with 50,000 barrels per day crude capacity, would have a capital cost of about $2 per barrel (like the one in Madras, India). A small refinery, with a capacity of 10,000 barrels per day, is assumed to have a capital cost of $3 per barrel (in line with the one being built in Peru). Finally, we have made similar rough assumptions about refinery operating costs, based on estimates made for the United Nations in early 1962.[4] A simple 50,000-barrel-per-day refinery might have direct operating costs (including fuel) of about 25 to 30 cents per barrel, compared to 45 cents per barrel for a 10,000-barrel-per-day refinery.[5] The next step is to compute the cost of importing the refined products which would be displaced by an indigenous refinery. For this two sets of data are needed: the prices for each product and the amounts of each product which can be obtained from the indigenous refinery. While there are variations in price and product demand pattern from country to country, we can gain some insight by using 1964 price data for products exported from Iran, and the output pattern of a "typical" simple refinery in an underdeveloped country. Round figures for export prices f.o.b. Iran (i.e., ignoring transport costs) would be about $4 per barrel each for gasolines and kerosenes, $3 per barrel for distillate fuels, and about $1.50 for residual fuel oils. Based on a product-mix for a simple refinery of 16 percent gasolines, 15 percent kerosenes, 20 percent distillate fuels, 43 percent fuel oil (and 6 percent refinery fuel), the value of the refinery's output would average about $2.50 per barrel.[6]

The above figures show that building a relatively large indigenous refinery, e.g., 50,000 barrels per day, is definitely economically attractive for an underdeveloped country. To see this, we need to compare the per barrel value of refinery output with the costs of the refinery: direct operating costs, crude oil costs, and capital costs. Annual capital costs equal the sum of the annual amount of depreciation plus the rate of return or profit on the capital investment. In the words of one authority:

> There is no fixed depreciation rate; theoretically it varies with the type of equipment (tankage having a longer life, for instance, than a reforming unit) but an over-all figure frequently adopted for interest and depreciation is ten per cent per year on the total investment. This is equivalent to complete depreciation in twelve years if an interest of 3 per cent is assumed, or in eighteen years if interest is 7 per cent per year.[7]

Using this 10 percent annual capital cost figure, based on the assumption that a simple refinery has a relatively long life,[8] and allowing a minimum return on the investment of 7 percent, would imply annual capital charges of 20 cents per barrel. With direct operating (and fuel) costs at 30 cents per barrel and the value of output at $2.50 per barrel, this leaves $2.00 per barrel to cover the f.o.b. price of crude oil. Since in the 1964 market the going price for the relevant Middle Eastern crude oil might be about $1.50 per barrel, this would leave an additional profit for the investor of some 50 cents per barrel. Alternatively, the real annual return on the investment would not be 3 or 7 percent but close to 35 percent.[9]

Looked at another way, the essence of the profitability of building an indigenous refinery lies in avoiding paying the high refinery profits which implicitly are being charged by refineries of the international oil companies located in the oil-exporting countries. Thus, while we have estimated refinery costs at $.50 per barrel, including a 7 percent return on investment, with oil products averaging $2.50 per barrel, deducting the cost of crude oil at $1.50 per barrel leaves an implicit gross refinery profit for the companies of $1.00 per barrel. This discrepancy arises largely because much of the refined products is exported at posted prices to affiliates of the international oil companies. Because these product sales to affiliates do not adequately reflect the major discounts available in crude oil prices, they allow the international oil companies exceptionally high refining profits.

The foreign exchange savings from such an indigenous refinery are even more attractive. It has been estimated that for a refinery of this size about 50 percent of the operating and fuel costs, and 25 percent of the capital costs would be in indigenous currency.[10] On this basis an initial foreign exchange capital investment of $1.50 per barrel would yield foreign exchange savings of 80 cents per barrel every year over the life of the refinery. This would mean a foreign exchange "payback period" of only two years.

The direct economic benefits to an underdeveloped country whose market can only support a much smaller refinery, e.g., 10,000 barrels per day, while not as dramatic, also appear impressive. With per barrel operating and fuel costs of 40 cents, capital charges of 30 cents, and crude oil at $1.50, this would still leave an "extra" profit of 30 cents per barrel. On a foreign exchange basis, with per barrel operating and fuel costs of 20 cents, capital charges of 25 cents, and crude oil at $1.50, this would involve a foreign exchange savings of 55 cents per barrel compared to importing refined products. Even with a $2.25 per barrel foreign exchange capital investment, this would involve a foreign exchange payback period of only four years.[11]

The above figures suggest the attractiveness to an underdeveloped country of building an indigenous refinery where there is sufficient demand, reasonably well balanced, to allow for at least a 10,000-barrel-per-day refinery. Once it has been decided, however, that an indigenous refinery should be set up, the question of ownership arises. There are four realistic alternatives here: total government ownership, private ownership either by the established majors or the newcomers, and joint ventures between the government and the foreign companies. In assessing the narrowly economic pros and cons of these four alternatives, the critical factor to be focused on is the potential impact on the cost of the imported crude oil. That is, it is safe to assume that in most cases the real cost of building and operating the refinery, other than the crude oil input, will be independent of ownership.[12]

The major disadvantage in considering the established majors for setting up an indigenous refinery is that, as we have seen, they have a vested profit interest in continuing to import refined products. Hence, they are most likely to engage in drawn-out negotiations. Even if the established majors agree to build an indigenous refinery, they are least likely to offer low prices for crude oil. This is because in most countries, either through government price control or competition triggered by independent refiners, refined product prices tend to equal crude prices plus refinery operating costs plus a relatively low refinery profit (low relative to crude oil profits). Hence a reduction in crude prices even within an integrated company will sooner or later lead to a reduction in product prices and thus in overall company profits. Further, because of their worldwide interests the established majors have the largest stake in maintaining high crude oil prices in each and every country.

The principal advantage to allowing established majors to build indigenous refineries is that it will cause the least trouble. Since the established majors already control the marketing operations within the country, there will be no need to attempt to force them to market someone else's refined products.

The international newcomers would appear to be more attractive candidates for an underdeveloped country in awarding indigenous refineries. Having made relatively large investments in oil exploration in recent years, they are usually particularly eager to cash in on their discoveries by finding outlets for this crude oil. This makes them tend to be more willing than the established majors to cut prices. This tendency is reinforced by the fact that newcomers normally do not have worldwide operations whose price structure will be endangered by a crude oil price cut in any given country. Another way of stating the same thing is that the potential outlet for crude oil in any given country is usually a much higher proportion of the newcomer's total overseas crude oil market than it is of an established major's.

The principal disadvantage in allowing a newcomer to build the refinery is that there may be difficulties in arranging marketing outlets for the products of the indigenous refinery. Thus, unless the newcomers are willing to spend sizable sums to build up a marketing operation or to buy out the established majors' marketing operations, it may be that the government itself will have to create a marketing organization to provide an outlet for the refined products.

The most obvious advantage of government ownership of the refinery is that it is free to seek its crude oil anywhere in the world, at the lowest price possible, in terms of indigenous currency and/or foreign exchange. At one extreme, the government may decide to use Soviet barter oil, which besides being low priced will have the fundamental advantage of virtually eliminating the foreign exchange cost. The enormous potential attractiveness to an underdeveloped country of this alternative can be seen from the fact that for a small refinery with capacity of 10,000 barrels per day, the total foreign exchange capital investment of $2.25 per barrel per year would be returned within a year and a half by replacing $1.50 per barrel oil with barter oil; for a larger refinery of 50,000-barrels-per-day capacity, the foreign exchange component of the capital investment would be returned in a year.

Even aside from barter oil, however, a government can undoubtedly obtain a better bargain in shopping for crude oil than can any refinery owned either by an international major or a newcomer. After all, the privately-owned refinery company is normally dealing with its parent in negotiating the crude oil price. When we consider that Middle Eastern production costs are about 10 cents per barrel and that a price of $1.50 per barrel still yields an after-tax profit to the companies of at least 55 cents per barrel (on $1.80 posted price oil), it is clear that there is still considerable room for downward pressure on crude oil prices. In addition the government will often be able to save the country a considerable sum by avoiding the overinflated transport prices charged to affiliate refineries by their parent companies. (See Chapter 12.)

A final advantage of government ownership is that this can help ensure that the refinery is located, built, and operated to maximize the social returns from the investment, which may differ from the private returns. For example, from the viewpoint of the economy as a whole, the best location for a new refinery may be inland, either to utilize internal crude oil or to develop the usually more backward hinterland of an underdeveloped country; from the profit viewpoint of the international companies, the best location will likely be on the coast in order to utilize their overseas oil, and the desired specific coastal locations may vary among the companies depending on their relative market positions within subdivisions of the country. Again, from the country's viewpoint it might be best to design and operate a refinery which would maximize production of low-priced fuel oil for industry and minimize production of high-priced motor gasoline for autos, while for the companies the goals would be the opposite.

A major disadvantage of government ownership of the refineries is that a large absolute amount of foreign exchange is initially required. In addition, marketing can be a problem in that the established majors may be completely unwilling to handle products refined by the government, unless they can at least provide the crude oil to the refineries.

This latter disadvantage touches on what is to our mind a less obvious, but in the long run more fundamental, advantage of government ownership. That is, since crude oil is the key not only to profitability for private companies but also to cost savings for the economy as a whole, the government should make a major effort toward exploring for and developing indigenous crude oil resources. For a variety of reasons discussed in the preceding chapter, it is not likely that the private companies will undertake this task. Moreover, normally no international oil company is going to build a refinery to process government crude oil, nor will a private company market government-owned refined products based on such crude oil. Therefore, in order to exploit the possibilities of indigenous crude oil production to the full, it virtually becomes incumbent upon the government to enter into both refining and marketing operations, i.e., for the government to develop an integrated oil operation within the country. In short, rational petroleum policy cannot be developed on a piecemeal basis for each level of the oil industry but must take into account the interactions at each level of the industry.

In this respect it might be noted that the joint-venture "compromise" will normally not provide a consistent solution to the overall problem. The only interest of the international oil companies, whether newcomers or established majors, in such joint ventures lies in finding outlets for their crude oil (barring the case where the company may be able to get a particularly high rate of return on the refinery investment through ignorance on the part of the government). The typical joint-venture deal, where the foreign company supplies the foreign exchange cost of a refinery, while the government provides indigenous exchange, is in basic cost terms simply a variant on the 100 percent foreign ownership situation. That is, the crucial crude oil cost is not likely to be markedly different or the company would not have entered the joint-venture deal in the first place. In fact, since the companies generally have an aversion to joint ventures, everything else being equal they would want a higher crude oil price under a joint venture than in a single ownership situation.

A good example of how fundamentally indifferent the oil companies are to the ownership of refineries in underdeveloped countries, as long as their crude oil outlets are preserved, can be seen from the deal which was worked for the first major oil refinery in Thailand:

> The refinery, backed by Shell, is to be operated for 10 years by the Thai Oil Refinery Company (TORC) after which it will be handed over, lock, stock and barrel, to the Thai government. . . . The refinery, linked as it is with major firms, has a high certainty of profit (Shell experts estimate profits conservatively at around US $2 million a year) . . .[13]

146

Since the estimated capital cost of the refinery was $28 million, a $2 million a year profit would only be a 7 percent per year return on this investment. The key to the majors' seeming largess, of course, lay in the crude oil profits to be derived from this 40,000-barrel-per-day refinery. Even taking a conservative figure of 55 cents per barrel as after-tax profit, the crude oil profits for this refinery would amount to about $7.5 million per year, or over a ten-year period almost three times the initial investment. The real annual rate of return on the investment would thus be over 30 percent per year rather than 7 percent. Moreover, after the ten-year period was up and the government owned the refinery, there would still be a hope of being allowed to continue as crude oil suppliers, particularly since the companies were the "donor" of the refinery.

In concluding, a word should be said about optimum refinery policy in those smaller countries in which, either because of a low absolute level of demand or a sharply unbalanced demand pattern, a "typical" 10,000-barrel-per-day refinery would not appear economic. Here we would make two basic points. First, in these countries the specifics of the situation should be examined very closely before ruling out an indigenous refinery. Under certain circumstances even a very small refinery can be economically attractive. For example, Israel has built a refinery with a capacity of only 4,500 barrels per day, at a cost of $1.5 million, or $1.02 per barrel.

The key to this very low cost is that it is an extremely simple refinery which will turn out gas oil and marine diesel oil, but mostly fuel oil. Where the product demand pattern of a country is such that fuel oil is the basic need, such a simple refinery may provide a neat solution. This is particularly true for countries seriously seeking rapid industrialization, since fuel oil can be the basic heat source for both industrial boilers and electrical energy. Moreover, a paper presented to the United Nations in late 1962, apparently based on extensive practical study by Jersey's researchers, strongly suggests that with proper design engineering "special" economic solutions may be far more commonly available than is generally believed. In the paper's own words:

> . . . it is reasonable to expect—and it has been demonstrated—that small refineries in the size range of 2500/5000 B/D can be built at a cost that compares favorably with the investment of refineries an order of magnitude larger. The operating and product costs will also reflect this investment reduction. It must be recognized that these small refineries are a different breed than the larger refineries and are custom-designed to meet the requirements of less developed countries. This philosophy and approach are essential for success.[14]

Finally, a better solution than simply continuing to rely on imported products may be provided by several small neighboring oil-consuming countries banding together to build a joint "regional" refinery sufficiently large to be economically attractive to the countries concerned. This could in fact be part of a plan to also undertake a joint regional exploration program, pooling the foreign exchange resources of each country in a search for oil, which search might not be worthwhile for any one country because its market demand is so low.

CHAPTER TWELVE

A Major Issue for the Future— Oil Transport Pricing

A GOOD UNDERSTANDING of the basic economics of oil transportation is vitally important for underdeveloped countries. This is because the cost of shipping oil from the producing countries to the oil-importing underdeveloped countries is a sizable component of their total oil import bill, amounting to several hundred million dollars per year, most of which must be paid for in scarce foreign exchange. The relative significance of transportation costs varies considerably among the underdeveloped countries, depending partly of course upon the distance from the major oil-exporting centers.[1] However, as we shall see, another important determinant is the method of pricing oil transport. In the following sections we shall briefly sketch out the basic factors determining the costs and prices of oil transportation and likely future trends in these determinants.

The principal method for transporting both crude oil and refined products between countries is by specialized ocean vessels, called "tankers." Since oil tankers are a highly mobile form of investment, tanker movements in relation to the underdeveloped countries must be analyzed as part of the worldwide tanker business. Over 80 percent of world oil exports move between the oil-exporting underdeveloped countries and the developed countries. As a result, since World War II the most important tanker routes[2] go from the Middle East west to Europe (generally through the Suez Canal) and east to Japan; from the Caribbean (Venezuela) north to the United States and east to Europe; and increasingly from North and West Africa to Western Europe (carrying Libyan, Algerian, and Nigerian crude oil).[3]

I. The Political Economy of International Oil

Physically, the international tanker industry consists of several thousand vessels of varying sizes, ranging from capacities of a few thousand deadweight tons (DWT—a measure of the maximum load which can be safely carried by a ship), up to several hundred thousand DWT. Financially, ownership of the tankers is distributed among the international oil companies, who own outright close to 40 percent of the world tanker fleet, and "independent" shippers (e.g., the Greek magnates such as Onassis and Niarchos), who own the remainder. However, at any one point in time, the international oil companies control most of the world's tanker fleet, through "chartering" (i.e., leasing) the independently owned tankers; chartering is done either for a single round-trip voyage, called a "spot" charter, or on a long-term basis for periods up to 20 years or more.[4]

The world tanker fleet has grown rapidly, increasing close to fivefold since 1939. One reason, of course, has been the long-run growth of oil consumption, particularly in Western Europe and Japan, which have virtually no indigenous oil. In addition, the need for increased tankerage capacity was heightened by the trend toward displacement of Venezuelan oil by Middle Eastern oil in the Western European market; for example, the nautical distance to Great Britain from the Persian Gulf, via Suez, is 6,400 miles compared to 4,100 miles from the Caribbean.[5]

At the same time that the demand for oil tanker services has been rising rapidly, several crucial developments have so increased tanker supply relative to demand that the price of tanker services (tanker rates) has tended to fall sharply over the last decade. The most dramatic event was the crisis stemming from closing of the Suez Canal in 1956–57. The immediate effect was to send oil tanker rates skyrocketing, owing to greatly increased demand for tankers needed to carry Middle Eastern oil to Europe using the much longer route around the southern tip of Africa.[6] However, one long-run effect of these increased spot rates was to trigger a tremendous shipyard demand for construction of new tankers, many of which were delivered well after the Canal had been cleared, thereby contributing to a relative surplus of tankers in recent years. The same long-run result may safely be predicted from the Suez closing in 1967. Thus, soon after the shutdown began *The Wall Street Journal* noted that ". . . the crisis has stimulated a supertanker-construction program that will add about 40% to the estimated hauling capacity of the world's tanker fleet within the next three years." [7]

In addition, the first shutdown of the Suez Canal brought home dramatically the vulnerability of the oil companies and Western Europe to Middle Eastern crises. It thus helped spark a search for more proximate oil, particularly in North and West Africa, and thereby ultimately contributed to reducing the demand for tanker capacity. Similarly, after the 1967 closing it was reported that:

> The crisis also has spurred development of new oil fields in lands such as Australia that hope to become significant oil producers. . . .
>
> Oil companies, seeking greater diversity of crude-oil supply to lessen the effect of any future Mideast disruptions, have accelerated development of new fields in such places as Alaska, Canada, West Africa, the Amazon Basin and Indonesia.[8]

The most fundamental development, however, affecting international tanker supply has been the major reductions in cost associated with building larger and larger tankers.[9] How rapidly the trend toward larger-sized ships has developed and even accelerated in recent years can be seen from the following reports, one in 1963 and one in 1967:

> Before 1939 a tanker of 16,000 d.w. capacity—this size was adopted for the war-built T-2 tankers—was designated as a "super tanker." Shortly after the end of the war, the first 24,000-ton tankers were laid down. Gradually the trend towards larger sizes has been intensified as evidenced by the commissioning of the turbine-tanker "Nissho Maru" (d.w. capacity: 132,334 tons) in October, 1962.[10]

* * *

> A design for a 500,000-ton oil tanker, twice the size of the largest now in operation, has just been completed by Lloyd's Register of Shipping.

In making its announcement, the society says that its detailed feasibility study has clearly demonstrated that tankers of such huge dimension are practical and present no problems of construction. . . .

The world's biggest oil tanker is the Japanese Idemitsu Maru, which has a deadweight of 205,000 tons and which was launched earlier this month. But the Gulf Oil Corporation of Pittsburgh has ordered construction of six tankers each of 300,-000 tons deadweight, which will go into operation in mid-1968.[11]

Along with the trend toward building larger tankers, increasing efficiency and automation in shipbuilding have led to a marked reduction of construction costs for even the smaller-sized ships. As a result, since the first Suez Canal closing construction costs have fallen by 50 to 70 percent. A medium-sized (45,000-ton) ship which cost $238 per ton in 1956 cost only $112 per ton in 1964.[12] Gulf Oil Company's order in 1966 for 300,000-ton vessels was estimated to be at a price of about $70 per ton or 30 percent of the per-ton cost of the medium-sized ship in 1956.[13]

The impact of these savings in construction costs on tanker rates should be very great, since it is estimated that capital charges (depreciation plus interest and/or profit on the investment) amount to about 40 percent of the cost of moving oil by tanker.[14] In addition, tanker operating costs have been reduced by higher speeds, improved unloading facilities, and lower bunker fuel prices.[15] Finally, increasing size and automation of tankers have tended to cut unit labor costs. Through automation, Shell, for example, expects to lower labor costs on a new 65,000-ton ship by 40–60 percent, with the automation costing less than 5 percent of the total investment.[16] As the ultimate, work is under way to design a completely unmanned ship.[17]

The effect of all these major cost-cutting innovations should have been to reduce drastically the price of oil tanker services to the underdeveloped countries, thereby significantly reducing their oil import bill. In fact, however, the underdeveloped countries have received only a part of the benefits accruing from these cost reductions. To understand the reasons for this we have to analyze the methods of pricing used in the oil tanker business.

Two basic rate concepts are utilized in the international tanker business. The first is known as "Intascale" (sometimes called "Scale"), whose purpose is to provide standard rate calculations for every possible voyage, against which actual market prices for that voyage could be expressed as a percentage "plus" or "minus." [18] For example, if the Scale base for moving oil from port A to port B was $10 per ton, and the actual market rate for that move was $7 per ton, the market price could be expressed as "Scale minus 30 percent." Scale or Intascale is thus simply a standard by which actual market rates can be expressed, just as centigrade is a heat scale by which actual temperatures can be expressed.

Of vital operative significance, however, is the "average freight rate assessment" (AFRA), which is computed quarterly by the same group which computes Scale (the London Tanker Brokers' Panel). In essence, the AFRA is an index which is supposed to estimate the implicit "average price" which was charged for shipping all oil during that three-month period. (The AFRA index is expressed in terms of scale, and is computed separately for tankers divided into different weight groups.) The crucial point about the AFRA index system is that it is an average obtained by using the prices paid in the last three months for carrying oil, *no matter when these prices were set*. As one expert, Professor Zannetos of M.I.T., has observed:

> The average freight rate assessment (AFRA) is supposed to show the average cost of a ton of oil delivered. Thus it is not a current index but a mixture of current and historic costs intended to show at any moment of time the cost of oil in transit.[19]

From the viewpoint of the underdeveloped countries, the fact that this AFRA index is used by the international oil companies for billing their affiliates in underdeveloped countries for transport costs is of vital significance.[20] The role that use of this historically oriented pricing system can play in inflating trans-port charges may best be illustrated by the following hypothetical but not completely unrealistic example.

Assume there are ten equal size boats in the world fleet, of which one operates in the spot market, five are on long-term charter, and four are owned by the international oil companies. Further assume that the Scale base equals $2.00 per ton of oil delivered; that the current spot market rate is Scale minus 60 percent (i.e., $0.80); and that the long-term charters, which had been contracted for over a 12-year period of secularly declining tanker rates, were made at Scale plus 60 percent, Scale plus 30

TABLE 12-1

AFRA INDEX
HYPOTHETICAL NUMBERS ILLUSTRATING METHOD
OF COMPUTATION

Data	Rate ($ per ton)	% ± Scale
Long-term charters		
1956–1969	$ 3.20	+60%
1959–1972	2.60	+30%
1962–1975	2.00	0
1965–1978	1.40	−30%
1968–1981	0.80	−60%
1968 sum of long-term charter rates	10.00	0
1968 average of 5 long-term charter rates ($10.00/5)	2.00	0
Spot charter 1968–1968	0.80	−60%

Computation Process

Add average rate for 5 long-term charters extant in 1968 ($2.00 weighted by 5)	10.00	
plus average rate for 4 company boats (taken same as long-term charters) weighted by 4	8.00	
plus spot rate (1 boat) weighted by 1	0.80	−60%
Total	$18.80	−60%
divided by the sum of weights (10) equals **AFRA**	$ 1.88	−6%

percent, Scale flat, Scale minus 30 percent, and Scale minus 60 percent, for an average of Scale flat (i.e., $2.00 per ton delivered). Using AFRA, the weighted average of these spot and long-term charters would be Scale minus 10 percent, or $1.80 per ton delivered; since the oil carried in the oil companies' own vessels would be priced at the average of long-term charters[21] or $2.00 per ton, the final AFRA index would be $1.88 per ton of oil delivered. (See Table 12-1.)

Under these assumptions, the affiliates of the international oil companies in the underdeveloped countries would then be billed for transportation of oil at $1.88 per ton. This price would be more than twice as great as the spot price of $0.80 per ton, the price an independent refiner could pay for transporting the oil. The reason for this discrepancy, of course, is that the AFRA index strongly reflects the "dead hand" of the past when oil transport costs and prices were much higher than they are today. As Zannetos, an expert on oil transport pricing has stated: "We may question the use of AFRA for purposes of marginal [i.e., competitive] economic analysis, since it is a weighted average rate of all existing contracts both spot and long term . . ."[22]

The use of AFRA by the international oil companies has been defended on pragmatic grounds. That is, since the AFRA index is largely based on historical data, it tends to change much more slowly than rates in the more volatile spot market.[23] In theory it would be true that, everything else being equal, stability of prices for any commodity, including oil, would be desirable. The heart of the problem with AFRA, however, is that everything else is not equal. AFRA contributes to price stability essentially by "averaging in" outdated high prices. Use of AFRA might be defensible if it were reasonable to assume that recent (before 1967 Suez conditions) charter prices for oil transport were merely cyclical lows which would be replaced later on by the high prices prevailing during the Suez crises, so that over the long-run future the AFRA index would be representative.

On the contrary, however, it is quite clear that, owing largely to the tremendous technological improvements reducing the cost of new shipping, future prices are not likely to rise much above their pre-1967 lows for any long-run period. Thus, the 1964 analysis of Mr. W. L. Newton of Petroleum Economics, Ltd., is just as true today:

> Surplus capacity, intense competition and shrinking profit margins have been as characteristic of the tanker business as of the international oil trade in recent years. That tanker owners must reconcile themselves to the persistence of low freight rates and must therefore continue to strive for maximum efficiency was the view expressed in an authoritative paper presented to the Institute of Petroleum in London last month. . . .
>
> Medium- and long-term charter rates are likely in the author's opinion to follow the downtrend of operating costs; the reason is to be found in his belief that no tanker shortage of more than a few months is likely in view of the large capacity of the world's shipyards and the striking reduction in building times. So far as the spot market is concerned, he would not exclude the possibility of sharp fluctuations, such as those witnessed during the past two winters.[24]

A recent confirmation of this analysis came in a late 1967 Shell study of its own tanker experience:

> By taking advantage of these economies of scale, Shell has kept the costs of its marine operations at about $560-million for many years, even though the carrying capacity of its owned and chartered fleet has increased from 200 T-2 equivalents (tankers of 16,750 deadweight tons) in 1947 to 550 a decade later and 1,100 in 1967. Looking into the future, Shell expects to lower its unit marine costs even further by "running down the number of expensive charters and replacing them by cheaper ones, but mainly by using bigger and more efficient ships." [25]

If then, as appears likely, over the foreseeable future freight rates in the long run are likely to be at or below pre-1967 lows, continued use of AFRA would simply impart "stability" by maintaining an artificially high price, which did not reflect the real supply-demand situation in international tankers.[26] Turning back to our hypothetical example, if new spot and long-term charters remained at Scale minus 60 percent, and assuming that one of the existing long-term charters would expire every three

TABLE 12-2

RELATIONSHIP BETWEEN AFRA AND MARKET RATES IN PERIOD OF DECLINING MARKET RATES

| | Year of Contract | | | | | | | | |
| | 1956 | 1959 | 1962 | 1965 | 1968 | 1971 | 1974 | 1977 | 1980 |
				Price expressed as % ± Scale					
Long-term charter									
Year of expiration 1969	+60%								
Year of expiration 1972		+30%							
Year of expiration 1975			0						
Year of expiration 1978				−30%					
Year of expiration 1981					−60%				
Year of expiration 1984						−60%			
Year of expiration 1987							−60%		
Year of expiration 1990								−60%	
Year of expiration 1993									−60%
Spot rate (1 boat)				−60%	−60%	−60%	−60%	−60%	−60%
Average rate for 5 long-term charters in contract during year					0	−24%	−42%	−55%	−60%
Average rate for 4 company boats at average long-term charter rate					0	−24%	−42%	−55%	−60%
AFRA average, 1 times spot rate plus [5 + 4] times average long-term charter rate, divided by 10					−6%	−28%	−44%	−56%	−60%

157

years, it would take twelve years before the AFRA index finally averaged down to the going rate of Scale minus 60 percent. On average over the next twelve-year period, the rate charged to affiliates would be Scale minus 34, or two-thirds more than the actual market rate. (See Table 12-2.)

In sum, use of AFRA by the international oil companies for affiliate billing prevents the underdeveloped countries from enjoying the benefit of the long-run worldwide reduction in shipping costs. It is analogous to providing these countries with price stability in crude oil by refusing to give them the going market discounts. It is a stability of price imposed upon the underdeveloped countries by the monopoly position of the companies in terms of the means of transport available to their captive affiliates. No independent refiner in recent years would normally pay the AFRA rate for his oil transportation; he could usually enter the spot or long-term charter market and get a much lower rate.

What then would be a "fair" basis for pricing oil shipped by the international oil companies to their affiliates? One solution might be to periodically compute an average of the "short-term" rates, e.g., of spot rates only or spot rates plus charters up to one year, for various size vessels in the international tanker fleet. The affiliates of the international oil companies could then be billed at the average short-term rate for the respective size tanker utilized by them. If it were feared that there might still be violent fluctuations in price, these could be eliminated by a government price stabilization fund. Thus, the companies could continue to price the oil within the country as if transport were billed at AFRA. If the computed short-term rates were higher than AFRA the companies could collect the difference from the fund, whereas if the short-term rates were lower the companies would pay the difference into the fund.

Such a system would have two advantages. First, since the computed average of short-term rates would reflect the true opportunity cost of transporting oil, i.e., the price which would be paid by an independent refiner, it would be a "fair" market price. Second, the government would be backing up its claim that over the long run AFRA rates lead to overpricing by a willingness to lose money if the claim is incorrect; conversely, if the claim is, as we believe, correct, then the potential overcharges instead will be channeled into a fund which ultimately could be used for economic development of the country.

In any event, it is clear that any such changes would be strenuously resisted by the international oil companies who profit greatly from the AFRA pricing system. In the words of *Petroleum Press Service*, described by an oil expert as "the major oil companies' journal," [27]

> But the point was made in discussion that the AFRA awards have, after close and critical examination, been widely accepted by governments for pricing purposes; and that their acceptance for this purpose has imparted a valuable element of stability to oil prices by averaging widely fluctuating freight rates through time. These are advantages that should not be lightly discarded, even if it is accepted that some revision of the AFRA basis may now be desirable.[28]

Again, it was reported in *Petroleum Intelligence Weekly*:

> Shell, traditionally the strongest supporter of the Afra system, is critical of the Newton analysis. Shell is still quoting contracts to big buyers with freights tied to and varying with Afra. The U.S. majors have never used the system as much, *at least not openly*.[29]

Undoubtedly more sophisticated plans than our suggestion could be constructed for retaining the virtue of price stability without the systematic bias toward overcharges which is embodied in AFRA. Moreover, the foregoing analysis has not dealt with some of the other complexities of affiliate transport pricing. For example, what is the optimum size boat to be used by an affiliate, and is there a conflict between the optimum for the affiliate and the optimum for the company as a whole? The principal point of our analysis is that the underdeveloped countries will have to pay much closer attention to the whole area of oil transport prices, giving it the same attention they have begun to give the question of the f.o.b. price of oil. For, as we shall show later, the overcharges stemming from use of AFRA can be as great as the hard-won discounts on f.o.b. prices of crude oil.

PART II

*India: A Case Study in the
Political Economy of International Oil*

CHAPTER THIRTEEN

Prologue: The Economic and Political Setting for the Petroleum Struggle in India

IN ITS TWENTY YEARS of existence as an independent nation, India has made little progress toward overall economic development. During this period total national income has grown about three percent per year but, with population increasing at two percent annually, per capita income has risen little more than one percent per year. Considering that per capita income in India is well under $100 a year, the magnitude of the failure is obviously enormous.

The principal reason for this low rate of progress has been the slow growth in the agricultural sector, which accounts for one-half of total national income. At the same time, fairly good progress has been made in the modern industrial sector, which now accounts for one-tenth of total income; the growth rate here has averaged close to eight percent per year.[1]

The faster growth rate for industry compared to agriculture has helped make the energy sector a rapidly growing and increasingly crucial area of the Indian economy. While agriculture accounts for close to one-half of total value of production, it consumes only about two percent of all commercial energy used in India.[2] Industry, on the other hand, while accounting for less than one-fifth of the total value of production, consumes about one-half of all commercial energy in India.[3]

Largely as a result of the rapid growth of industry in India following independence, the total amount of commercial energy consumed by all sectors has increased over twice as fast as national income. Not only does industry utilize far more energy per unit of output than agriculture, but also the rate of increase of energy consumed within industry in India has been almost twice as fast as the rate of growth of industrial output itself. This has been due to the more rapid growth in India of heavy industry, such as steel and chemicals, compared to light industry such as textiles; everywhere in the world technology is such that heavy industry consumes more energy per unit of value output than light industry.

The differing growth rates within the Indian economy have been associated with major structural changes in the energy consumption pattern. While coal is still the most important source of commercial energy, its share[4] has been declining steadily, from 82% in 1950 to 69% in 1965. Hydroelectric power has been the fastest growing energy source, and its share in these years has risen from 4% to 9%. Petroleum fuels consumption has also been rising very rapidly, going from 21 million barrels in 1950 to 88 million barrels in 1965, with its share increasing from 14% to 22%.

While the industrialization of the Indian economy is clearly the basic cause of the rapid growth in total energy demand, it is not obvious why it would lead to accelerated growth in petroleum consumption, nor is it obvious why the particular pattern of ownership and control in India has emerged. To understand these developments it is necessary to look at the various other economic and political factors affecting energy and petroleum policy within India.

To our mind the key to analyzing and understanding the development of energy patterns and petroleum policies within India (or any underdeveloped country) involves a careful analysis of the balance of forces among the various groups, both domestic and international, concerned with petroleum and energy. While this balance of forces changes over time, the change is not continuous but rather involves distinct stages reflecting major changes in one or more groups' position. Hence, the basic approach of this analysis of India will be to focus on critical turning points in the fortunes of each group concerned with petroleum, as seen from an examination of the major policy questions. In essence we will attempt to combine a dynamic historical analysis with a topical approach while always attempting to trace the underlying power forces which have caused major policy shifts.

In order to do this, it is necessary to have in mind a clear idea of the crucial groups involved in the struggle over energy and oil policy development. In Part I we analyzed the general positions and goals of the major international forces. In a country like India, where the state plays a key role in the development of the economy, the central government is the focal point of these international forces. Thus, in addition to being a major force in its own right, precisely because it makes crucial decisions about the energy sector the government naturally becomes the focal point of efforts by the other major groups to attain their own goals. Within the whole period under discussion the pressures on the Indian government by the international forces have been particularly significant because the central government itself has had basic divisions on petroleum policy. In turn, these divisions can be seen as a reflection of basic ideological differences within the government and the country as a whole, corresponding to the divisions all over the world between the Left, Right, and Center.

The Left in the Indian government has consisted of the more socialistically oriented members of the ruling Congress Party, along with their supporters from the more extreme left-wing parties (particularly the Communist Party). Its basic strength lay in what was known as the "Menon-Malaviya" wing of the Congress Party. In terms of energy policy, the Left has generally favored government ownership over private ownership and development of indigenous energy resources over relying on foreign energy resources. Where foreign assistance was deemed essential, the Left has preferred dealing with the East rather than the West. From a historical viewpoint, the Left probably reached its apex of power in the early 1960s, just before the Sino-Indian military clash. The Left was particularly powerful in the energy sector up to that juncture because one of its leaders, K. D. Malaviya, was Minister of Mines and Fuels, with responsibility for the coal, oil, and natural gas industries.

II. India: A Case Study

The Right in the Indian government has consisted of the members of the Congress Party who are more oriented to private enterprise, along with their supporters from the more extreme right-wing parties. In terms of energy policy the Right favored private ownership over public ownership, reliance on foreign energy resources as opposed to developing indigenous ones where the latter step was considered too costly, and dealing with the West as opposed to the East. The Right was personified by a Congress Party leader, Moraji Desai, who was Minister of Finance in the Nehru government and later Deputy Prime Minister under Mrs. Gandhi. The center of the Right's strength in the government was the Ministry of Finance, which was particularly concerned with the balance of payments problem, and desirous of attracting large-scale foreign private capital investment and Western governmental aid to help overcome that problem.

The Center, which during most of the Nehru regime was the dominant group, never maintained a clear-cut position, but tried to balance the various considerations and the conflicting forces. Thus, while it may be said that it ideologically leaned toward public ownership over private, and development of indigenous energy resources over relying on foreign ones, the Center was also interested in attracting private foreign investment and gaining economic aid from the West as well as the East. Moreover, the Center itself contained a mixture of views and changed in response to its changing members and to changing external forces, both in the national and international arena. Leadership of this shifting Center resided in Prime Minister Nehru himself until his death.

However, nationalist feeling sometimes cut across economic ideologies and interests within the government. Thus, for example, the Left could often win support from the Center, and sometimes even the Right on the question of developing indigenous petroleum, by appealing to national pride and interests; moreover, some elements of the Left would ally themselves with the Right when it came to supporting Indian entrepreneurs over foreign corporations. In addition, bureaucratic interests within the government could be of some significance: for example, the atomic energy technicians might favor reliance on the West because they could obtain more nuclear assistance from there, or the Ministry of Irrigation and Water Power might favor expansion of the public energy sector simply because hydroelectric power falls in the public sector. This is simply to indicate that, particularly in the late 1950s and early 1960s when the balance of forces appeared most fluid, the situation analyzed in the following chapters was even more complex than is indicated by the groups and elements mentioned above.

CHAPTER FOURTEEN

The Seeds of Conflict: Evolution of the Indian Oil Industry in the 1950s

PRIOR TO AND EVEN FOLLOWING upon India's gaining of independence in 1947, control of the oil industry in India lay overwhelmingly with private foreign companies, largely utilizing foreign physical resources. Of the 20 to 25 million barrels of oil consumed in the early 1950s, over 80 percent was imported refined products derived from crude oil produced and refined primarily in the Middle East.

As befitted the legacy of British colonial rule, the Indian oil industry was dominated by British corporations. Burmah-Shell, whose shares are equally divided between Burmah Oil Company (of Scotland)[1] and Shell Transport and Trading (a branch of the Royal Dutch Shell group) sold well over half of all petroleum products consumed in India. Additionally, Burmah Oil, which produced and refined a small amount of crude oil in Eastern India, controlled about 10 to 15 percent of the product market. The other major companies were Standard Vacuum Oil Company (a 50–50 partnership of Jersey and Mobil) and Caltex (a 50–50 venture of Standard of California and Texaco).

During the early years of the Indian First Five Year Plan (1951–55), oil consumption and consequently imports of refined petroleum products increased rapidly. By 1954, when consumption equaled 31 million barrels, the cost of imported petroleum products amounted to almost $200 million, or 15 percent of India's total import bill; in fact, petroleum imports were three-fourths as great as food imports.

Mainly because of the large and growing drain on foreign exchange, the Indian government successfully pressured the three largest marketers into building refineries in India, thereby reducing the oil import bill through substitution of imported crude oil for the more expensive refined products. Standard Vacuum built a refinery with a capacity of 9 million barrels a year (1954), Burmah-Shell a 15-million-barrel refinery (1954), and Caltex a 5-million-barrel refinery (1957).

Building these indigenous refineries basically represented an involuntary or undesired step for the companies. The advent of the refineries did not change product prices in India, since these were set as equal to product prices in the Middle East plus freight, insurance, and port charges necessary to bring the products into India (i.e., "import parity"). Hence, the only change in profitability for the companies would be the addition of profits derived from their Indian refineries minus the reduction in profits for their Middle-Eastern or Indonesian refineries. Since, as we have seen, the larger Middle-Eastern and Indonesian refineries would be more efficient than the smaller ones built in India, it is clear that the net impact of the change would be to reduce the companies' profits. In addition, the companies had to make a sizable investment of close to $100 million in India, a country at that time considered susceptible to revolution and nationalization.

Why then did the companies agree to build these refineries? There are essentially three related answers to this. First, the companies saw India as a promising growth market for oil and that, given the foreign exchange problem posed by importing refined products, ultimately they might jeopardize their whole position in India if they did not accede to the government's desires. Second, as long as they could still import their own crude oil, the reduction in profits emanating from the change was still worth accepting, when weighed against the possibility of being excluded from the Indian market.[2] The companies would still make about $30 million in after-tax profits on the sale of the crude oil moving through the three refineries.[3] Thus, while they might expect to earn only about 10 percent per year on their refinery investment from refining operations as such, when the refinery is viewed as necessary for getting the crude oil profits, its profitability rises to about 40 percent per year, which is quite attractive indeed. Third, the companies had the legal protection of refinery agreements negotiated between the Indian government and themselves. The essence of these long-term agreements provided that each company had the right of importing and refining oil from its own sources and that the companies could not be nationalized.[4]

In any event, the completion of major refineries by Burmah-Shell and Standard Vacuum during the First Five Year Plan involved a significant change in the character of the Indian oil industry. Whereas in the early 1950s only 10 to 15 percent of all products sold in India were refined indigenously, by 1955 the proportion was almost three-fourths, even though yearly demand had increased from 20 to 25 million barrels to 35 million barrels. Also, foreign investment in petroleum facilities in India jumped from $47 million in 1948 to $218 million in 1955; petroleum's share of total foreign investment in India increased in this period from 8 percent to 24 percent and actually surpassed foreign investment in plantations (such as tea), which had historically been the major foreign sector in India.[5]

During the course of the Second Five Year Plan (1955–60), Indian government pressures on the international oil companies intensified. Oil's negative impact on the foreign exchange situation was an increasingly serious underlying problem, for two reasons. First, India had succeeded in overfulfilling the targets of the First Five Year Plan.[6] Buoyed by this success, the Second Five Year Plan envisaged much more ambitious targets and, in the course of pursuing this accelerated growth, imports increased rapidly, leading to a major deterioration in India's foreign exchange position; between 1955 and 1960, imports rose from $1.4 billion to $2.3 billion, while India's gold and foreign exchange reserves plunged from $1.9 billion to $0.7 billion. Second, increasing industrialization and urbanization fueled a rising demand for petroleum, with consumption increasing from 35 million barrels in 1955 to 54 million barrels in 1960.

A second cause of government pressure arose from what is known in international business circles as the problem of "local equity." Because this problem has also played an important role in sharpening antagonism between national governments and foreign oil companies in many other underdeveloped countries, it is worth analyzing in some detail the developments in India.

The general background to the problem may be summarized as follows. Governments of underdeveloped countries, particularly ones which like India have recently emerged from colonial status, frequently sought to promote the indigenous private business sector. This has been a goal both because the indigenous businessman was often a key supporter and/or member of the new governments, and because such promotion strengthened indigenous control of the economy. Providing governmental assistance to native entrepreneurs was not only a way of rewarding government supporters, but it frequently served to unite Left, Right, and Center within the country in pursuit of the goal of increasing indigenous control of the national economy. Even the Left in a country could be expected to favor a change from private foreign control to indigenous private control of resources (if these were seen as the only alternatives).

Hence, it is a widespread phenomenon in underdeveloped countries that nationalist governments have frequently sought to further the role of indigenous businessmen at the expense of foreign investors. Much of this favoritism involves behind-the-scenes pressure, since any nation which, like India, seeks substantial amounts of foreign aid and/or foreign private capital, cannot discriminate too openly. Nevertheless, the pressures are no less real for being subsurface.

The specific goal of national government pressures has usually been to induce foreign private capital to enter into partnership with local businessmen. Preferably local businessmen would have the majority of equity stock, but at least they should be 50–50 partners or at the very minimum have a significant minority equity position. The theory of the national governments is that the foreign companies will provide capital, particularly scarce foreign exchange, and technical know-how, while local businessmen will contribute indigenous currency capital along with knowledge of local conditions; both groups will share in the profits of the business. A common long-range goal of the governments is that this type of partnership will help to develop a larger indigenous business group and technical class.

Occasionally, this policy is enunciated fairly explicitly by a government:

> On August 15, 1958, the Indian Government explained that . . . While majority Indian participation is generally favored . . . the Government stated that in suitable cases it will allow foreign investment as the bulk of the capital structure of a new enterprise provided Indian participants can exercise effective influence on the company to develop Indian "know-how." [7]

Whether the Indian government's policy was explicit or implicit, however, foreign investors could hardly have been unaware of it, as indicated by an official United States publication on "Investment in India":

> Investment Policy: . . . As a rule, the major interest in ownership and effective control of an undertaking should be in Indian hands. . . . the Government urged Indian business to seek collaboration with foreign companies able to supply capital equipment as equity participation in a project.[8]

The major weapon which India like other underdeveloped countries had in pressuring private foreign corporations to enter into joint ventures is the requirement that virtually all significant new capital investments or enterprises need governmental approval. The government's refusal to grant a license or to provide "reasonable" terms for the license, or simply its stalling on the licensing negotiations, can effectively forestall any foreign investment project. As a result, foreign companies seeking to invest in India have generally found it the better part of valor to make joint venture arrangements with Indian businessmen or, if necessary, with the government.

The established international oil companies, however, in India as elsewhere in the underdeveloped countries, have usually been the most tenacious holdouts to any policy of allowing any local equity in their operations. Indian government interest in obtaining local equity in the oil industry dates at least from the early 1950s when the companies were building their indigenous refineries, which clearly would be a crucial sector of the economy. Moreover, the need for specific refinery licensing agreements between the companies and the government gave the Indian government bargaining leverage.

In any event, whether due to government pressure or their own desires, in the course of negotiating their refinery agreements, each of the established majors agreed to allow the Indian public to buy *preferred* stock in the new refining companies. Standard Vacuum early issued $1.6 million worth of such preferred stock to the Indian public, but neither Burmah-Shell nor Caltex followed suit. One reason was that in later years the Indian government was bringing pressure for common stock rather than preferred stock for the Indian public. This was because preferred stock is vastly different from common stock equity. The preferred shareholder is promised a fixed return on his investment, e.g., 7% per year, but gets no more than that and does not share in the general profitability (or losses) of the company. Moreover, ownership of preferred stock gives the investor no voice or participation in the actual operations of the company. In these crucial respects preferred equity is closer to debt capital than it is to common equity; one gets a relatively safe fixed income, but no power nor chance of large gain.

It should be understood that the question of whether to allow local investors to share through common stock in the various phases of the established majors' operations in India was continually under the companies' active consideration. Yet, while foreign companies operating in other industries in India were frequently giving a majority of the common equity to local people as early as the 1950s, to this day not one of the three majors has allowed any local common equity in its Indian refining or marketing operations. The reasons for the different position taken by foreign oil companies as opposed to foreign companies in other industries are worth examining.

For foreign investors the major practical advantages from allowing local equity are the following: (1) Local equity is often necessary in order to get government approval for the initial investment or expansion. (2) Local equity may bring in "local know-how"—either knowledge of the country's market or, frequently more important, indigenous political influence. (3) Local support and political influence generated by local equity may reduce the chances of nationalization. (4) Local equity reduces the amount required to be invested by the foreign company.

The basic disadvantages of local equity are: (1) It reduces the profits which can be made by the foreign company—if the investment is considered to be a "good thing," there is usually no desire to give up a share of it. (2) Local owners, even if they do not have a majority of the equity, can have a voice in the operations of the company.

In the case of the international oil companies, all of the advantages listed above were foreseen, with the first being the most important consideration. The first disadvantage of giving up part of the profits was not insurmountable, since the refining and marketing operations were not inordinately profitable; moreover, part of a good thing is better than none of a good thing. The most important factor, then, in leading the companies to reject local equity was the second disadvantage—the possibility of local interference with the company's operations.

Specifically, the greatest fear was that minority stockholders would raise questions about the prices paid by the indigenous refining companies to their parent companies, particularly for crude oil. After all, local investors would not participate in the profits of the parent companies from their overseas crude oil. Therefore, the profit-maximizing interests of local investors would lie in importing the lowest-cost crude oil possible, from anywhere in the world. In pursuit of this, Indian investors might pressure management as to why their company was paying such high prices for its crude oil, compared to prices paid by independent refining companies in Japan or Western Europe, or they might even demand that the companies import the low-priced oil offered by the Soviet Union. Local investors might also raise questions about the freight charges billed by the parent companies to their Indian affiliates for transporting oil in tankers controlled by the parent companies. Possible minority stockholders' court suits on these questions might be very embarrassing; even if successfully defended against, such suits might lead to general ill feeling against the companies as well as the necessity of disclosing pricing and cost data. Finally, even if none of these problems ultimately developed in India, an offer of local equity there would generate pressure for local equity in other countries in which these problems might arise.

In sum, our basic thesis is that the fundamental objection of the established international oil companies to granting local equity in refineries was fear of interference with the most profitable and most sensitive sector of their business, namely, the sale of their overseas crude oil. It is interesting to compare this interpretation with that presented in a leading book on development economics:

> Solid business reasons may also militate against joint ventures. The major oil companies of the Western world resist joint venturing mainly for two reasons: 1) because oil exploration is a risky business attended by many failures which a large organization with world-wide affiliations and capital reserves can sustain better than local interests concentrated on one venture only; and 2) because the requirements of established world-wide operations and markets cannot be taken into account in joint ventures in individual, less developed countries.[9]

This second point, while rather vague, may be interpreted as an oblique statement that worldwide profit maximization may be incompatible with joint ventures. Thus it may be a euphemistic way of alluding to the various conflicts of interest, particularly in transfer pricing, between parent and affiliate in international oil companies. However, the first point, which implies that the international oil companies refrain from admitting local equity into oil exploration in order to protect the local investors from risk, is completely wrong. Such an approach would be totally inconsistent with the profit maximization goals of the international oil companies, which, everything else being equal, seek to monopolize riskless investments and share risky ones.

Moreover, in the case of India at least, the facts are inconsistent with this interpretation. As we shall see later, the *only* local common equity participation granted by the international oil companies in India was precisely in the area of joint ventures for oil exploration. The logical interpretation of this, we think, is that since oil exploration is the relatively risky part of the oil business, there is greater willingness on the part of the foreign companies to share the equity with local capital. This is particularly true since the established companies have relatively little to gain from such exploration. On the other hand, in India as in most underdeveloped countries where the demand for oil is growing rapidly and where there are shortages of oil products and/or imports of refined products which can be replaced by domestically refined products, an indigenous refinery is a relatively riskless business (at least with regard to market demand fluctuations). Moreover, as we have seen, it is often a necessity for realizing crude oil profits.

Thus, despite the fact that throughout the 1950s it was well known that the Indian government desired local common equity for its citizens in the Indian refining (and marketing) companies, no such equity arrangement was ever made.[10] Moreover, there is no doubt that Indian investors would be eager to have such a share in the Indian oil industry. It was believed that a common stock offering to the Indian public for either the marketing or the refining branch of an international oil company in India would be oversubscribed many-fold. Support for this view can be seen from the response to Jersey's 1963 offer of equity stock in its Malayan refining and marketing company:

> Esso Standard Malaya Ltd., Esso Eastern's affiliate in the Federation of Malaya, concluded in April its first public sale of stock. The company's offer of approximately $5.8 million in common stock and $2.3 million in redeemable debenture stock was immediately bought up by the public in Malaya, Singapore, and the British Borneo territories. . . .
>
> The offering, easily the biggest ever made in the area . . . remained open for one hour only. The response was overwhelming. More than 150,000 applicants from all parts of Malaya, Singapore, and Borneo subscribed in excess of $97 million. This

surpassed the total raised by the three largest of all previous stock issues by any other company.[11]

In this period, the unwillingness of the international oil companies operating in India to provide local common equity thus alienated both the Indian Government and the Indian business community.[12] While not a decisive point of conflict between these companies and the government, it undoubtedly contributed to a background of tense relations. These relations were to come to the fraying point over the question of "Soviet oil," which we turn to in the next chapter.

CHAPTER FIFTEEN

India and Soviet Oil

By THE END of the Second Five Year Plan government-company relations became inextricably bound up with the whole problem of "Soviet oil." By 1960 the Indian government had largely succeeded in getting the established majors to import crude oil rather than refined products, thus saving considerable foreign exchange. At the same time, however, the rising demand for oil in India threatened to more than offset this saving, thereby increasing the total oil import bill. Moreover, as we have seen, India's foreign exchange position had been deteriorating rapidly.

Faced with this problem, the government had five basic alternatives: (1) Attempt to suppress the growth in oil demand, either through the price mechanism or through limiting imports; (2) seek to find crude oil within India; (3) force a reduction in the prices paid for imported oil by the established majors to their overseas affiliates; (4) import oil on a barter basis, involving no outlay of hard foreign currencies; (5) obtain Western foreign aid to purchase oil. At one point or another the Indian government worked along each of these lines. Considerable effort, however, was devoted to (3) and (4), which, as we shall see, soon became closely interrelated. Because the Soviet Union was the only practical source of barter oil, the pressures of the Indian government were brought to bear on the companies to import Soviet oil.

The predicament of the established majors came into the open in mid-1960, when it was revealed that the Soviet Union had offered to supply India with large quantities of low-priced crude oil on a barter basis. As reported in an Indian newspaper at the time:

The Soviet offer is said to be at a price which is twenty-five per cent below what the British and American companies charge. . . .

The Soviet offer has other solid advantages. As in the case of earlier transactions of different kinds, the Soviet Union is agreeable to accept payment in rupees and to spend it in purchasing Indian goods. India cannot possibly expect more favourable terms at a time when her urgent need is to save foreign exchange.[1]

The Soviet offer was undoubtedly tempting to the Indian government. Specifically, the Russians offered to supply some 18 million barrels of crude oil per year or about half of India's crude imports at that time. Despite the long haul from the Black Sea to India, the Soviet crude oil was offered on a delivered basis at $1.81 per barrel CIF Bombay or $0.25 a barrel cheaper than the companies' CIF price.[2] Thus the Soviets were offering to sell oil to India at a price which would immediately save India almost $5 million per year. Even more important, since the Soviet oil was to be paid for completely with rupees (ultimately by barter trade), the foreign exchange saving would amount to the full cost of the displaced Western oil or some $33 million per year. Despite the fact that the Soviet offer was clearly an attractive one for India, to this day not one drop of Soviet crude oil has ever moved through an Indian refinery.

At about the same time, the Russians also offered to sell refined products to India at a delivered CIF price substantially below that of the companies.[3] In July, 1960, following the companies' refusal to handle Soviet crude oil, the government-owned Indian Oil Company signed a contract to import Soviet products on a barter basis (particularly kerosene and automotive diesel oil, which were in short supply).[4]

In the light of the vast potential of this refined product trade for India, the actual amounts of Soviet products imported into India have been relatively small. Between 1961 and 1965 India imported a total of 10 million tons of refined oil fuels, of which Soviet products accounted for less than one-fourth.[5]

The principal reason has been that the established majors,

who controlled most of the marketing and distribution facilities in India, have adamantly refused to handle any Soviet products in their marketing networks; this would be even worse than handling Soviet crude oil, since it would not only eliminate the companies' crude oil profits but also their refining profits. In addition the Indian government seems to have moved very slowly to develop its own marketing facilities. Thus, the Third Five Year Plan (covering 1961–65) allocated only $21 million for this, or about a tenth of what the private companies have invested.

The inability of the Indian government to take full advantage of Soviet refined products was potentially even more costly than its inability to utilize Soviet crude oil. In value terms, about three-fifths of India's oil imports in the early 1960s have consisted of refined products; in the 1961–65 period India imported close to $500 million worth of refined oil products. Yet, until recently, most of these refined product imports had to be paid for with scarce hard currency.[6]

The Soviet crude oil offer brought into action virtually all of the forces involved in oil and energy policy considerations. Because it came at a particularly crucial point in Indian and world history, the offer and its rejection make a fascinating case study in the political economy of international oil. Superficially the Soviet crude oil offer involved simply the proposal of a deal between two sovereign governments. Below the surface, however, it triggered and involved continuing political and economic struggles among nations and institutions. It is important for a full understanding of the events that followed to recall some aspects of the historical setting within which this struggle took place.

First, following the upsurge in worldwide oil exploration and discovery triggered by the Suez crisis, the period after 1958 was one of declining oil prices all over the world. Second, mid-1960 was just prior to the beginning of the Third Five Year Plan in India, with India in a position of severe foreign exchange shortage and increasing dependence on foreign aid for its industrial programs. Third, the Soviet barter oil offer to India came almost precisely at the same time as a similar offer to Cuba—an offer which was rejected by the Western oil companies operating there and which led directly to their nationalization by the Castro government. Fourth, the offer came at a time of particularly great tension in the Cold War between the Soviet Union and the United States; in May 1960 the shooting down of a United States U-2 plane over the Soviet Union led to the termination of the Paris summit conference between President Eisenhower and Premier Khrushchev. Fifth and finally, within India the position of the ruling Congress Party had been weakening, owing partly to the lack of success in the almost completed Second Five Year Plan.[7] These factors should be borne in mind as we examine the problems and opportunities posed by the Soviet barter oil offer for each of the major groups involved.

The Soviet Union may well have had some direct economic motivation for its offer to barter oil for Indian goods. Nevertheless the background to the Soviet offer indicates that it was at least partly aimed at increasing Soviet influence in India. The first Soviet action potentially affecting Indian oil was the conclusion of a Soviet-Indian trade agreement in 1953, which included crude oil and refined products as possible imports into India; however, such trade was moot at the time, since there were neither indigenous refineries nor state marketing facilities to handle Soviet oil. Thus, the first really significant Soviet action came in 1956 with the granting of a long-term credit to India of about $125 million to purchase Soviet industrial equipment. Included in this credit was the money for building the second Indian government-owned refinery in Bihar; Romania provided the credit for the first one, in Assam. Not only were the Russians to provide funds for all necessary equipment, but in addition they were to provide technical assistance for constructing the refinery and for initially running it while simultaneously training Indian nationals to ultimately take over operations. Finally, during this pre-1960 period the Russians (and the Romanians) were providing the Indian government major assistance for its oil exploration program.

Set against the background of this Soviet oil assistance,

which contrasted strikingly with the established majors' continuing struggles with the government over such issues as local equity, the Soviet barter oil offer placed the companies in a very embarrassing position. The only available facilities for processing Soviet crude oil in India were their refineries. However, while the companies were not anxious to come into open conflict with the Indian government, there was very little likelihood of their ever agreeing to handle the Soviet oil. The basic reasons for this should be amply clear by now. They stem from the fundamental economics of the international oil industry wherein most of the profitability lies in the crude oil, rather than in the refining or marketing process. Moreover, even if the oil companies were willing to accept the partial loss of profits from using some Soviet oil in India, such a step could have dangerous worldwide repercussions; it would open the door for many other countries, similarly hard pressed for foreign exchange, to demand that the companies use Soviet crude oil in their indigenous refineries.

What would be the possible gains to the companies from agreeing to handle Soviet oil? An obvious one would be elimination of the threat of nationalization of their refineries and marketing facilities, the ultimate weapon that could be used by the Indian government to force them to handle the Soviet crude. A second gain would be that the government might grant their requests to expand their existing refineries or build new ones. On the other hand, with the real profitability coming from refining one's own crude oil, there would be no great benefit to expanding refining capacity which would use "profitless" Soviet crude oil. Moreover, even if the ten percent per year type of profit likely from refining activities alone were acceptable to the companies (at least when combined with the hope that in the future they might be allowed to import more of their own crude oil), investing more money in additional refinery capacity would make the companies even more vulnerable to the threat of nationalization; there would be a grave danger of simply throwing "good money after bad."

Thus, the basic gain to the companies from accepting Soviet crude oil ultimately boiled down to avoidance of the possibility of nationalization. As we shall see, the oil companies had good reason to believe that in India, at least, nationalization would not follow their refusal. However, even if nationalization were a distinctly possible response, it is very likely that the companies still would have acted as they did. After all, while the total investment of foreign oil companies in all phases of the petroleum business in India was about $300 million in 1960, all of this investment was "sunk," either in physical facilities or in the minimum rupee holdings required for business and/or political reasons. Hence, from the companies' viewpoint the practical loss from nationalization would be the value of that part of the earnings stream generated by internal operations (i.e., excluding crude oil profits) which might realistically be remitted abroad over the years. Reported total internal earnings had been declining, going from $35 million in 1956 to $23 million in 1959.[8] Stacked up against this would be the minimum profitability to the companies from importing their own crude oil, about $30 million per year, plus $15 million per year profits from importing their own refined products. Thus, leaving aside the potential losses from the "demonstration effect" of accepting Soviet oil, even if it were considered that the probability of nationalization was as high as fifty-fifty, rational profit-making calculations by the companies would lead to a decision to reject the Soviet offer. Since the profits of the international companies from their Indian operations were only a small part of their worldwide profits, adding in the danger of establishing a precedent makes it clear that rejection of the Soviet offer was a relatively easy decision for the companies.

Additional proof of this can be seen from the fact that in the same month the same three foreign oil companies involved in India—Jersey, Shell, and Texaco—rejected a demand of the Cuban government to process Soviet oil. The amounts at stake in Cuba were similar to those in India, and the Cuban government's response to the companies' refusal was nationalization. (See Chapter 24.) Despite this, the companies continued to refuse to handle Soviet oil in India.

Since the established majors would not agree to handle imported Soviet crude oil (or products) they were left with two basic alternatives. They could do nothing at all, thereby forcing a naked showdown with the Indian government, or they could offer some other proposal to help save foreign exchange for India, thereby softening the impact of their refusal. The most basic step which they could take and did was to lower the prices charged for the crude oil they imported into India, to meet the delivered price offered by the Russians.[9]

At the same time that these price moves were made, the companies had to take additional steps to meet the Soviet threat. A first prerequisite was that there be a united front in dealing with the Soviet challenge in India. Otherwise, the Indian government might play off one company against another, for example, by offering to allow major expansion to that company willing to process some Soviet crude oil. The problem was considerably simplified by the fact that each of the three majors in India faced a similar situation; their basic profits derived from supplying their own crude oil, and all had worldwide interests which would be threatened by the potential repercussions of yielding to Soviet oil within India. Thus, it may well have been sufficient for each company to make its own decision to realize that the others were likely to take a similar position. Nevertheless, coordination was certainly in the minds of the oil companies, and it is not unlikely that they jointly worked out a common policy. Suffice it to say that in fact all three companies adopted the same essential policy vis-à-vis the Soviet Union's crude oil offer in India.

A second prong of the oil companies' effort was to attack Soviet crude oil as "political oil." This was part of the previously discussed continuing worldwide effort to discourage countries from importing Soviet oil. As we have seen, the heart of the companies' argument was that Soviet barter oil was "subsidized" for political reasons, i.e., to "penetrate" the underdeveloped countries and make them dependent on the Soviet Union for this crucial commodity. The implication was that the Soviets would be a dangerously unreliable source of oil supply as compared to the private oil companies which could be relied on to supply oil regardless of political considerations. At this point it suffices to quote one Indian newspaper to clearly indicate how little weight this argument carried in India:

> The oil companies may give the stock answer that the Soviet offer has political strings, visible and invisible. But, according to expert British opinion, these companies themselves no longer believe in that answer. The bulk of Soviet oil exports has gone to West European markets. Soviet political strategists must be more than ordinarily dense and naive if they believe that cheap oil would win political friends for them in West Germany, Sweden, France, or Italy. Even in Cuba, where there is room for the play of political motives, the Soviet Union is getting much needed sugar in return for oil. From West European countries the Soviet Union is getting equally urgently needed capital equipment. . . .[10]

The same newspaper also stated:

> One may concede that the British and American companies in India have acted from purely commercial considerations without political or ideological overtones. But whatever the motives, India must judge both the Soviet offer and the British and American attitude from the point of view of her own national interests.[11]

A third part of the companies' effort was to convince India of the large amounts of foreign exchange the oil companies had already saved India, not only by their price cuts but particularly through their building of indigenous refineries. The companies stressed the additional foreign exchange savings which could be gained from allowing them to increase their refining capacity in India. Finally, a fourth and more significant line of attack for the oil companies was to seek the aid of their home governments and major international institutions such as the World Bank in helping to persuade the Indian government not to "force" Soviet oil upon the companies.

There is no question but that the United States and British governments had good economic as well as obvious political reasons for wanting to aid their respective oil companies in blocking the Soviet oil drive in India, as well as in other countries.

The Soviet oil offensive was economically harmful to these governments because it was hurting the interests and profits of its important national companies, which could also have serious negative effects on the balance of payments. Moreover, insofar as the Soviet oil drive in India brought closer relations between India and Russia, it tended to further the public sector at the expense of the private sector; thereby, it hindered the possibilities for all Western companies, not just those in the oil industry, to invest in India.

At the same time, the Soviet oil drive presented a difficult problem for the Western governments. While they had important levers of power in India, particularly through their foreign aid programs, the situation was delicate. After all, the Soviets were also offering foreign aid to India. Too much pressure by the Western governments, or the wrong timing, might backfire and strengthen the hand of the Left in India, thereby pushing the country toward the Soviet Union and away from the West.

Such an event would have been a serious blow to the United States particularly, which foresaw great potential for foreign investments in India at this time:

> According to U.S. Ambassador Ellsworth Bunker: now is the time of opportunity for private industry to combine opportunities for legitimate profit with a challenge of building a greater industrial nation.

This quotation, appropriately enough, comes from a magazine first published in mid-1960 by the International Relations Department of the United States Chamber of Commerce, entitled, *World Challenge: A Bulletin on the Communist Offensive as It Affects American Business in World Affairs*. *World Challenge* spells out the United States' view in more detail:

> The climate for participation in India's industrial development was never more favorable, according to a recent U.S. Trade Mission to Bombay and Western India.
>
> Among the many attractions which India now holds for foreign investors are:
>
> —A hugh [*sic*] reservoir of easily trainable manpower.

—A large body of private industry with good managerial skills and initiative.

—Immense resources of fertile land not yet intensively cultivated and awaiting only the application of water, fertilizer, and modern agricultural methods.

—Large resources of iron ore and coal situated in close proximity.

—Rich natural resources of manganese, thorium, water power.

Investment incentives in India include these factors:

—India has honored its commitments on the repatriation of profits and capital arising from foreign investment; Indian laws provide equal protection for foreign investors.

—Foreign companies are now permitted to own more than 51% of the capital of new companies.

—Approval of private investment in sectors of the economy hitherto reserved for development by the public sector.

—Favorable tax depreciation incentives.

—Signing of an investment guaranty pact with the U.S., conclusion of an agreement for the avoidance of double taxation, and availability of the Export-Import Bank line of credit.[12]

One proposed solution to the problems raised by the Soviet barter oil offer in India was for Western government (primarily United States) financing of oil imports (crude and/or products) into India. This idea was expressed in almost identical words in two Indian newspapers:

It is a hard doctrine to accept that three foreign companies, by refusing to refine Soviet crude oil, should be in the position to impose a big burden on the country's balance of payments. It is suggested that the easiest way out would be for the West to match the Russian offer by a medium-term loan covering this country's oil imports during the next few years of critical foreign exchange shortage.[13]

*   *   *

It has been suggested that a way out would be for the western countries to match the Soviet offer by a medium-term loan covering India's oil imports during the next few years of critical foreign exchange shortage. Failing some such gesture, India's needs are bound to come into conflict with western commercial interests more and more sharply as the years pass.[14]

There was ample precedent for this under the Marshall Plan (ECA) when United States loans were made to Europe to import American crude oil and products. However, this experience had led to considerable friction between the United States government and the private oil companies, particularly over oil pricing.[15] In the light of this, the oil companies were fearful that government financing of their oil imports into India might lead to pressure by their home governments either to reduce their prices or to buy crude oil from outside their own affiliates (excluding of course the Soviet Union as a possible supply source).[16] Hence, at a minimum the companies wanted modifications in the foreign aid financing regulations which would allow them to continue importing oil from their own sources at their own prices.[17] Even this, however, was considered a last-ditch desperation measure.

The United States government, on the other hand, also appeared reluctant to get into the financing of oil imports into India. One problem was that aid to India was generally conceived as providing assistance for specific capital projects, rather than for raw materials or other supplies necessary to run capital projects. Perhaps more important, officials of the United States foreign aid agencies feared getting embroiled in an oil-pricing controversy.[18] Since relations between the Indian government and the oil companies were quite strained, this could endanger the whole U.S. foreign aid program in India and the overall relations between the United States and India.

For these various reasons, it is not surprising that both the companies and the United States government never had any great enthusiasm for using foreign aid money for importing oil into India. In any event, no such money has ever been so used, despite the fact that increasing amounts of Western aid to India have been used for importing raw materials as opposed to equipment for projects.

Aside from Western governments, one of the most important international forces affecting the Indian government and the Indian economy is the World Bank. About the time of the Soviet crude oil offer, World Bank loans to India equaled about $0.5 billion, or almost four-fifths of India's total holdings of gold and foreign exchange. Moreover, India was then drawing up plans for her Third Five Year Plan (1961–65) which envisaged about $5 billion in foreign aid from the Western countries, with the World Bank expected to play a significant role both as lender and as coordinator for the Western countries. Hence, on any such critical issue as importing Soviet crude oil, the Bank would have to be influential—at a minimum in the sense that the Indian government's knowledge of the Bank's position would be a silent but powerful operative force.

Since the question of Soviet oil in India was closely related to the whole question of oil pricing and the role of private foreign enterprise in the Indian oil industry, there can be little doubt that the Bank would stand strongly opposed to forced introduction of Soviet crude oil. This would be so if for no other reason than that it would weaken the position of private enterprise as a whole in India. At the worst, if events led to another Cuba, the Bank's own enormous stake in India would be jeopardized. In any event, knowledgeable sources within the oil industry reported criticism by the World Bank and by Western aid-giving nations of the Indian government's growing role in the oil field. (In fact, as we shall see, this was probably one of the factors which led the Indian government to renew negotiations at this time with foreign oil companies for exploration rights in India.)

Finally, the Indian government, the focal point of these international pressures, undoubtedly was divided on the question of what to do about the Soviet crude oil offer. While we have little direct information on this question, it seems likely that all three ideological wings of the government would have wanted the oil companies to quietly accept the Soviet offer; after all, the Right and Center could see in it significant economic advantages for India, in terms of foreign exchange saved and expanded markets for India's exports. The main differences within the government must therefore have been on the questions of how hard to pressure the companies to handle the Soviet oil, and what response to take if they refused.

For the Indian business community, a crucial element of the Congress Party, there must have been considerable ambivalence. On the one hand, Indian businessmen would have a consumer's point of view toward oil, desiring the lowest-cost energy possible. Moreover, since during this period there were frequently production-halting shortages of energy, e.g., in the textile industry, the Soviet oil offer would also be attractive insofar as it offered the potential of increasing the energy supply. And, Soviet barter oil could release foreign exchange for use in importing other materials and capital goods for Indian industry. Finally, the Indian business community could have little affection for the foreign oil companies, since oil prices were considered high and in addition Indian investors had been excluded from participation in the profits of the Indian oil industry.

On the other hand, the Indian business community had to fear the negative effects of any attempt to force the oil companies into accepting Soviet crude. The dangers of a sharply adverse Western governmental reaction to such a move could not be underestimated, particularly in the light of the Cuban events. The fear of provoking a break with the West, and/or causing a sharp internal swing to the Left, undoubtedly outweighed the possible attractions of forcing the issue.

The tone of the Indian press at the time of the companies' rejection of the Soviet crude oil offer suggests grievance and helpless frustration:

> In refusing to handle Soviet crude oil, the three western oil refineries in Bombay may be justified by the letter of their contract with the Government of India, but they have thereby lost a large part of the good will they once had in India. . . . They ought to know that there is a world surplus of oil, that Russian competition, which is already strong, is bound to grow against them and make them more and more vulnerable, and that they can afford to do nothing that antagonises public opinion anywhere. Above all they ought to know that India in the interest

of her development plans is anxious to be on the best of terms with the Government and business interests everywhere. . . .

It is said that following the Soviet offer, the western oil companies have indicated their willingness to reduce their price for crude oil to something like the Soviet price. But they are prepared to go no further. . . . Their present price policy imposes on India's balance of payments a burden of about fifty million sterling [i.e., about $140 million per year] by way of cost of oil imports. This is what India will save by buying oil from the Soviet Union. If the three western companies now agree to accept a lower price in order to meet Soviet competition, they cannot escape the charge that till now they have been exacting from India a larger price than was justified on commercial grounds, and that they were taking full advantage both of their position as a world monopoly and of India's helplessness as against their combined power to dictate terms. What is really disturbing is that but for the Soviet offer, India would never have realised how heavily she is and has been overpaying for British and American oil. The western companies must be aware that their present refusal to co-operate with the Government of India means a delay of some two years in the Indian plan to increase supplies of oil.[19]

The extreme Left in India favored nationalization of the companies for their refusal to handle Soviet crude oil.[20] But, whether voluntarily or because outvoted by the Right and Center, a key Left leader in the Congress Party and the government announced the government's rejection of this approach:

Oil Minister Keshiva Deva Malaviya confirmed India plans no Cuban-style reprisals against three Western oil companies which have refused to process Russian crude oil in their India-based refineries.

Mr. Malaviya said India fully intends to live up to its obligation under a 20-year agreement signed with Caltex, Standard Vacuum, and Burmah Shell. . . .

Under an agreement signed in 1948 the companies have the privilege of importing and refining oil from their own sources.[21]

Instead, the reaction of the Indian government was to try to retaliate against and outflank the oil companies by moving on several fronts: (1) investigating the pricing policies of the companies; (2) providing increased support for an integrated state oil industry—with government petroleum exploration, building of oil refineries and a petroleum marketing organization (which could handle Soviet refined products); (3) refusing to allow the expansion of existing refineries or the building of new refineries by the three established majors; (4) seeking to bring newcomers into the oil industry in India—ranging from U.S. independents like Phillips Petroleum to state oil companies like ENI. These four sets of developments encompass most of the principal events in the 1960s. The analysis of each of these will thus bring us to the present status of the oil industry in India. Before proceeding to this analysis, we conclude this chapter with our estimate of the main effects of the Soviet crude oil offer and its rejection.

First, while rejection was a logical step for the oil companies in terms of both their short-run Indian profits and their long-run worldwide interests, it caused considerable negative reaction from the Indian government and the public. Second, the rejection of the Soviet crude oil offer by the companies placed the Indian government in a difficult position. Because it was essentially committed to a process of economic development through large-scale reliance on foreign assistance, particularly from Western governments, the Indian government was unwilling and/or unable to take extreme measures in retaliation for the oil companies' action. On the other hand, it tried to achieve some of the economic benefits which would have been obtained from using Soviet oil through a series of smaller retaliations which did not involve a dramatic head-on confrontation.

Third, Western pressure, whether explicit or implicit, undoubtedly played a major role in the Indian governmental decision. For, even countries with much stronger bargaining positions than India have yielded, as indicated in a 1962 petroleum trade journal report:

On the brighter side, Bolivia has refused a Soviet offer of $150 m. [million] in credits for financing purchases of machinery for oil and tin production. And the Japanese government, under pressure from Washington, has resisted tempting offers from Moscow which would have linked 60 m. bbl. [60 million barrels] of cut-price oil a year with Japanese supplies of oil pipeline.[22]

Nevertheless, it should be noted that the rejection of the Soviet crude oil offer in 1960 did not end the matter. In fact, the standing Soviet offer to provide cut-rate barter oil was a continuing irritant to government-oil company relations, through the early 1960s at least.

CHAPTER SIXTEEN

Oil Prices: Indian Government Pressure in the 1960s

THE SOVIET CRUDE OIL OFFER of mid-1960 and its rejection by the companies was the basic cause of the formation of an Indian government committee to investigate the problem of oil pricing:

> . . . in August 1960 the Indian Government appointed an official committee to look into oil pricing. The Committee, which was known as the Damle Committee after its chairman, was to determine the future basis of oil prices and to look into the manner in which the companies were conducting the business of refining.[1]

The findings of the Damle Committee and the debate between the companies and the government over them are worth analyzing in some detail. The Committee tried to come to grips with all the major aspects of pricing which still trouble government-company relations in India and other underdeveloped countries today. In addition, the implementation of the Committee's findings shed a good deal of light on the limits of Indian governmental power with regard to the companies.

The ultimate findings and recommendations of the Damle Committee[2] covered three main areas: (1) crude oil supplies for Indian refineries; (2) prices of refined products; (3) allowable expenses for marketing and distributing products in India.

In the first area, the Committee found that crude oil prices were being sharply discounted by many private companies in sales to other countries. In addition, it found that crude oil was available on an even more advantageous basis from the Russians. The Committee therefore "recommended" that the companies: (1) buy crude oil from the lowest price suppliers, or (2) force their own suppliers to cut crude oil prices to the lowest competitive price, including that of the Russians, (3) eliminate the intermediate purchasing companies between the producing end of an international oil company and its refining affiliate in India, i.e., have the refinery in India buy directly from the producer in the Middle East.[3]

The response of the oil companies to each of these recommendations regarding crude oil purchasing practices was negative. Essentially the companies argued that the discounts given by their producing affiliates to their Indian refineries were "competitive." This was because the only cheaper oil available was "politically motivated" Soviet oil or "small" amounts of "commercial oil" both of which ostensibly were available only "temporarily" rather than on a long-run basis.[4] That is, the oil companies argued that Soviet oil was not comparable to "free world" commercial oil because it was an "unreliable" supply; since heavily discounted "commercial oil" was available only on a temporary basis, the companies did not feel obligated to meet these lower prices either.[5]

What were the facts of the matter? Generally speaking, there was no large-scale "discounting" of world crude oil prices until the reopening of the Suez Canal in mid-1957. Subsequent price developments in the 1957–61 period have been ably summarized by a Venezuelan oil economist in 1961 as follows:

> During the second half of 1957 and 1958 there were a number of downward adjustments in US oil prices. . . .
> The wave of crude price reductions which had started in Texas spread to the Middle East via the Caribbean. . . .
> The competitive price cuts in themselves did little to stem the volume of discounts being offered. . . .
> . . . A growing rivalry between companies was apparent which was further intensified by the appearance of the international "minors." . . . the small companies have proceeded to penetrate the market by the only method available to them— price discounts. The independent buyer in the interests of obtaining still lower prices has fanned the flames of rivalry by taking advantage of the imperfect state of communications within the industry. . . .
> In August 1960, there was a further round of price cuts.

Esso the price leader on this occasion, reduced prices by 4 to 14 US cents and alleged that discount sales were now so prevalent that they could no longer be considered marginal. . . .

From published sources discounts of from 15 to 30 cents a barrell are said to be available to most independent buyers in Middle East.[6]

From the above (and the author's own experience) it seems clear that the Indian refineries have generally not received the benefits of the discounts being given in other countries to independent buyers of Middle Eastern crude. The first discount off posted prices given to Indian refineries was about $0.15 per barrel of Arabian crude oil, and did not come until June 1960. Yet, all during 1959 and up to mid-1960 a discount of $0.11 per barrel was available to independent buyers of this same crude oil. It is true that the granting of discounts in India in mid-1960 brought Indian prices generally in line with going market discounts. However, from 1961 on, through 1962 at least, while the posted price for Arabian crude oil was $1.80 per barrel, it was available to independent buyers at $1.45, while the price to the Indian refineries was $1.62. Thus, even while the Indian refinery was receiving an $0.18 per barrel discount, the market discount was about $0.35; the Indian refinery was thus paying $0.17 more per barrel than independent buyers, compared to $0.11 more in 1959, indicating a widening gap.

Over the whole 1959–61 period, the average difference between the price paid by independent buyers and Indian refineries for Saudi Arabian crude oil was about $0.14 per barrel. During this period about 30 million barrels of oil were imported from Saudi Arabia, indicating that had the Indian refineries gotten the prices paid by independent buyers, India would have saved foreign exchange amounting to $4 million. Assuming the same situation existed for the 70 million barrels of crude oil imported in 1959–61 from other Middle Eastern sources (and Indonesia), the saving would have been $14 million during these three years. If, as appears likely, since 1962 Indian crude oil has been imported at a price 5–10 percent greater than independent buyers were paying, the additional foreign exchange cost of this crude oil has averaged about \$3.5–\$7.0 million per year.[7] Thus, the total additional cost from 1959 through 1966 would be \$30–\$50 million or enough to pay the complete cost of a modern 50,000-barrel-per-day refinery.

The Damle Committee also concluded that discounts on refined products produced in the Persian Gulf should also be available to the Indian marketing branches. It therefore recommended that the Middle East f.o.b. product prices used in computing the Indian product prices be reduced by 3 to 10 percent.[8] The conclusions of the Damle Committee were based on four factors:[9] (a) the belief that ultimately refined product prices were based on crude oil prices (plus markups for refining and transportation costs); hence, if crude oil were available at a discount, products should also be available at a discount; (b) the fact that discounts on some products had been received by at least one Indian importer from a Middle East supplier; (c) the fact that the Russians were supplying refined products at sharp discounts; (d) the fact that the established majors on occasion had underbid even Soviet prices on refined products.

The established majors' position was that none of these points supported the conclusion that products were readily available at discounted prices. As to the Damle Committee's implicit theory of refined product prices, they argued that it may be theoretically true in the long run, but did not necessarily determine product prices in the short run. They further argued that crude oil discounts rose because, since 1960, under the pressure of Middle Eastern governments (whose revenues are based on crude oil posted prices) the companies had not been free to adjust these posted crude oil prices to market conditions. On the other hand, they had been free to adjust posted product prices, so that there has been no need for discounting to meet competitive market conditions.

The fact that a Middle East supplier (CFP) had given a small Indian importer discounts on product prices was dismissed as evidence only of "marginal sales of a distress character." [10] The companies argued that Russian refined products, like Russian crude oil, could not be considered relevant since it too was "political" oil. Finally, the fact that the established majors in India had on occasion reduced their product prices below Russian prices was justified as a normal competitive procedure to hold on to existing customers.

Thus, the heart of the difference between the Damle Committee and the companies was a factual question as to whether or not discounts off posted prices on refined products in the Persian Gulf were "readily available." (The companies did not deny that Russian refined products were available at sharply discounted prices, but simply ruled them out of the relevant domain of discourse.) The facts of the matter are that while discounts off product prices were not as freely available as discounts off crude oil prices, they were available to knowledgeable independent buyers.

The reason for the lesser availability of discounts off posted product prices was only partly the fact that pressures of Middle Eastern governments prevented cuts in crude oil posted prices, thereby encouraging crude oil discounts. Perhaps more fundamentally, the fact was that crude oil was in relatively much greater oversupply than refined products; for one thing, at that time refineries in the Persian Gulf were operating at close to maximum capacity. More important, the fundamental and enduring pressure on crude oil prices did not arise simply from the discovery of prolific Middle Eastern oil fields. For even prior to this crude oil had usually been in relative oversupply in the sense that in a truly competitive market, i.e., one containing a large number of independent suppliers, crude oil prices would have dropped sharply.

The crucial new fact was that a significant part of the crude oil was discovered by newcomers who found themselves with large amounts of Middle Eastern crude oil but little or no refining operations outside the United States. Hence the newcomers could not themselves turn this crude oil into refined products, but had to move it into the market through independent refining companies, which necessitated giving major discounts. In most countries of Asia and Africa indigenous refining capacity was being fully utilized. This meant that the international majors could continue to import refined products for their marketing affiliates without giving these affiliates the discounts which they had to give their refineries in order to meet the sharp competition in the crude oil market. The relatively stronger position of the international majors in product pricing also was buttressed by their low refining costs in their large-scale Middle Eastern refineries.[11]

The major exception to this general situation was the highly competitive Japanese market, in which a great part of refining and marketing facilities was not controlled by the international majors.[12] It was in this market that great discounting took place, both in crude oil prices and in product prices. For example, in 1963, the international majors were giving independent refiners in Japan discounts off posted prices for large volumes of refined products; these discounts typically ranged from 5 to 15 percent. Hence, the indisputable facts were that for the knowledgeable independent Japanese buyers, not only the newcomers but the international majors themselves were giving sizable discounts off posted product prices.[13]

The key to the difference between a country like Japan as opposed to India undoubtedly lay in three crucial factors: (a) many more independent refiners and marketers in Japan; (b) greater knowledgeability in Japan, particularly at the governmental level, about oil pricing; (c) the greater economic strength and independence of Japan, which enabled it to apply greater pressure on the oil companies' prices. There seems little doubt that the Indian government demonstrated considerable lack of knowledge about oil pricing. Widespread discounting had been going on for almost two years before it was at all introduced into India. As one Indian paper noted in July 1960:

> What is really disturbing is that but for the Soviet offer, India would never have realised how heavily she is and has been over-paying for British and American oil.[14]

The final major area in which the Damle Committee made recommendations was the allowable expenses and profits for the oil companies' marketing operations. The need for determining "allowable expenses" and "allowable profits" stemmed from the fact that the government set ceiling prices at both the refining and marketing levels. Moreover, the marketer's ceiling price in India, prior to the Damle Committee report, was essentially based on the cost of the refined products he purchased plus a markup of 10 percent.[15]

The Damle Committee recommended basing market prices on allowed refinery prices plus a before-tax profit of 12 percent on "capital employed" in marketing and distribution. The Committee also recommended "disallowing" some of the items which the oil companies wanted included in "costs" (in arriving at profits) and in "capital employed." Specifically, the Committee recommended disallowing the following items as components of cost: donations, bonuses to employees, interest on debt, and losses from bad debt or on sales of assets;[16] the Committee also recommended excluding short-term bank loans as an element of capital employed.

We do not intend to analyze the pros and cons of these various recommendations, most of which were strongly opposed by the companies.[17] It is our view that the differences between the views of the companies and the Damle Committee (or the Indian government) stemmed fundamentally from a basic conflict of interest; as such, they could not be solved by tortured exegesis of theoretical accounting principles, nor appeals to "equity" or "justice." Take, for example, the question of whether or not short-term bank loans which are continually renewed should be treated as long-term loans, and therefore be included in the capital base. Part of the problem is that different concepts of "risk" are implicitly embodied in the opposing positions. From the companies' viewpoint, if one is operating in a situation where immediate nationalization is possible, not only should they be compensated for the long-term capital they have tied up in the country, but also for the short-term capital which could also be seized. From the government's viewpoint, the question is whether it should compensate the companies for risks against governmental actions; at a minimum the Indian government may well have felt that the companies had an exaggerated view of their risks.

Again, there was a disagreement over whether or not Standard Vacuum's branch in India should be allowed to include in its cost of importing oil a "pro rata" share of the parent company's headquarters' expenses (in New York and London). Related to the question of the magnitude of the costs—the Committee stated that the amounts paid were "exorbitant" [18]—there are clearly differing views as to the function of the company's overseas headquarters. The company's position is that the headquarters provides valuable technical and financial services at a low cost to its affiliates.[19] The government, implicitly at least, views part of the headquarters work as involving actions not necessarily benefiting India; for example, some headquarters employees will devote their efforts toward developing proposals for negotiating with and trying to win concessions from the Indian government itself. The problem might be dramatized this way: should the headquarters' employees who devoted many hours to drawing up a reply to the Damle Committee report have their salaries charged to the operations of the Indian company, and thereby be subsidized by the Indian government?

The principal conclusion we draw from the debate over the recommendations of the Damle Committee is that generally there is no "right" or "objective" criteria for evaluating such recommendations. To the extent that the basic interests of the oil companies and government diverge, their views on these questions naturally tend to differ. Conflicts stemming from divergent interests are not usually solved by appeals to "equity," but rather are settled by the relative bargaining power of the parties.

Thus, the "recommendations" by the Damle Committee that the companies cut their crude oil prices, or buy crude from other sources, or even change their whole international structure were in one sense quite naïve. As profit-maximizing organizations the companies were not going to take any of these steps unless they felt compelled to do so, i.e., felt that their worldwide long-run profitability would be reduced by not taking these measures. The companies had already introduced price discounting into India in an attempt to offset the impact of the Soviet crude oil offer. Finally, while the power of a government to compel concessions from the companies depends partly upon its knowledge-ability about the oil business, even more important is its willingness to use strong pressures, and if necessary force, to obtain its goals. Since the Indian government had already shown in 1960 that it was not willing to force a showdown with the companies, the latter had little reason to fear that rejection of the Damle Committee recommendations would now be disastrous.

The basic truth of this is borne out by the government's use of the Committee's recommendations. Fundamentally, the government accepted all the findings of the Committee. However, since the established majors steadfastly refused to import products at discounted prices, the government chose to act in this area by reducing the maximum selling price of all refined products (imported or not) in line with the Committee's estimates of available product discounts. The principal defect of this "solution," from the governmental viewpoint, was that there would be no reduction in the foreign exchange outlay of the Indian affiliates for importing oil; instead, the companies' internal rupee profits would be reduced. (Since the companies generally prefer to repatriate money as quickly as possible, from their viewpoint this aspect of the solution was a virtue.) Finally, while the Committee submitted its report to the government in July 1961, the companies were not given a copy until September 27, 1961; on that day they were also told that the government's decisions would be implemented beginning October 1, 1961. This "peremptoriness" of the government probably stemmed from sheer frustration. The government must have had a renewed feeling of impotence that it could not obtain the foreign exchange savings which would result from the companies either taking Soviet oil or reducing prices on their own imported oil. Hence, because the government felt it could do nothing about imported oil prices, it tended to waste no time doing something about the internal prices that it could more easily control.

In the years following the Damle Committee Report, the oil companies have been very successful in maintaining their prices on oil imports into India. Despite the formation in 1964 of another governmental group to study oil prices in India (the

"Talukdar Committee"),[20] the government has been unable to obtain major reductions in crude oil prices. Thus, more than three years after the Damle Committee Report, "in response to Government requests" the established majors, after initial resistance, agreed to reduce prices on their imported crude oil by two cents per barrel.[21]

A more substantial reduction came in July 1965, when Burmah-Shell reduced the price of imported Iranian oil by $0.07 per barrel, and of Kuwaiti oil by $0.04. This was almost immediately followed by similar price cuts by Caltex and finally Jersey, which cut its price on an Arabian crude oil mixture by $0.09 per barrel.[22] Several things are noteworthy about this round of crude oil price cuts. First, it was led by Burmah-Shell, which is not surprising in view of the fact that its half-owner, Burmah Oil Company, has the least interest in importing high-priced crude oil. Second, trade sources clearly indicated that the prices charged to the affiliates of the established majors in India continued to be higher than the price paid by independent refiners:

> Independent refiners in Japan are digesting these [1965] price cuts and preparing to argue that they too, should get lower prices. Japanese buyers tell PIW in Tokyo, for example, that while they've made no proposals to their suppliers yet, they believe that supplying companies "will want to maintain the same price differential" that has existed between prices paid by Japanese refiners and those paid by affiliates of the suppliers in other east-of-Suez markets.
>
> Suppliers can't be expected to buy that argument easily. They'll undoubtedly tell the Japanese that the move in India was designed to eliminate the "differential" and keep prices charged their affiliates in India on a parity with prices in other markets east-of-Suez such as Japan.[23]

It is worth noting that even if the suppliers' argument was true, it implicitly admits that for some time at least there was a differential between prices paid by affiliates and independent refiners.

Finally, the oil companies sought to give the impression that their price cuts had nothing to do with formation of the

Talukdar Committee or any other governmental pressure to cut prices. Thus, *Petroleum Intelligence Weekly* reported that:

> For its part, Shell acted, a source close to the company's London headquarters says, only after a continuous appraisal convinced the company lower prices were needed to keep its Indian affiliate on a par with other Far East affiliates. That, and not local pressures, he added, is the real reason for Shell's decision.[24]

On the other hand, *The New York Times* stated:

> Observers noted, however, that Burmah Shell and the other two private oil companies here are operating under the shadow of two implementing orders that give the Indian Government the right to seize the companies' stocks [of oil].[25]

The background developments here indicate that once again the companies chose to use crude oil price cuts to deflect pressures for handling Soviet oil. In April 1965, the government-owned Indian Oil Company announced plans to import 2.3 million tons of refined products from the Soviet Union in 1965, or about 20 percent of total market demand.[26] Further, the government prohibited all imports of kerosene and automotive diesel oil from other than the Soviet Union. However, the government's limited marketing facilities, combined with the complete unwillingness of the companies to facilitate in any way the importing of Soviet oil, quickly led to product shortages.[27] This in turn led to another confrontation between the government and the established majors. In June 1965, the government took steps, under the Defense of India Act, to be in a position to compel the companies to distribute refined products as ordered:

> The government has assumed control over the distribution of all petroleum products here because, say Government officials, the private oil companies are causing "artificial shortages" by withholding kerosene and high speed diesel oil from consumers.[28]

The government and the companies traded charges as to who was to blame for the shortages, but the petroleum press saw clearly that the companies' refusal to handle Soviet products was the crucial issue:

> The underlying point of contention between the companies and the Government concerns the companies' unwillingness to distribute Russian products. Up to now the private concerns have steadfastly refused to handle Russian imports because of the repercussions such a step could have in other countries with foreign exchange difficulties.[29]

As part of the developing confrontation the oil workers union in India called for a nationwide strike of the established majors, aimed at forcing them to assist in distributing Soviet refined products.[30] However, the scheduled strike of oil workers was called off at the request of the Indian government itself.[31] In any event, this further confrontation between the established majors and the government undoubtedly served to worsen existing relationships between the two groups, particularly since the shortages of refined products continued.

These developments probably increased the pressure for additional price reductions in imported oil. In August 1965, the Talukdar Committee submitted its report to the Indian government, calling for additional reductions of four to seventeen percent in crude oil prices and three to fifteen percent in product prices.[32] India's Oil Ministry, however, thought that the price cuts should be even greater.[33] As a result, the Oil Ministry commissioned the Indian Oil Corporation to undertake its own survey of crude and product prices by visiting several European countries.[34]

Finally, another measure recommended by the Talukdar Committee and considered by the government was to (belatedly) emulate the Japanese government—by reducing the foreign exchange allowed for importing crude oil as a way of forcing reductions in crude oil prices.[35] Since, as we shall see in the next chapter, at this point in time (1965) the established majors appeared clearly barred by the government from further major refinery expansion, this plan seems poorly conceived. Without the "carrot" of refinery expansion, the likely response of the companies to any cut in their foreign exchange allocation would be to cut their quantity of imports rather than their prices.[36]

Two additional aspects of the overall pricing conflict in India, which are inherent in the existence of oil affiliates in under-developed countries, are worth noting. Both reemphasize the apparent governmental ignorance and/or weakness which handicaps the securing of low-cost energy supplies.

First, when Jersey announced its crude oil price reduction in July 1965, it said that part of the cut came from importing into India a different crude oil mix, containing a higher proportion of heavy, less expensive crude oils. There was nothing to indicate that Jersey could not have provided this lower priced crude mixture in previous years, thereby saving India considerable foreign exchange (but of course reducing its own profits). Moreover, the Talukdar Committee called attention to the fact that the Indian government had no idea exactly what crude oils were in the mixture being imported into India:

> It is not possible to judge, unless the crudes are identified and composition of the mixture is known, whether the average price is in accordance with the world market price. . . . In order to satisfy Government that purchase is being made at world market prices, Esso should indicate the constituents and the proportions thereof.[37]

It might be noted that even if the individual crude oils were being imported at the recommended prices, the total cost of the mixture still might be higher than necessary. The companies can deliberately import a high proportion of high-priced crude oils in order to maximize overall company profits.

A second aspect of the companies' oil pricing which is even more important is the freight rates charged Indian affiliates for the crude and products they import. Little governmental attention has been paid to this problem, particularly compared to the efforts made to get a reduction in the f.o.b. prices of oil. Yet, the amount of "overcharging" due to the companies' use of the out-of-date AFRA rates as opposed to the more realistic "spot" rates (discussed in Chapter 12) has been of the same order of magnitude as the hard-won crude price cuts.

For example, at the time of the Soviet crude oil offer to India (mid-1960), the AFRA tanker rate being used by the international oil companies was "Scale minus 12 percent," which for shipments from the Persian Gulf to Bombay meant a freight charge of 24 cents per barrel. Use of the spot rate, about Scale minus 47 percent, would have meant a reduction in crude oil transport costs of $0.10 per barrel, or almost $4 million per year.[38] At this rate, the cumulative cost to India from the use of AFRA pricing in the 1960s would have been tens of millions of dollars for crude oil alone.[39] Additional millions were undoubtedly lost due to use of AFRA for pricing refined product shipments.

Moreover, the Indian government seemed largely unaware of the magnitude of the stakes involved in the freight pricing problem as well as the issues involved. Thus, the Damle Committee concluded a brief analysis of the AFRA system, used by all three established majors in India, as follows:

> The adoption of the spot charter rate is likely to introduce an element of uncertainty, which is not worth risking, particularly when the current low rates for spot charters are automatically reflected in the over-all level of AFRA. We, therefore, recommend that ocean freight may be calculated at the AFRA rate for the time being.[40]

The Talukdar Committee apparently agreed, for it made no recommendation for change. This indicates that the Indian government has failed to appreciate both the basic long-run trends in tanker rates and the way they would affect the relationship between AFRA and spot rates. (See Chapter 12.)[41]

In sum, various problems of pricing which are inherent in the affiliate relationship have been, and are likely to continue in the future, as sources of conflict between the oil companies and the government in India. During the course of this struggle over pricing there has been a parallel struggle over the question of who was to own the oil refineries in India. As we shall see in the next chapter, the pricing conflict has been an important factor in helping to answer this crucial question.

CHAPTER SEVENTEEN

The Struggle over Oil Refineries in India: Public versus Private Sector

EVER SINCE THE EARLY 1950S, when the international oil companies recognized the political necessity of building indigenous refineries, the question of who was to own them has been one of the most burning issues facing the Indian government. It has involved not only those in the energy field, but also top officials in finance, foreign affairs, and defense. The basic question has been whether petroleum refining would be owned by the government or the private sector. A secondary issue has been which specific oil companies should be chosen for any role allowed to the private sector.

The Soviet Union has had a major influence on the answer to the key question of public versus private ownership. For one thing the Soviet Union (and Romania) provided in the late 1950s the technical and economic assistance necessary for the Indian government to mount its own oil exploration effort. More directly, the Eastern European countries provided similar technical and economic assistance for the Indian government to build three government-owned oil refineries.

The first two government refineries were established in Eastern India to process crude oil discovered in Assam. The first, at Nunmati, was designed with initial capacity of 0.75 million tons per year, and was built at a cost of $37 million, of which the Romanians provided the foreign exchange component of $12 million. The second, at Barauni, had initial capacity of 2 million tons per year, and cost $88 million, of which the Soviet Union provided the foreign exchange component of $37 million. Included in these cost figures were the expenses of building a complete town at each refinery location. The third government refinery was set up in Western India, at Koyali, to process the crude oil found in that region by the Soviet-aided government oil exploration program. The initial capacity of this refinery was also 2 million tons, with an estimated cost of $63 million, the foreign exchange component of which was again provided by the Soviet Union. All of these refineries were designed so as to be readily expanded by close to 50 percent, which expansion has in fact taken place. Thus, while the total initial design capacities was about 36 million barrels per year, by 1966 the three refineries had a capacity of 51 million barrels per year.[1]

While the government refinery building program was impressive in terms of aggregate capacity, its short-run impact on the Indian energy picture was weakened by the long timelag between the agreement to build a refinery and the realization of full production. For example, while the decision to build the refinery at Nunmati was made in 1958, it was not until January 1, 1962, that the refinery actually went into operation. Moreover, through most of 1962 the refinery was dogged with various operating problems.[2] As a result actual production during the inaugural year was only one-third of the designed capacity. Similar problems beset the Barauni and Koyali refineries, both initially and during their expansion stages.[3]

At the same time as the government's refinery building program was having "teething" troubles, demand for refined oil products continued to rise rapidly. As a short-run solution for this imbalance, the government could either increase imports of refined products or allow the established majors to (relatively quickly) expand their existing refineries. In the long run, to meet increasing demand the government could either build more government refineries, allow private companies to build new refineries, or enter into joint ventures with the companies. In the following sections we analyze the short-run developments in relation to their impact on long-run policy decisions in this area.

Before the government-owned refineries came on-stream in the early 1960s, the imbalance was largely met by allowing the established majors to increase capacity at their original refineries. For example, Standard Vacuum's had an initial design capacity of 1.2 million tons per year, but by 1961 throughput had been expanded to 1.9 million tons. Similarly, annual capacity at Burmah-Shell's refinery was increased from 2.0 million tons to 2.8 million, and at Caltex's from 0.65 million to 0.90 million tons. Total private sector refinery capacity increased 40 percent between 1957 and 1961, from 4.33 million tons to 6.05 million tons.[4]

The period of the late 1950s and early 1960s was marked by an intense struggle over the respective roles of the government and private enterprise in petroleum refining; this same struggle continues today in India as in many other underdeveloped countries. Crucial events took place during this earlier period which set the framework for the current debate and struggle.

On one side of the debate stood the Left wing of the Indian Congress party, and particularly Mr. Malaviya, the Minister of Mines and Fuels. Malaviya's aim was to make the Indian government the dominant factor in all levels of the Indian petroleum industry. Up to a point he was highly successful, getting the government to mount a successful oil exploration program, build three wholly owned government refineries, and enter into marketing and distribution of oil products. Ultimately, however, the basis of his success lay not in a governmental consensus that the State should play the dominant role in the Indian oil industry, but rather in the fact that he had been able to undertake this program at virtually no cost (particularly in foreign exchange) to the Indian economy—by utilizing Soviet and Romanian foreign aid. Through this mechanism, he had been able to overcome internal opposition, particularly in the Finance Ministry, the stronghold of the Right in the Indian government.

By and large, Malaviya had utilized Communist Bloc assistance in a skillful manner to attain his goals. His program appealed not only to the economic self-interest of India, but also to the spirit of nationalism. His (successful) government oil exploration program made it logical to build government refineries to process this crude oil; in fact, it made it a virtual necessity, since the private oil companies would have no interest in building refineries in India to process government crude oil.

When demand for petroleum products continued to grow in the period during which the government-owned refineries had not yet come on-stream Malaviya helped to soften the foreign exchange impact of the gap by importing Soviet refined products; this simultaneously reduced the pressure for private sector expansion. Since the oil companies refused to distribute the Soviet products through their marketing apparatus, this logically gave an impetus to developing a government-owned oil marketing network. In addition, their refusal to handle Soviet crude oil or products tended to make Indian public opinion run strongly against the companies.

The weak link in Malaviya's program was that, for a variety of reasons, the Indian government was only willing to make a drive for dominance in the oil industry if it could do it relatively cheaply, particularly in terms of foreign exchange costs. While the Russians were willing to provide large quantities of crude oil at no foreign exchange costs to India, they probably were unwilling and/or unable to provide capital funds and equipment for underwriting indefinite expansion of the government-owned refinery program. Again, while the Russians were willing to sell refined products to India at no foreign exchange cost, the fact that it takes much longer to build up a marketing organization than a refining or exploration group meant that importation of Soviet refined products would not solve the short-run supply-demand imbalance. Thus, the combination of rising demand, a severe foreign exchange shortage, the Indian government's inability or unwillingness to force the companies to handle Soviet crude oil or refined products, and the limits to foreign assistance for government operations in the oil industry served to undercut Malaviya's program.

An illustration of the basic weakness of Malaviya's position can be seen from the skirmish which took place in 1961–62 over the question of replacing the refinery agreements under which the three international oil companies operated in India. These agreements, reached shortly after India's independence, appear more like treaties between sovereign states than like the usual commercial licenses under which private companies operate in foreign countries. None of the many other foreign companies which have set up operations in India have such special agreements; all operate under ordinary industrial licenses. That the refinery agreements infringe on ordinary government prerogatives can be seen from provisions which prohibit the government from placing an import duty on crude oil, guarantee the refineries from nationalization for 25 years after they begin operations, and give the refineries complete freedom to choose their own sources of crude oil supply.

In an attempt to put pressure on the oil companies to revise these refinery agreements (described by one Indian newspaper as "no longer such good bargains as they had seemed at the time"),[5] Mr. Malaviya in October 1961 forced the established majors to cut back refinery production to their 1960 level. This involved cutbacks in annual throughput of crude oil of about 9 million barrels. The additional foreign exchange costs for importing refined petroleum products instead of this would equal close to $10 million annually. Since this cutback was directly hurting India as well as the companies, it would seem to stem from a position of weakness rather than strength.

In any event, this test of strength was cut short nine months later, in June 1962, upon the beginning of military hostilities between Communist China and India. This date was clearly a critical turning point in Indian energy policy history, as well as the history of the country, since the conflict with China greatly weakened the Left in India.[6] Nevertheless, even prior to this date, Mr. Malaviya had drawn considerable criticism for his policy of curtailing the crude oil runs of the private oil companies. As a postscript, it might be noted that the best concession he could get in June 1962 was the willingness of the companies to "discuss" with the government a change of the refinery agreements into ordinary licensing agreements. To the date of this writing no such change has been made by any of the companies.

Naturally, the strongest opponents of the Indian government's role in petroleum refining were the three established majors. They were intensely eager to have the growing demand for petroleum products in India met by increasing the scope of their own refinery operations in India. The companies' motivation, of course, derived from the basic economics of the international oil industry which vests enormous profitability in crude oil—provided that one can find an outlet for it. Thus, even with $1.80 per barrel posted price crude oil selling at $1.60 per barrel (f.o.b. the Middle East), the crude oil profits alone to the international oil companies from additional outlets would be about 65¢ per barrel (after all taxes), or $12 million per year on a 50,000-barrel-per-day refinery.

Clearly then, for any of the international oil companies with "surplus" crude oil the rewards from obtaining additional refining capacity in India would be great. The three established majors, however, had even greater incentives than the other international oil companies. For one, the established majors had already existing refineries which could be expanded relatively cheaply, giving an exceptionally high rate of return on refinery investment. In addition, since they already had existing marketing networks there would be no problem of finding an outlet for the additional refined products; presumably there would be a very high rate of return on the relatively low-cost expansion of these facilities too.

With these strong incentives the established majors continually pressed the government for permission to expand their refinery capacity, particularly after the 1962 conflict with China had led to the end of production restrictions on their existing refinery capacity. This effort was made despite the fact that the Third Five Year Plan, drawn up in 1961, stated flatly that "developments in the field of refined petroleum products are visualized entirely in the public sector." [7] Early in 1963 all of the established majors made proposals to the Indian government for expanding their existing refineries. Thus, the issue was formally joined between Malaviya and the Left wing of the Indian government (both of whom wanted petroleum refining to be kept in the public sector) and the established majors (who desired to expand their existing refineries); the battle raged also within the Indian government itself.

The Sino-Indian conflict which weakened the Left in the

Indian government, while also reducing India's foreign exchange reserves and making her even more dependent on foreign, particularly Western, assistance, had made the prospects for large-scale private sector refinery expansion appear to be excellent. Nevertheless, in January of 1963 Mr. Malaviya announced a major expansion program for the government sector, stating that there was no need for private sector expansion. Malaviya said that the three existing government refineries would be expanded by an additional 18 million barrels per year capacity. In addition, a "publicly owned" refinery with an 18 million-barrel-per-year capacity was to be built in southern India (at Madras). Thus, for the time being at least, the drive of the established majors for addition 1 refining capacity was blocked. It is important to analyze how Malaviya was able to carry this policy within the Indian government.

Malaviya's real success came from using his historically strongest card, Soviet foreign aid. Thus in February 1963 it was announced that the Soviet Union had agreed to provide the foreign exchange component of the cost of expanding the three government refineries. The other way that Malaviya promoted the growth of the public sector was more semantic than real. An oil refinery was now to be considered "in the public sector" if the government owned at least half of the equity stock of a joint-venture refinery.

Phillips Petroleum, a United States newcomer to the international oil industry, had prepared in 1962 an offer to set up such a joint-venture refinery. The Indian government was to have 51 percent of the equity capital, Phillips Petroleum 25 percent, and Indian private investors the other 24 percent. The 50,000-barrel-per-day refinery was to be built in Southern India. The Indian government was to control the crude oil supply (excluding Soviet oil) and the products made by the refinery; the latter were to be distributed through the new government marketing organization. Phillips was to get a fee for technical services in running the refinery. Finally, the bulk of the total investment would be provided by loans from equipment suppliers and other sources; Phillips' own equity investment was to be guaranteed by the United States government against expropriation and currency inconvertibility.

Phillips had strong incentives to offer such a joint-venture deal. As a newcomer Phillips did not have to worry about whether such a deal would bring government pressure for joint ventures in already existing refineries in India or other underdeveloped countries. Since Phillips did not have an established marketing operation in India, the ability of the government to market the refinery's products would also be a positive feature.

Most important, the deal finally worked out between Phillips and the Indian government was extremely profitable for Phillips, even as estimated by a petroleum trade journal:

> The Government, which controls product prices, has assured Cochin that the refinery's average income on a barrel of products sold will be $1.35 per barrel higher than the delivered cost of its crude oil supplies. Industry sources estimate most refiners abroad operate on a much lower margin, say, between 60¢ to 75¢ per barrel.[8]

Thus, the refinery probably would make extra gross profits amounting to between 60¢ and 75¢ per barrel of crude oil processed, or from $11 to $14 million per year. Assuming a 50 percent tax rate, Phillips would probably earn extra profits in the neighborhood of $1.5 million on its $5 million equity investment or about 30 percent per year.

This represents only the directly visible profits. In addition, in the deal Phillips was made the agent for negotiating a 15-year crude supply contract for the refinery, which contract it awarded to Standard of California. While there was considerable uncertainty as to how much Phillips would directly make on this supply contract,[9] it is hard to believe that Phillips will not indirectly profit from this crude oil contract. After all, Phillips negotiated a price of $1.57 per barrel f.o.b. Iran, which discount of 21¢ per barrel off the posted price was in line with that being given by the established majors to their affiliates in India. Even if it is debatable whether Phillips could have in fact obtained a lower price from non-Soviet sources, it seems clear that at that price there would have been many eager suppliers. The enormous profits from such a crude oil supply contract, cumulatively equal to close to $150 million for Standard of California, might conceivably place the recipient under some obligation for a reciprocal "favor."

Again, the fact that Phillips itself received the construction contract, must have been a source of indirect profit for Phillips. Finally, Phillips was to receive a technical service fee which might have provided additional profits. All in all, it seems clear that the winning of this refinery deal must have been a real bonanza for Phillips.

At the same time, other aspects of the deal would tend to make it acceptable to various important groups in India. The fact that the refinery was to be "in the public sector" (and excluded the established majors) would appeal to the Left and Indian nationalists. Since Indians for the first time in history were to be allowed a real equity share in the country's oil industry, and a highly profitable share to boot, the deal was undoubtedly popular with the Indian business community. Finally, the fact that the Indian government had to provide little if any foreign exchange for this refinery project undoubtedly appealed to "Centrists" as well as the Finance Ministry.

In any event, it is clear that this decision of the government did not put to rest the question of expansion of the established majors.[10] Their hopes undoubtedly increased in June 1963, when Mr. Malaviya was "compelled to resign on charges of improper conduct in office," and India's *Eastern Economist* commented:

> It is to be hoped that the Government's commitment to its present oil policy does not mean that no steps will be taken, even after Mr. Malaviya's exit, to give the quietus to those sterile controversies and in their place work out a constructive dialogue of co-operation between the Government and the international oil companies. Here there are two specific issues on which a healthy reorientation of policy seems certainly to be most desirable. First there is the question of the further expansion of the Burmah Shell and Esso refineries in Bombay. Mr. Malaviya has been adamant in refusing to consider this expansion. He has been

giving a number of reasons for the stand taken by him, but it is no secret that there is no agreement within the Government on the soundness of his attitude.[11]

In late 1963 it appeared that the established majors had virtually achieved their goal. The *Far Eastern Economic Review* reported, "Increases in capacity in the refineries of Burmah Shell, Esso and Caltex have been recommended by the Indian Planning Commission, which, if approved, will add about 3 million tons to annual capacity."[12] The same month *World Petroleum* reported from New Delhi, "Expansion of Indian refining facilities by Esso, Shell, and Caltex will obtain government approval providing the three companies agree to permit the government to share in the equity capital, according to O. V. Alagesan, the Oil Minister."[13] Unfortunately for the companies, Alagesan was quickly overruled: "Prime minister Nehru has informed the new oil minister Alagesan that the oil policies of India's former minister Malaviya must be continued—with no expansion in private-sector refineries unless future conditions force such a move."[14]

Among other things, the question of the refinery agreements still vexed relations between the established majors and the government even after Mr. Malaviya's departure. *Petroleum Intelligence Weekly* summarized the problem as follows:

INDIA—No early decision is expected to end current refinery agreements held by Burmah-Shell, Esso, and Caltex, although discussions between Shell and the Government continue. The Government would like to end the agreements and put the oil companies under license deals now applying to most other private business ventures.

The companies seem disposed to give up their agreements provided the Government guarantees them: the right of free choice in supply crude oil, equal treatment in tax and marketing rights with public sector refineries, and the right to repatriate profits.[15]

While still hoping for expansion of their existing refineries, Jersey and Burmah-Shell apparently decided to try and "ride with this trend [toward joint ventures] rather than buck it."[16]

Thus in early 1964 each submitted a proposal for a joint-venture refinery, with the company to be a minority partner with the government for the next refinery to be built at Madras.

However, since the 1963 government decision to expand public sector refining capacity by joint ventures with private companies, there has been an intense struggle among established majors and newcomers for the right to build new refineries and provide the all-important crude oil. By early 1964, the Indian government had been approached by nine different groups of companies, each willing to take a minority partnership in a joint venture with the government for the Madras refinery: the three established majors, Mobil, Phillips, Continental, a Japanese consortium, ENI, and a partnership between Standard of Indiana (AMOCO) and Iran's state oil company, NIOC (Standard Vacuum's 1962 world-wide breakup had removed Mobil from India).[17]

Initially the apparent front runner in the race was ENI,[18] which had long been interested in the Indian oil market. Thus, back in 1961, Malaviya had negotiated a $100 million broad-gauge petroleum loan with ENI.[19] For ENI, aggressively exploring for crude oil in many countries, the refinery deal probably represented a chance to get a foothold in the Indian market, which might ultimately provide an outlet for any newly found crude oil.[20] In any event it was reported that ENI's request for "6% plus" as the rate of interest on its credit for building the refinery was a stumbling block to a deal.[21]

The ultimate winner of the Madras refinery award, which was packaged with associated plants for manufacturing lube oils, petrochemicals, and fertilizers, was the AMOCO–NIOC partnership. The main features of the final agreement were reported to be as follows:

(1) ownership of the refinery would be 51 per cent for the government, 25 per cent for the two companies, and the remaining 24 per cent for Indian businessmen.

(2) the companies were to provide all of the foreign exchange investment, estimated at $65 million, partly by arranging loans to be repaid over a sixteen year period, and with interest charges at "somewhat less than 6 per cent."

(3) the companies had the right to supply India with crude oil at a price between \$1.35 to \$1.37 per barrel, FOB Iran; this was reported to be "8 cents per barrel below current prices on Indian imports."

(4) the Indian government would purchase about 300 million barrels of crude oil from the two companies, over a 20-year period, with the right to use the oil in Madras or any government-owned refinery; the companies would give India short-term supply credits for purchasing the crude oil.[22]

The principal attraction of the deal to the two companies undoubtedly was the huge crude oil supply contract, particularly since they had recently discovered large amounts of oil in their Iranian offshore fields (Darius and Cyrus). NIOC, as a state oil entity, was especially interested in finding outlets for Iranian crude oil to reduce the country's dependence on the international consortium which controlled most of Iran's oil. According to a press report, "in the eyes of the Government, the winning bid topped the others in crude price, supply credit, foreign exchange assistance and loan amortization." [23]

In this deal the Indian government retreated from its apparent desire to keep supply contracts for crude oil on a short-term basis—a desire based on the belief that crude oil prices might decline further in the future.[24] An added attraction which might have helped overcome the government's reluctance for long-term crude contracts was that the Iranian government might reciprocate by opening the door to Indian government exploration in offshore Iranian areas.[25]

The latest "public sector" refinery scheduled to be built in India is at Haldia, near Calcutta in eastern India. The pattern of competition to win this award was essentially a duplicate of that for the Madras refinery, except that with the increasing world oversupply of oil the offers became even more favorable to India. As an indicator of the pressing drive to find outlets for crude oil, Mobil was willing to provide not only a \$65 million loan to finance the 50,000-barrel-a-day refinery (plus a lubricating plant and a fertilizer plant) but in addition would give the Indian government 100 percent ownership of the refinery (providing technical services for a fee). In exchange, Mobil wanted a long-term supply contract for 250 to 300 million barrels of crude oil (and a 49 percent share in the lubricating and fertilizer plants).[26]

Ultimately, however, the Indian government kept the Haldia refinery as a 100 percent government-owned one without losing complete control of the crude oil supply. It was able to do so by combining Romanian and French government aid. Romania agreed to barter $12 million of lubricating plant equipment for Indian iron ore. The French government agreed to loan India $20 million to buy refinery plant equipment from CFP. In addition, CFP received a crude oil supply contract for 45 million barrels, or less than one-fifth of that which Mobil sought; moreover, it was reported that "The crude will be delivered from Iran and will be priced lower than other suppliers were willing to go." [27] Finally, "Both CFP and Romania will play 'assistance' roles in running the government plant," [28] which is scheduled to be on-stream by 1970.

However, despite the fact that the established majors have been shut out of any new refineries in India, they have continued to grow in India by expanding their old ones. Between late 1963, when Nehru ruled out their expansion "unless future conditions force such a move," and the end of 1967, the established majors have with government approval quietly increased their capacity by about one-half, or an amount equivalent to one of the newly licensed refineries. This situation is another stark measure of the relatively weak position of the Indian government vis-à-vis the international oil companies.

Epilogue: Refinery Expansion and Demand Forecasting

An interesting and instructive footnote to the struggle for refinery expansion in India was the role played by forecasts of future demand for petroleum products. The first instance came in late 1962, when a "numbers game" struggle began between the established majors and Mr. Malaviya. India's Third Five Year Plan (1961) had projected expansion of refinery capacity to 11 million tons by the end of 1965–66. Between 1960 and 1962, however, demand grew extremely rapidly, increasing by 10 percent in 1961 and leaping by 15 percent in 1962. As a result, a subcommittee of the Indian government's Oil Advisory Committee (OAC) was told to again forecast the likely product demand in 1965–66.

The subcommittee, dominated by representatives of the established majors in India, quickly drew up a forecast of demand through 1966 increasing at about 15 percent per year;[29] this very high growth rate was nominally based on an (superficial) "end use" analysis, but in actuality was simply developed to justify a case for expanding the existing refining capacity of the established majors. Malaviya, on the other hand, developed lower demand figures, and in addition made optimistic forecasts of the dates by which government oil refineries would come on-stream. Thus, while the high demand forecasts of the OAC's subcommittee ultimately were largely adopted by the Indian government for planning purposes, Mr. Malaviya was able to reduce the pressure within the Indian government for immediate major expansion of the existing refineries of the established majors. While the government raised its targeted total refinery capacity for India from the figure of 11 million tons developed in the Third Plan to 17 million tons, this effort of the industry-dominated subcommittee did not succeed in getting major expansion of the existing refineries.

A similar attempt by the established majors either to get expansion of their existing refineries or be allowed to build additional refineries seems to have been made in 1964:

> India's decision to go ahead with the new refinery [at Madras] may not come any too soon, if latest industry forecasts of the country's demand growth prove accurate. A new study by the private oil companies in India indicates that in 1966, demand will reach 18.1-million tons a year. Right now, some 10-million tons of capacity is on stream, including the first stage of the state's new Barauni refinery. . . .[30]

This 1964 forecast of the established majors implied a doubling in demand between 1963 and 1966, or an enormous growth rate of over 25 percent per year. In fact, petroleum demand in 1966 turned out to be less than 14 million tons, or far below the companies' publicized prediction of 18 million tons.

It should be noted that this was not the first time (nor likely to be the last) that different petroleum demand forecasts were put forth to further self-interest. For example, we have the following report:

Long-term forecast: Why the experts differ. In November, 1963, Sir Maurice Bridgeman, Chairman of BP in a paper entitled "World-Wide Production—Its Prospects and Problems" predicted that in the year 1990 world crude production (excluding the Soviet bloc) would reach 56 million barrels per day. Thus begins an article published in OPEC's "Inter-OPEC Newsletter."

The periodical goes on to say that recently, Mr. J. H. Loudon, Managing Director of Shell, in a paper delivered to the Institute of Petroleum, predicted that world consumption (including the Soviet bloc) would reach 100 million barrels per day.

Why the startling difference in predicted growth rates between the two sources, both of them eminently qualified?

In the BP paper, an attempt was made to show that the world would have sufficient petroleum reserve in the long-run to meet demand. Obviously, a low figure for demand in 1990 would facilitate this task.

In the Shell paper, an effort was made to stagger the imagination with the sums of money ($350 billion over the next 25 years) that the industry would need to meet growing demand. Obviously, a high figure for production requirements in 1990 would make this more credible.

Conclusion: The companies can and do "bend" their long-term forecast curves for public consumption.[31]

CHAPTER EIGHTEEN

Exploration for Crude Oil in India

AN IMPORTANT ASPECT of the Indian government's attempts to gain a dominant position in the oil industry was its activities in crude oil exploration. In addition to its significance for developing an integrated public sector oil industry, the whole history of oil exploration in India is worth analyzing as a case study in the pros and cons of natural resource development by government versus private enterprise.

Until the early 1950s Burmah Oil was the only sizable company exploring for oil in India.[1] In fact, prior to the entry of the Indian government into oil exploration in 1957, a Burmah Oil subsidiary had discovered the only three productive fields in India (Digboi, 1889; Nahorkatiya, 1953; and Moran, 1956—all in Northeast India). Up to that time the only exploration effort made by the international oil companies in India was that of Standard Vacuum (a 50–50 partnership of Jersey and Mobil). In 1951–52 it carried out an aereomagnetic and ground survey of the complete West Bengal basin, stretching across eastern India and East Pakistan. Following this, in 1953 Standard Vacuum entered into a joint venture with the Indian government for exploration in this area. The partnership was based on a 75 percent share for Standard Vacuum and 25 percent for the Indian government. The venture, known as the Indo-Stanvac project, drilled ten test wells and found no oil or gas;[2] in 1960 the project was closed down after $17 million had been spent.[3]

The joint-venture agreement provided the following major concessions to Standard Vacuum: (1) the project was given a 10 thousand square mile concession area; (2) managerial control was totally in the hands of Standard Vacuum; (3) all of the losses of Standard Vacuum, over and above those of the government, would be deductible against Standard Vacuum's marketing income; (4) a depletion allowance was given which would provide effectively a 50 percent tax rate on the oil produced; (5) if oil were found, it was to be priced at import parity.

From the terms of this joint venture it seems obvious that the Indian government was far more eager than Standard Vacuum to have the oil exploration effort undertaken. Typically, in underdeveloped countries where an oil company is strongly interested in exploring for oil it pays bonuses to the government for the right to drill in concession areas; in this case there was no such bonus to the Indian government. Moreover, the provision that Standard Vacuum's losses could be deducted from its marketing income in effect shifted one-half of the exploration cost to the Indian government, even though it was to get only one-quarter of the profits.[4] As a result, when the project was completed the government would not agree to similar agreements for other possible exploration areas. The government was particularly opposed to the provisions pertaining to depletion and those allowing the company to write off exploration losses and retain complete managerial control.

The government's first major venture on its own resulted from a 1955 tour by an Indian delegation to study the oil industry in various countries and determine what assistance India could obtain for its oil industry. The delegation, which visited the Soviet Union, Romania, Great Britain, the United States, and Latin America, was headed by Mr. Malaviya, then Minister of Natural Resources. As a result of this tour, the Indian Oil and Natural Gas Commission (ONGC) was established in 1956, and a group of oil experts from the Soviet Union and Western countries came to India. However, the work of the Soviet team of experts, which provided a detailed integrated plan for oil exploration, became the basis of the Indian governmental program. In the words of the Indian government:

> When it was decided that local exploration should be undertaken, principally (but not exclusively) as a Government enterprise, an approach was made to a large number of countries for

technical and financial assistance. Initially, only the USSR and Rumania offered to help; for this reason a major part of the exploration and production activity in the country today is carried on with the support of these countries.[5]

Thus, the Russians provided the bulk of the foreign exchange funds for importing drilling equipment as well as the skilled technicians necessary to start the program and train Indian nationals to carry it on later.[6] The first major event of the ONGC drilling program, which began in 1957, was the finding of the Cambay gas field in 1958. The first major oil find by the ONGC came in 1960, with the discovery of the Ankleshwar field (part of the Cambay Basin), north of Bombay.[7] Some other significant discoveries reported to date by the ONGC include fields at Kalol, Nawagan, and Lakwa.[8]

During the period that the ONGC was launching its exploration program, the Indian government also entered into joint-venture oil exploration with Burmah Oil Company. The joint-venture company, Oil India, Ltd., was set up in 1958 to produce and explore in the already proven Nahorkatiya and Moran Fields (and also to construct a 720-mile crude oil pipeline from these fields to the government refineries at Gauhati and Barauni). The 1958 agreement, in which the government received a one-third interest in Oil India, in effect gave the government a share in proven oil fields in exchange for which the new company got licenses for producing in the area and an outlet for this crude oil. In 1961 the government's equity in Oil India was increased to 50 percent, in return for which the company was given a concession for oil exploration in areas adjacent to the proven fields. In addition, Oil India was guaranteed a net return (after all taxes) of 9 to 13 percent on its equity base.[9]

In ten years (1956–66) of pursuing this two-pronged approach to finding crude oil, the Indian government has spent close to $400 million on exploration and production. While in the last few years India's oil reserves have become classified data, it is clear that the effort resulted in discoveries totaling about a billion barrels of crude oil, plus natural gas equivalent to several hundred million barrels of oil, or a finding cost of about $.35 per

II. India: A Case Study

TABLE 18-1

INDIAN GOVERNMENT EXPENDITURES AND RETURNS
FROM OIL EXPLORATION INVESTMENT, 1956–1966
BASES FOR DCF CALCULATIONS

Data used	ONGC	Government (50%) share of Oil India, Ltd.
Exploration expenditures ($ million)		
1956–1961	55	30
1961–1966	252	45
Returns (million barrels)		
Oil discovered, 1956–1966	750	185
Gas discovered, 1956–1966 (crude oil equivalent)	78	113
Total, 1956–1966	828	298
Projections of annual production from discovered reserves (million barrels)		
1966	18	9
1971	47	13
1972	50	18
1983	50	18

SOURCES:

Exploration expenditures.

ONGC: 1956–1961, *Third Five Year Plan*, p. 514; 1961–1966, by subtraction from 1956–1966 figure in *The Oil and Gas Journal*, November 6, 1967.

Oil India, Ltd.: Maximum approximation (including cost of pipeline construction) based on data in *The Indian Petroleum Handbook: 1966* (New Delhi: Petroleum Information Service, Statistics Division), p. 12.

Returns.

Oil: Estimated from data in India, Press Information Bureau, "Towards Self-sufficiency in Oil and Oil Products in India" (Press Release, August 15, 1966).

Gas: Estimated from data in India, Ministry of Information and Broadcasting, "Oil Industry in India" (December 24, 1964).

Future projections of annual production.

1966, actuals; 1971, official estimates in *The Oil and Gas Journal*, November 6, 1967; 1972–1983, figures projected constant to approximately exhaust all discovered reserves over 20-year period.

barrel. (See Table 18-1.) Thus, while large volumes of oil were discovered under this two-pronged government oil exploration effort, large amounts of money and resources were also spent in the process. How then do we judge whether or not the expenditure proved to be worthwhile?

As argued earlier, the best way to evaluate success of a government oil exploration program is to submit it to the same criterion used for evaluating private investments: the DCF rate of return on the investment program. For government investments this rate of return should be computed both on an indigenous currency basis and on a "foreign exchange basis," i.e., using "shadow prices" to reflect the fact that the existing currency-exchange ratios may be unrealistic. Ideally, the calculations should also take into account all the economic, political, and social ramifications of the investment program. For example, the value of a skilled manpower group developed in the course of an oil exploration effort, or of the increased political independence emerging from finding indigenous oil should ultimately be included in the "returns" from the oil exploration investment. Clearly, we cannot put quantitative values on these factors whose evaluation must largely lie with government planners on a judgmental basis. Hence, the following evaluation of the Indian government's oil exploration program is based solely on comparing the monetary investment in the oil exploration program with the monetary value of the oil discovered.

Even after limiting ourselves to this more narrowly defined economic data, there are still difficulties in estimating the rate of return on the Indian oil exploration program. The principal problem is that the government's program is a relatively new one. Since there is inevitably a sizable time lag between the outlay for oil exploration and the flow of returns from oil production, it is necessary to make some forecast of the timing of future Indian crude oil production from these already discovered reserves. The measurement problem is further complicated because government data lump together both genuine new "exploratory" expenditures and "development" expenditures, i.e., drilling in already proven fields.[10]

II. India: A Case Study

Despite these problems, it seems clear that the Indian government oil exploration effort has been a major economic success. Examining the ONGC first, available data show that from 1956 through 1961 total expenditures were only $55 million and from 1961 through 1966, $250 million. On the basis of this expenditure of roughly $300 million (about one-half in foreign exchange), as of 1966 it was estimated that the ONGC discovered at least 800 million barrels of oil (including natural gas), which, valued at import parity, would amount to close to $1.5 billion worth of oil.

A simple-minded analysis might conclude that the rate of return on this investment was 500 percent. This would be untrue for two basic reasons. First, it would fail to take into account the fact that it will take many years for this oil to be produced, and oil produced ten years from now has less value than oil produced today (essentially because the money saved today could be productively ploughed back). Second, and to a lesser extent, it will take additional expenditures to actually produce this discovered oil.

Nevertheless, two approximate DCF calculations of the rate of return on investment, which take these factors into account, were made on the following basis. (See Table 18-2, sources for which are found in Table 18-1.) First, the future time pattern of production from estimated ONGC reserves discovered by 1966 was forecasted, conservatively. Second, the oil to be produced from these discoveries was given two values—an upper limit of $1.50 per barrel and a lower limit of $1.00 per barrel.[11] Under these assumptions the DCF rate of return on ONGC investment through 1966 would range between 9 and 14 percent per year, compounded annually through 1983. (The rate of return might be even more, since perhaps half of the investment required foreign exchange, and all of the returns are completely foreign exchange because the discovered oil substitutes directly for imported crude oil; the DCF rate of return using a higher value for foreign exchange would clearly be even greater. In addition, oil exploration can thus be seen as a good hedge against devaluation for India.)

TABLE 18-2

DCF RATE OF RETURN ON ONGC OIL EXPLORATION INVESTMENT, 1956–1966

Year	Investment in millions of dollars (a)	Oil production est. in millions of bbl. (b)	Oil production in millions of dollars $1.50/bbl. (c)	Oil production in millions of dollars $1.00/bbl. (d)	Net return in millions of dollars $1.50 (c − a) (e)	Net return in millions of dollars $1.00 (d − a) (f)
1956						
1957						
1958						
1959	$ 55	o	$ o	$ o	−55	−55
1960		o	o	o	o	o
1961		o	o	o	o	o
1962		2	3	2	3	2
1963		5	8	5	8	5
1964	$252	5	8	5	−244	−247
1965		10	15	10	15	10
1966		18	27	18	27	18
1967		24	36	24	36	24
1968		30	45	30	45	30
1969		36	54	36	54	36
1970		42	63	42	63	42
1971		47	71	47	71	47
1972		50	75	50	75	50
1973		50	75	50	75	50
1974		50	75	50	75	50
1975		50	75	50	75	50
1976		50	75	50	75	50
1977		50	75	50	75	50
1978		50	75	50	75	50
1979		50	75	50	75	50
1980		50	75	50	75	50
1981		50	75	50	75	50
1982		50	75	50	75	50
1983		50	75	50	75	50

DFC rate of return (compounded annually)   14%   9%

SOURCES: See Table 18-1.

II. India: A Case Study

The precise interpretation of these 9 to 14 percent DCF figures is to say that, if the Indian government had taken the money which it invested in oil exploration and put it in a bank which paid 9 to 14 percent per year compound interest on deposits, it would have ended up in the same monetary position as if it had made the oil investment. Another way of viewing this is to say that the "oil exploration subsector" of the Indian economy would have had a real growth rate of 9 to 14 percent per year. Clearly then, in an economy which has historically grown at well under 5 percent per year, the 9 to 14 percent rate of return on oil exploration investment marks the ONGC effort as a major success for the Indian government.

Similarly, the government's investment in a 50 percent share in Oil India, Ltd., has also paid a handsome return. The government's oil exploration outlay in the 1958–60 period was $30 million and in 1961–66 $45 million; the government's half-share of the discovered oil and gas likely amounted to at least 300 million barrels. Using similar assumptions as before about future production and the future value of oil to the economy, the rate of return to the government on its oil exploration investment through Oil India, Ltd., would range between 13 and 18 percent per year. (See Table 18-3 and, for its sources, Table 18-1.) (The greater return on the investment in Oil India compared to ONGC probably reflects the fact that much of Oil India's exploration was in relatively proven areas, whereas that of the ONGC was wildcatting in virgin territories.)

A more precise estimate of the rate of return on the Indian government's oil exploration program cannot be made until the program has a longer history. Nevertheless, it has become increasingly clear even to skeptics that the governmental oil exploration program is already a proven success. As early as 1961 the conservative *Eastern Economist* made the following comments about Mr. Malaviya:

> The striking of oil in Cambay and Sibsagar and the proving of a very substantial oil-field in Ankleshwar, capable of supporting a 2 million tons refinery, placed his entrepreneurship on the only sure basis of success, which is success itself. It is right that

TABLE 18-3

DCF RATE OF RETURN ON INDIAN GOVERNMENT'S SHARE OF OIL INDIA, LTD.
INVESTMENT IN OIL EXPLORATION, 1958–1966

Year	Government investment in millions of dollars (a)	Government share of oil produced in millions of bbl. (b)	Oil production in millions of dollars $1.50/bbl. (c)	$1.00/bbl. (d)	Net return in millions of dollars $1.50 (c − a) (e)	$1.00 (d − a) (f)
1958						
1959	$30	0	$ 0	$ 0	−30	−30
1960		0	0	0	0	0
1961		1	1	1	1	1
1962		2	3	2	3	2
1963		3	5	3	5	3
1964	$45	4	6	4	−39	−41
1965		5	8	5	8	5
1966		9	13	9	13	9
1967		10	15	10	15	10
1968		11	16	11	16	11
1969		12	18	12	18	12
1970		12	18	12	18	12
1971		13	20	13	20	13
1972		18	27	18	27	18
1973		18	27	18	27	18
1974		18	27	18	27	18
1975		18	27	18	27	18
1976		18	27	18	27	18
1977		18	27	18	27	18
1978		18	27	18	27	18
1979		18	27	18	27	18
1980		18	27	18	27	18
1981		18	27	18	27	18
1982		18	27	18	27	18
1983		18	27	18	27	18
DCF rate of return (compounded annually)					18%	13%

SOURCES: See Table 18-1.

those of us—and this includes *The Eastern Economist*—who have often wondered whether the public sector was justified in taking these risks, should admit that, in the result, Mr. Malaviya has been justified by the events.[12]

Upon the departure of Mr. Malaviya from the Cabinet in 1963, the same journal commented as follows:

There is also another aspect in which this assurance of continuance of the existing policy will be found agreeable. It is that the country's own potential of oil and gas should be energetically explored and wherever found commercially profitable, vigorously developed. Here again, Mr. Malaviya has considerable achievements to his credit. At a time and over a period when there was general scepticism about Indian oil reserves, he stuck with courage and conviction to his course of geo-physical surveys and exploratory drillings which have ultimately paid a fair dividend. There is no doubt that this programme of indigenous oil development must continue to be pursued with the vigour and confidence with which Mr. Malaviya has hitherto presided over it.[13]

Interestingly, this praise came prior to the highly successful government exploration efforts in 1964 which had contributed to the *Far Eastern Economic Review*'s view: "1964 was India's oil year. . . ." [14]

It should be made clear that there has been strong opposition to any government oil exploration program in India. The opposition was partly based on the purported lack of oil in India. Thus, despite the paucity of exploration efforts:

. . . a notion has been sedulously cultivated over a century that except at Digboi, in Assam, which too was nearing exhaustion, there were no large oil reserves in India. This myth, however, has been exploded by the Indian Government's Oil and Natural Gas Commission which, though set up only six years back, has found three oil-fields in Gujarat in as many years.

But despite the universally acknowledged fact that gulfs are potentially good oil-bearing areas all the world over, the Cambay basin and adjoining areas in Gujarat, in which the present oil-fields have been located, were left unexplored to date, although basic information about the existence of tertiary rocks in the Cambay basin was available during the British regime in India since 1925–30. In Assam itself, though Digboi was discovered 70 years ago, the adjoining vast alluvial tracts of the Brahmaputra valley remained unexplored till as late as 1952, when the very first well in Nahorkatiya proved a producer.[15]

This argument was linked with the one of the alleged "great risk" in oil exploration. Under the subheading "inspired criticism," one writer noted in 1961:

During recent months a mounting campaign has been going on in the country, both inside the Parliament and outside in the Press, against the official oil policy. The main points of this criticism are:

(1) The supply of oil in the world is concentrated in a few centres, specially regions around the margin of seas that fill the depressions between continents where these approach each other most closely. The Indian oil area forms a very weak link in the general belt of oil fields. The search for oil in India is, therefore, essentially a wild goose chase. It is one of the riskiest ventures which may lead the Government to unthinkable financial disaster. . . .

That there are insufficient reserves of oil in India and that except for Assam the rest of the country is almost dry in respect of this vital mineral has been the key-note of propaganda in interested quarters.[16]

Some Indians, while acknowledging that the government exploration program has had successes, have accused the government of hurting India's economy by "refusing" to allow a larger role for private oil exploration in India. Thus, the previously-cited editorial in *The Eastern Economist* lauding the exploration achievements of Mr. Malaviya upon leaving the Cabinet, also argued as follows:

But where Mr. Malaviya went wrong was that he seemed to take delight in creating controversies of quite a different kind —controversies which served no useful purpose but on the contrary were quite harmful since they denied to the country the resources of the international oil companies in expanding the domestic industry as rapidly and economically as may be possible. . . .

. . . A new approach seems also desirable on the question of enlisting the resources of the international oil companies more extensively for surveying and drilling for oil. While the exploration of possible domestic oil deposits is certainly very desirable, this country does not have so much spare capital that it should prefer to use its own funds for this highly speculative programme to securing resources on reasonable terms from all other sources as may be open to us. It is possible that the international oil companies had not been as active in the past in exploring and drilling as we might have liked, but it is highly probable that the responsibility for this state of affairs is not wholly one-sided. Just because of his crusading zeal so admirable in many respects, Mr. Malaviya has been apt to pursue his oil policy with a dogmatism which has quite frequently come in the way of his own purpose of building up the national oil industry rapidly, economically and efficiently.[17]

Later (in 1965) *The Eastern Economist* spelled out its position in detail:

We should be able to make a deeper dent on our crude import bill if we streamline our oil exploration policy. But are we doing so? The evidence is to the contrary. The clarification by an official spokesman of the statement of the Minister for Petroleum and Chemicals, Mr. Humayun Kabir, at his news conference on December 17 that all on-land oil exploration work in the country was reserved for the ONGC (to be undertaken by it either directly or through contractors) runs counter to the policy enunciated in the exploration rules. (The Minister's observation was that all oil exploration in the future was proposed to be undertaken only by the public sector.) The exploration rules provide for minority participation by foreign oil explorers in joint ventures for exploration on land also. Any firm Government decision reserving all on-land oil exploration for the ONGC, despite the fact that its operations hitherto have not been very efficient and it cannot be expected to explore all the 400,000 square miles of the potential oil-bearing sedimentary areas on land even in the next two decades, will therefore be a pure dog-in-the-manger policy.

The stepping up of our crude output at a rate faster than that planned for the fourth Plan requires that, along with expanding the operations of the ONGC, we should make an effort to interest foreign oil explorers in oil prospecting in our country. . . .

The three main factors that inhibit foreign oil explorers from taking active interest in oil prospecting in our country, apparently, are the desire of the Government to have a majority interest in the joint ventures to be floated for the purpose, heavy taxation and frequent changes in our fiscal policy. . . .

In oil exploration, what, in fact, our policy for the fourth Plan should be is that the ONGC should be asked to concentrate on developing the oilfields already discovered by it and on areas where its operations are at an advanced stage. Diffusion of its energies over wider areas is neither in the interests of the ONGC nor in the overall interests of the country. We would even suggest that the ONGC should withdraw from Assam and let exploration work there be undertaken by Oil India—that well-established partnership between the Government of India and the Burmah Oil Company. In lieu of the oil reserves already discovered by it, the Government's share in Oil India could be raised suitably. For exploring new areas or areas where the operations of the ONGC are only in the preliminary stages, foreign oil explorers should be invited.[18]

Not only did domestic critics seek to limit the Indian government's oil exploration program, but similar efforts were made by important Western forces. Thus, the Levy report to the World Bank, which was circulated to governments of underdeveloped countries in 1961 argued that in light of India's huge capital needs for its Third Five Year Plan, the government could not afford to undertake a major oil exploration program completely on its own. Thus, particularly in light of the "risk" associated with oil exploration it would be better to call on the oil companies for help in this area; any losses from oil exploration would be borne by the companies while gains would be shared between them and the government.[19]

Again, in 1963, *Petroleum Intelligence Weekly* reported the following:

II. India: A Case Study

Temporarily shattered: India's dreams of foreign oil exploration fostered by former oil minister Malaviya. Faced with the realities of foreign exchange problems, India now plans no overseas ventures in the near future, reports in New Delhi indicate. In the midst of an ambitious domestic oil development program under the Third Plan (1962–1966), India has decided all available resources should be utilized at home. *The World Bank is understood to concur with this view. In a report to the Indian Government*, the Bank has pointed out that the cost of oil investment under the Third Plan would hit $514.5-million, with nearly 40% required in foreign exchange. Drilling equipment and rigs have not yet been produced in India. *Thus, says the Bank, India could not enter into foreign exploration with its present limited resources of men, material and money.*[20]

Earlier, the same source carried the following report:

Plans for India to make an Iranian oil search now seem remote with the recent veto of the Planning Commission. Despite hopes of the Oil and Gas Commission to minimize foreign capital outlay by joining an American partner in the venture, the Planning Commission said the Iranian move would invite criticism from foreign-aid sources as an unnecessary expenditure.[21]

Today foreign critics of the Indian government's oil exploration program, forced to recognize its success, have shifted their gears to proclaiming that India has a great oil potential; but, it is said, this potential cannot be realized without the "help" of the international oil companies. Thus, *The Oil and Gas Journal*, in a 1967 article entitled "Lone-wolf oil policy hurts India," argued as follows:

India is determined to develop a thriving, self-sufficient oil industry but is unwilling to accept outside help.

A short-sighted government policy based on the "India first" approach is retarding the very growth the country is seeking. . . .

It's not that desire is lacking. India is eager to build her oil and fertilizer industries. But she is unwilling to meet a partner halfway. And she hasn't the money, the techniques, talent, or equipment to do the job on her own. . . .

To boost local reserves, ONGC has been carrying on an aggressive exploration campaign for the past 10 years, but results have not been world-shaking. . . .

In its one joint Indian venture with a foreign company, ONGC has fared very well. But the lesson seems lost on the government. Oil India Ltd. (O.I.L.), a government-Burmah Oil Co. venture, has increased its reserves in recent years by 17 to 18% through careful selection of drilling sites and intense geologic study. . . .

Geologically, the prospects for new oil discovery are very bright. There are some 400,000 sq. miles of sedimentary basin in the country, plus excellent drilling targets on the continental shelves.[22]

It is one of the ironies of the international oil industry that in India, as in other countries with government oil entities (e.g., Mexico), successful government oil programs have not stilled the critics, but instead have led them to urge limiting future activities. Some reasons for this should be readily apparent from our previous analysis, while others will be discussed later. At this point, however, we turn to the question of the extent to which the limited role of the oil companies in India's oil exploration may be attributed to governmental policies.

The role of the oil companies as regards oil exploration in India is a classic case of the behavior which could be predicted *a priori* from a good understanding of the institutions and the framework within which they operate. Specifically, if a knowledgeable and honest student of the international oil industry had been asked in 1960, "What would be the reaction of the various oil companies to the changes in India's government rules[23] aimed at encouraging private exploration for oil in India?" he probably would have forecast the following:

First, the established majors would have little direct incentive to explore for oil in India under any reasonable set of conditions. This simply reflects the fact that they all had large quantities of already discovered, low-cost crude oil available in the Middle East, which would be backed out by any oil discovered in India. (In relation to their own annual production of crude oil in the Eastern hemisphere, crude oil reserves of Royal Dutch Shell would last 57 years, those of Jersey 79 years, Standard of California 82 years, and Texaco 82 years.)[24] With crude oil available to these companies at a real cost of about $0.10 per barrel f.o.b. the Middle East, plus shipping costs of no more than $0.25 per barrel to bring the oil to India, the cost of finding and producing crude oil in India would have to be no more than $0.35 per barrel simply for them to break even on the deal.[25] Clearly, since the proposed oil exploration program would be "wildcatting," it would be a highly unattractive proposition under these conditions. (Note our estimate of per barrel finding costs alone for the ONGC of $.37 in the 1956–66 period.)

The only real positive incentive to an established major for exploring for oil would be some kind of "tie-in" deal, particularly the right to build new refineries which would utilize the discovered crude oil and hence not back out its present Middle Eastern supplies. The only other significant motive for exploring would be a negative one; that is, if it was felt that unless the company went ahead with oil exploration, the government or competitors would do so and would ultimately find oil which would back out the established major's present Middle Eastern supplies.

Except for the last point, the preceding conclusions would also hold for the remaining international majors and the international minors who had recently discovered large quantities of cheap Middle Eastern crude oil. Basically, neither of these groups of companies needed crude oil; what they needed were outlets for crude oil.

A second prediction would be that the only companies with any strong incentive to explore for crude oil without "tie-in" arrangements would be those which lacked cheap external crude, but already had a refining or marketing position within India. Burmah Oil is the only sizable company which fits in this category. Burmah Oil in 1959 was in the unique position of having the only indigenous crude oil production (which was slowly declining) along with its own small refinery in Assam. Furthermore, while Burmah Oil marketed these products in Assam, through

Burmah-Shell it also had a half ownership in the largest refinery and marketing operation in India. While Shell, with large supplies of Middle Eastern crude, would have as little incentive to explore in India as the established American majors, Burmah Oil would have considerable incentive. It could either move any discovered crude oil through the existing Burmah-Shell refinery, or the refineries of the other established majors, or be allowed to build new refineries. In any event, the crucial point is that Burmah Oil had no significant amounts of external crude oil which could be backed out by indigenous discoveries.

The actual history of oil exploration in India fits this predicted model quite closely. Up through the 1950s, the only productive field had been discovered by Burmah Oil Company. The only other significant exploration endeavor had been a small-scale joint venture of Standard Vacuum with the government but this project had been undertaken in the early 1950s, when crude oil was not in such great oversupply and was partly underwritten by the Indian government through tax subsidy.

Despite the changes in government rules in 1959, no exploration efforts have been undertaken by any of the established majors. Only Burmah Oil has made any significant efforts, operating with the government through Oil India. In analyzing this joint venture, the United Nations commented:

> By purchasing an additional 15.75 per cent of the shares of Oil India Ltd. at par value from the Burmah Oil Company for cash, the Government increased its holding in this company to 50 per cent, the same as that of Burmah Oil. In future, each financial partner will bear its share of the costs, the foreign currency being provided by the Burmah Oil Company. In calculating profits, a deduction will be made for the past and future exploration and development costs of Oil India. This is thus neither an agreement with the usual fifty-fifty sharing of profits nor an association such as ENI has entered into with foreign governments. . . .[26]

The UN concluded:

> The agreement with the Burmah Oil Company is regarded

by the Indian Government as a model which it hopes will be adopted by other enterprises.[27]

It is not surprising, in the light of our previous analysis, that this hope has not been realized.

While the logic of the situation should suffice to indicate that oil exploration in India is not an attractive investment for the established majors, the behavior of Standard Vacuum following the government's 1959 invitation should prove the point. In early 1960, Standard Vacuum prepared an economic study of the potential profitability of exploration in the Jaisalmer region of the State of Rajasthan in Northern India. The study was based on the belief that there was a "good chance" that oil would be found, and for calculation purposes the discovery of a 500 million barrel field was assumed. In addition, because the proposed exploration area was right across the border from Standard Vacuum's natural gas fields in West Pakistan, a separate study was made premised on a "good possibility" for finding natural gas even if no oil were discovered; in this case a discovery of 3,000 billion cubic feet of natural gas was assumed. Depending on various other assumptions, on a total exploration investment of $5 million, if crude oil were found the DCF rate of return on an ultimate investment (including development expenditures) in the neighborhood of $100 million was estimated to range between 21 and 30 percent per year over a 25-year period; if only gas were found, on an ultimate investment of $10 million the DCF rate of return would be 15 percent per year.

Clearly, Standard Vacuum's potential rate of return on the investment if crude oil were found would be high by normal standards. Moreover, during this 1960–61 period the Indian government was under considerable pressure from the World Bank and the Western aid-giving nations to allow the oil companies into oil exploration; as a result it resumed negotiations with a number of companies besides Standard Vacuum, including some newcomers. Nevertheless, Standard Vacuum never was able to reach an oil exploration agreement with the Indian government.

A fundamental reason would appear to be that the profit

rates as calculated were for Standard Vacuum as a separate entity, rather than for its parent-owners, Jersey and Mobil, who would have to deduct the lost profits from their backed out external crude oil. The effect of this deduction would be to sharply reduce their overall potential profit rate on the Indian exploration investment. Only the fear that others would find substantial amounts of oil in India, thereby backing out Jersey and Mobil's crude oil without Standard Vacuum making any offsetting crude oil production profits, gave Standard Vacuum's owners any strong incentive to negotiate with the Indian government; in addition it was hoped that undertaking oil exploration would improve the company's public and government relations in India, thereby paying off in future profitability in other projects and areas within the country.

From Standard Vacuum's viewpoint, the oil exploration negotiations with the Indian government ultimately broke down because the Indian government was insistent on a 50:50 joint venture project, while the company wanted it to be a wholly owned one. Clearly, however, the reluctance of the established majors to yield to the government on questions such as joint participation in oil exploration springs from their outside interests which give them little direct economic incentive to explore for oil in India. Rather, oil exploration by them is at best a defensive measure, aimed at limiting the efforts of government or of possible competitors. It is essentially governmental pressures which propel the established majors toward oil exploration in India, rather than the intrinsic economics of the situation itself. As an Indian newspaper commented in March 1962:

> By the time the Damle Committee reported last July, the position of the oil companies had been weakened. Russian supplies were coming through, more oil had been discovered in India, public sector refineries were being set up and a competitive sales and distribution organisation was in the process of being completed. The oil companies have since made some clumsy efforts to regain lost ground. For instance, when an agreement was signed with the French Petroleum Institute for oil exploration in the Jaisalmer area of Rajasthan, Standard Esso

tried to revive negotiations in which it had earlier been involved with the Indian government over this area without success.[28]

A good conclusion for this analysis of oil exploration in India, and the struggle over public versus private control, is the following lyrical but essentially accurate Indian view:

The saga of India's search for mineral oil springs from her will to fight a myth—that geologically the country was incapable of yielding any oil.

There were other myths too: that she did not have the technical knowhow to enter a venture requiring a high degree of skill; that even if she desired it was not proper for an underdeveloped country to fritter away her resources on ventures which were full of risks; and that the oil companies alone, who commanded resources, experience and technical skill, could do it.

But within a short span of less than five years these myths have been exploded. The country has discovered a sedimentary area of nearly 400,000 square miles spreading over the Gangetic delta and running down to the Cauvery basin in the south. A zealous band of trained geologists, geophysicists and drillers is well on the job. By experience the country has learnt that the oil venture is not so risky as it is made out to be. And the indications of oil and gas discovered in areas where oil companies had earlier showed no interest have also established that the public sector alone could have undertaken the search for mineral oil in India.[29]

CHAPTER NINETEEN

Oil Byproducts: Fertilizers in India

No ANALYSIS of the oil industry in India would be complete without some examination of the country's petrochemical industry. (See Chapter 9 for a general discussion of the relationships between the oil and petrochemical industries.) The branch of petrochemicals examined here, fertilizers, is undoubtedly the most significant one for India, as well as for many other underdeveloped countries.

The potential significance of fertilizers for India is particularly great because her consumption is so low. At the end of the Second Five Year Plan, 1960–61, India was using about 2 kilograms of fertilizer plant nutrients (primarily nitrogen) per hectare of arable land, compared to 10 for the Soviet Union, 38 for the United States, 304 for Japan, and 456 for the Netherlands.[1] The previously cited FAO study of the response of agricultural output to fertilizer use in various countries suggests that given India's low level of fertilizer consumption, an additional ton of fertilizer (ammonium sulphate) could generate about 12 additional tons of rice production. With rice selling in India for $91 per ton, this would mean additional production of $1,100 worth of rice, compared to the Indian price for a ton of ammonium sulphate of only $376, or a return of $3 worth of additional agricultural production for a $1 investment in fertilizer.[2] These figures indicate the great desirability of increasing fertilizer consumption in India.

Fertilizer consumption can be increased either through imports or building indigenous fertilizer factories. India's shortage of foreign exchange has been the principal deterrent to increasing fertilizer consumption, since large amounts of foreign exchange are required for either course. Thus, if consumption of fertilizers in India has grown rapidly in percentage terms in recent years, it is largely due to the fact that earlier usage was practically nil. In the last year before the First Five Year Plan, 1950–51, total consumption of fertilizer (in quantities of plant nutrient) was only about 70,000 tons, or less than the output of a large fertilizer factory. During the First Five Year Plan total consumption doubled, during the Second Plan it more than doubled, and during the Third Plan consumption almost tripled, reaching a level of 840 thousand tons. Thus, over the 15-year period, fertilizer consumption in India increased by an average growth rate of over 20 percent per year.[3]

To put this apparent rapid growth in further perspective, it should be noted that the fertilizer production goals set by the Indian government have been consistently missed by large margins. For example, the Second Plan set a target for 1960–61 fertilizer production of 410 thousand tons while the actual achievement was only 153 thousand tons, or a shortfall of 63 percent; again, the Third Plan 1965–66 production target was 1.4 million tons, while actual production was only 0.35 million tons, or a gap of 75 percent.[4] As a result of these production shortfalls, the increasing consumption has been met by rapidly rising imports. In 1955–56 and 1960–61 imports accounted for one-half of total fertilizer consumption, while in 1965–66 they equaled almost 60 percent of the total.

Thus, Indian food production has become increasingly dependent on fertilizer imports, which by 1965–66 amounted to $80 million, or 3 percent of India's total imports.[5] The relatively enormous role of foreign aid here can be seen from the fact that in January 1966 the United States agreed to provide India a loan of $50 million to "facilitate the urgent purchases of fertilizers." [6]

It is important to understand the reasons for India's failure to meet production targets in the fertilizer field. Under the Industrial Policy Resolution of 1956 which delineated the basic division of industries between the public and the private sector, fertilizer, unlike petroleum, was classified as a "Schedule B" industry, to be developed both by the private sector and the public sector:

> . . . Schedule B industries will be progressively State-owned, but private enterprise will be expected to supplement the efforts of the State in these fields.[7]

Up through 1960–61, production of nitrogenous fertilizers, which had increased by only 19,000 tons during the Second Five Year Plan, lay entirely in the public sector, deriving from three government plants.[8] In an effort to increase nitrogenous fertilizer production by 1965–66 to ten times the 1960–61 level, the government licensed the private sector to have a capacity of 0.62 million tons by 1965–66, with the public sector targeted to have capacity of 0.68 million tons. It was believed that this capacity of 1.3 million tons would allow production of 1.0 million tons by that year.[9] In fact, actual capacity reached only 0.5 million tons, and production fell even further short of its target, reaching only about half of actual capacity.[10] The primary reason was that by and large the private sector did not use the licenses given to it by the government.[11] The causes for this are instructive and worth analyzing in some detail.

Most of the major planned fertilizer projects in the private sector have involved collaboration between Indian businessmen, foreign private firms (primarily United States), and in some cases the government (central or state) in India. The primary reason for bringing in the foreign collaborators of course has been for their supplies both of foreign exchange and technical know-how. Two types of foreign firms have been interested in participating in an underdeveloped country's fertilizer business: the chemical companies which have a long experience in the fertilizer business, and the oil companies which in recent years have been diversifying into petrochemicals.

In general, oil companies in the United States have had several reasons for getting into petrochemicals and fertilizers. First, through their favorable tax treatment, stemming from large depletion allowances, they have tended to generate enormous sums of capital for which in recent years there have been insufficient opportunities for investment in the oil industry proper; in some cases, diversification of risk was also a motive. Second, since fertilizers and petrochemicals involve similar types of technological processes, the companies have felt it relatively easy to make the transition. Third, the oil companies themselves produce the naphtha and/or natural gas which is a basic raw material in the fertilizer-making process.

In recent years the established majors have had a particularly great incentive to get into the fertilizer business in India. This is because the demand pattern for refined petroleum products in India, as in many underdeveloped countries, conflicts with the technologically limited product supply pattern from a typical oil refinery. Unlike in the United States where about half of the total demand for petroleum is for the "top of the barrel"—gasoline—the "underdeveloped" demand pattern has been largely for the "middle distillates"—diesel oil (for industry, trucking, and railways) and kerosene for heating and lighting—or for the "bottom" of the barrel—heavy fuel oils and residual asphalts.[12] Since gasoline is essentially treated naphtha, it follows that as more and more refineries have been built in underdeveloped countries, there has been a tendency to develop a naphtha surplus relative to market demand within the country. A natural use for this surplus naphtha would be as a feedstock for fertilizer plants constructed near the petroleum refineries.

For the established majors, a naphtha surplus was not a significant problem in India until the beginning of the 1960s. Prior to that time, to the extent that their refineries generated more naphtha than could be used internally for gasolines, the remaining amount could be turned into gasoline and exported to their affiliates in gasoline-short countries, e.g., Australia.[13] With the drive by most oil-importing countries to eliminate refined product imports by building indigenous refining capacity, and with most of the Asian-African countries having the same type of product supply/demand imbalance as India, it was clear to the companies that a major expansion of their refinery capacity in

India would leave them only two real alternatives:[14] (a) to find internal outlets for naphtha, such as fertilizers, lubricating oils and petrochemicals; (b) to burn off the excess naphtha, thereby deriving no profit at all from it.

The critical point for the established majors in India was that they would face a serious naphtha surplus *only if* they were allowed large-scale refinery expansion. Since their primary interest in fertilizer projects was as an outlet for potentially surplus naphtha, their offers to build fertilizer projects were effectively tied to obtaining the right to build additional refinery capacity. Similarly, for the newcomers to refining in India, the interest in fertilizers was usually closely related to its potential for profitably absorbing surplus naphtha.

For the foreign chemical companies, which cannot benefit from petroleum outlets, interest in fertilizer investment in the underdeveloped countries is based on the high direct profit potential of the projects themselves. It is likely that there will be a rapidly growing demand for fertilizers for years to come. The market for indigenously produced fertilizers is even more assured, since the governments will make every effort to replace imports.[15] Moreover, the capital-intensive chemical companies also generate huge cash flows from their profits-plus-depreciation and they are usually seeking profitable outlets for investment. Finally, the chemical companies' control of and experience with the complex technology embodied in the large-scale plants that they desired to build in a country like India would help to ensure a high rate of return on these investments.

Of course, the profitability of any petrochemical project in an underdeveloped country where governments control product prices depends to a great extent on the prices allowed by the government. In this respect, India would potentially appear to be particularly attractive to foreign companies, since the price for fertilizer charged to farmers there is among the highest in the world; in 1961–62 for a ton of ammonium sulphate the price paid by farmers "at the farm gate" was $376 in India, $278 in Japan, and $240 in Thailand.[16] However, as we shall see, the question

247

of the prices to be allowed the companies for fertilizer in India has been one of the major sources of conflict between the government and potential foreign investors.

Just as in the area of oil refineries, the most fundamental struggle over fertilizers has centered around the question of public versus private ownership. During the early 1960s, both public and private sector fertilizer plant expansion lagged. However, unlike the situation in refineries, the established majors have not felt any strong pressure in recent years to enter the fertilizer business. For one thing, the established majors have sometimes been able to contract to supply naphtha to fertilizer plants being built in India by others. Most important, they have not been allowed the major expansion in refining capacity which they desired and which would have led to their having a sizable oversupply of naphtha. As a result, with the food and fertilizer situation in India becoming increasingly desperate, the Indian government has had to virtually abandon its goal of public sector control of fertilizer plants. Instead the government has been ever more intensely trying to attract foreign investment, particularly by United States companies.

A signal event marking the general swing away from the public sector toward the private foreign sector was the holding of a week-long conference in New Delhi in April 1964 between representatives of major United States corporations and top Indian government officials. The significance of the conference was neatly summarized in a *Business Week* article aptly entitled "India Opens Wider to Foreign Funds":

> Recently, T. T. Krishnamachari, wealthy former businessman who took over as India's Finance Minister last August, has been campaigning to woo foreign private capital. He has been redefining the concept of "socialism"—or mixed economy—to which all Indian politicians pay at least lip service, and asserting that socialism is a system that "seeks to stimulate and reward individual initiative and enterprise." T.T.K., as he is called, declares that private enterprise, which still accounts for nearly 90% of India's over-all production, must carry much of the load in speeding economic development.

. . . This month the Indian government is taking action to back up its words. Some 28 top U.S. businessmen from such companies as Union Carbide Corp., Firestone Tire and Rubber Co., International Minerals and Chemical Corp., Deere and Co., Standard Oil Co. (N.J.), and Westinghouse Electric International Co. are flying to New Delhi to look at investment possibilities in oil refining, chemicals, machinery manufacture, and other basic industries.

The group is sponsored by the Business Council for International Understanding, an organization of major U.S. companies with foreign interests. The delegates will be accompanied by Under Secy. of Commerce Franklin D. Roosevelt, Jr.

They will meet with Indian government officials and businessmen, to discuss the impact of such measures as India's recent reduction of corporate taxes, ending of price controls, and exempting foreign technicians from income tax. As an added inducement, the government is playing up the fact that U.S. manufacturing investment in India has an average after-tax return of more than 20%—one of the highest in the world.[17]

Some of the changes in India which were the background to this conference were also well described in the same article:

India's wooing of foreign companies is a significant shift toward greater reliance on private investment as a means of achieving its economic targets.

Behind the shift is a serious slowdown in economic growth under India's current Third Five-Year Plan (1961–66). . . .

Partly, the slowdown was the result of Communist China's attack on India's northeast frontier in 1962, which forced the country to divert scarce investment funds to a defense buildup. Partly, it was due to poor crop weather and bottlenecks in electric power and transportation. Leftist Oil Minister K. D. Malaviya halted expansion of private refineries and launched the government into refinery construction projects that are running behind schedule. Ever since India won its independence in 1947, the government's attitude has been colored by Prime Minister Jawaharlal Nehru's suspicion of private business.

Now, however, new men are coming to leadership in India as a result of the Chinese invasion and Nehru's illness this year. The left wing of the ruling Congress Party was weakened by the political fallout of the attack, and leftist Ministers Krishna Menon and Malaviya were ousted. Rising leaders such as T.T.K. and Lal Bahadur Shastri, Minister Without Portfolio who has been Acting Deputy Prime Minister since Nehru's illness, are middle-roaders both in politics and in economic policy.

T.T.K. is especially interested in attracting more outside investment because of India's desperate shortage of foreign exchange. Payments of principal and interest on past loans are taking a bigger and bigger share of India's export earnings . . .

. . . the big projects are in refineries, petrochemicals, and fertilizers. International oil companies, anxious to assure an outlet for their crude production, are eager to expand existing refineries or participate in new ones. Besides the government refinery in which Phillips will have a share, oil and chemical companies have presented more than a dozen proposals for refinery construction and petrochemical plants that would convert the surplus of naphtha and other light refinery products into chemical fertilizer plants to boost India's stagnant farm output.[18]

The conference brought together not only leaders of major United States corporations, but also many of the top Indian government officials (and members of their staffs), including the Minister of Finance, the Deputy Chairman of the Planning Commission, the Minister of Steel, Mines, and Heavy Engineering, the Minister of Petroleum and Chemicals, the Chairman of the State Bank of India, etc.; in addition, leading representatives of Indian industry attended. A summary of discussions[19] at the conference indicated the Indian government's increased eagerness to attract United States investment. For example, there was this interchange between an Indian Cabinet Minister and a United States businessman:

The Minister commented that there had been some misunderstanding of what India meant when it spoke of socialism and that perhaps he should provide a definition to dispell such misunderstandings. In the Indian context, socialism meant only that development must percolate to all levels of the population. [The businessman] replied that "that is what we mean by capitalism." [20]

Despite the general interest of the Indian government and industrialists in attracting United States investment, there were clearly limitations which produced difficulties. For one thing, the Indian government desired to channel United States investment into specific industries.[21] On the other side, the U.S. businessmen were worried about whether India's limited foreign exchange would allow future repatriation of profits earned by their investments.[22] Even more crucial, it became clear during the conference that there was a huge gap between the Indian government's notion of a "fair" rate of return on investment and that of the United States companies. For example, in the panel discussing fertilizer and petrochemical industries, the following exchange took place:

> The GOI [Government of India] then posed the question as to why American capital appeared to be reluctant to invest more in the fertilizer industry in India. The American spokesman responded by saying there were uncertainties and doubts which they hoped could be resolved through these discussions, in connection with (1) the market, (2) methods of distribution, (3) prices, (4) sources of feed stock, and (5) the issue of Indian collaboration. . . .
>
> The GOI said that prices were based on production costs, depreciation, overheads and a reasonable profit was also taken into account. . . .
>
> The U.S. asked whether the GOI could define what it meant by a "reasonable profit."
>
> The answer was that a general rule of thumb was $12\frac{1}{2}$ percent before tax on total capital employed. The GOI observed that Americans with their efficiency and know-how ought to be able to have a better profit margin than public sector producers and that, therefore, the guarantee that they would receive the same price as the public sector plant, operated in their favor because it opened up the possibility of the Americans having a higher profit margin in their operation.[23]

The government's target rate of return of $12\frac{1}{2}$ percent before taxes, equivalent to about 6 percent after taxes, is far below what any United States firm seeks from overseas investment.

This gap was brought out in an understated way at the conference itself:

> Question: . . . "What does American business consider a fair return on foreign investment?"

> Answer: [A United States' chemical company executive] Each company has its own opinion on this point. The return desired depends on other available investment opportunities, the risk, the tax structure of the foreign government, and the time required to put the plant on a profitable footing. More specifically, a return of 10 percent is considered an acceptable minimum, but 15 percent would be the goal of most investors.[24]

In fact, the foreign profit goals of United States companies, particularly chemical and oil companies and especially in a "risky" country like India, would be more in the range of 15 to 25 percent per year, after taxes.[25] It is the companies' knowledge that foreign firms generally will not invest in India without these high rates of return which contributes to their concern about the problem of repatriating their own profits from India.

Finally, another area of divergency between the government and the companies was highlighted in a statement by an Indian government official:

> "We are more anxious to get things out of your heads than out of your pockets. Indians have the ability to learn if they have someone to teach them."
> The [official] said, "If you are only looking at profitability for the next 2 to 3 years, the prospects are not too good. But if you look in terms of 10 to 15 years, they are better."[26]

The problem for the companies is that they have no desire to wait 10 to 15 years for high profits. Moreover, it is precisely the companies' control of "things in their head," i.e., technology, which helps give them leverage against the Indian government, both in terms of getting a higher rate of return on their investment and also as some defense against nationalization.

In any event, the results of the Indian government's swing toward private foreign capital as indicated by the conference were not long in forthcoming. In January 1965 *The New York Times* reported:

> The Indian Government has instituted a crash program to build new fertilizer plants that would raise production capacity to 1.2 million tons.
>
> The Bechtel Corporation of San Francisco is presently studying the feasibility of building five plants for the manufacturing of a million tons of fertilizer.
>
> The program of the Government's Planning Commission calls for the new industry to fall into the mixed private-public sector.
>
> To attract capital, India has put out a new welcome mat adorned with tax incentives. The Government now seems willing to allow foreign interests to hold majority interests in new plants, particularly in the petrochemical industry.[27]

However, all was not yet clear sailing for United States private investment. Thus, the Bechtel proposal was rejected by the Indian government in early 1965, according to the *Far Eastern Economic Review* "because it stipulated high prices and demanded control over marketing and distribution."[28] (*Petroleum Intelligence Weekly* reported that another reason for the breakdown was the old issue of control of the oil feedstock supply.[29]) Nevertheless, this event proved to be merely a temporary setback to foreign private capital.

The key to the situation appeared to be an agricultural crisis in India, combined with increasing United States government pressure on behalf of private foreign capital in India. In December 1965 *The New York Times* reported from New Delhi:

> Chidambara Subramaniam, India's Minister of Food and Agriculture, left today for Washington, where he will confer with United States officials on steps to meet India's acute food crisis.
>
> India has told the United States that she will need between 10 million and 14 million tons of food grains next year to fill the deficit caused by her worst drought of the century. Most of this must come from the United States, the only country with a large enough surplus.

II. India: A Case Study

Since last June the United States has declined to sign a new year-long Food for Peace program, insisting that India must first take some hard decisions about food policy. . . .

Even if he gets the assurances from Mr. Subramaniam, Mr. Johnson may prefer to hold up the signing of a new year-long Food for Peace agreement until he meets with Prime Minister Lal Bahadur Shastri in February.

It was probably no coincidence that the Indian government announced several important steps long sought by the United States.

At a news conference Prof. Humayun Kabir, Minister for Petroleum and Chemicals, said that all fertilizer plants licensed up to March 31, 1967, would be able to set their own prices and organize their own distribution networks for a period of seven years.

He also announced the establishment of a high-level three man committee to expedite negotiations with foreign investors in the fertilizer field.

The Government's decision to free prices and distribution in fertilizers was seen as an important breakthrough in efforts to attract foreign capital.

The United States still would like to see the Indian Government drop demands for 51 per cent ownership of joint ventures in the fertilizer field.[30]

This latter United States governmental goal was partially met by mid-1966:

The Indian Government signed a collaboration agreement today with the American International Oil Company for the establishment at Madras of India's largest fertilizer plant. . . . [The agreement is] an important breakthrough in India's dealings with foreign companies interested in investing in the Indian fertilizer industry.

The United States government and the International Bank for Reconstruction and Development have insisted that India provide easier terms for foreign private investment in fertilizer plants as one condition of resumed economic aid. . . .

However, to get the agreement, the Indian Government has made certain concessions on management control of the

254

joint venture that may increase Prime Minister Indira Gandhi's political troubles.

The left wing of Mrs. Gandhi's Congress party has been attacking the Government's new fertilizer policy, calling it a "sellout" to American economic pressures.

These critics are already contending that the Government has given American International too much management control for a company that has only 49 per cent of the equity in the joint venture. The Indian Government retains 51 per cent of the equity.

The Government and officials of American International contend that the agreement provides only a "50–50 split" of management control.

They concede that this is slightly more than American International is technically entitled to according to the equity shares, but they say private banks in the United States would not approve loans for a project in which the American collaborator did not have a major voice in management. . . .[31]

The New York Times also bluntly called attention to India's impotence in this situation:

Many of the trends now evident had their origin in the last months of Prime Minister Lal Bahadur Shastri's life.

Much of what is happening now is the result of steady pressure from the United States and the International Bank for Reconstruction and Development, which for the last year have been urging a substantial freeing of the Indian economy and a greater scope for private enterprise.

The United States pressure, in particular, has been highly effective here because the United States provides by far the largest part of the foreign exchange needed to finance India's development and keep the wheels of industry turning.

Call them "strings," call them "conditions," or whatever one likes, India has little choice now but to agree to many of the terms that the United States, through the World Bank, is putting on its aid. For India simply has nowhere else to turn.[32]

As the preceding analysis has demonstrated, this bold United States government policy has been quite successful, in the short-run at least. Questions have been raised, however, about the long-run efficacy of these tactics. Thus, *The New York Times* stated in early 1967:

> The freeze had the desired effect of forcing India to beg, buy or borrow food grain from any country willing to come to her aid and some who were not. But it also left a legacy of humiliation in a nation that had come to regard the United States as its stanchest friend.
>
> What effect this "squeeze" on the Indian Government ultimately may have on the self-help measures sought by the United States is conjectural. But there is no question about the embarrassment of those elements of the Indian Government most closely identified with this country.[33]

It remained, however, for a radical United States journalist, I. F. Stone, to clearly spell out the ultimate dangers of this policy to long-run United States governmental interests:

> Wider attention should be called to President Johnson's plan to change the name Food for Peace to Food for Freedom. To link food to "freedom" as Johnson interprets it is to use hunger as a lever in favor of "free enterprise." Behind his strange and cruel reluctance to commit himself on the additional grain India desperately needs for a famine already begun is Johnson's insistence that India deal with U.S. oil companies for new fertilizer plants instead of building them under public ownership. The reality is that India has too much of this "freedom" already; we cannot remember an Indian famine which was not aggravated by the freedom its speculators and money-lenders enjoy. Johnson had better begin to worry more about the contrast with China. Isolated and without foreign aid, U.S. or Soviet (India has both), China has met its food needs in the same drought years without famine. That is really "self-help" but not the kind the White House likes. That contrast may make a new generation in India feel that "freedom" is only freedom to starve. That's not the whole story, but distended bellies are indifferent to debate.[34]

CHAPTER TWENTY

India: Conclusions and Lessons for the Future

OVER THE LAST TWENTY YEARS oil policy in India has been the focus of a continuing series of struggles. The struggles have taken place between the Indian government and the foreign oil companies, as well as within the government, and have brought into play the power of foreign governments and international economic institutions. The struggles have involved every level of the oil industry from exploration to marketing to petrochemicals. The most fundamental issue has been the respective roles of government versus private foreign business in the Indian oil industry.

For the established international oil companies, the profitability of India has always been as an outlet for external low-cost crude oil or refined products. Thus, for the companies the key goals have been maximum prices for their oil and control of the refining and marketing facilities. On the other hand, the concerns of the Indian government have been to minimize the foreign exchange burden in the oil sector and to ensure indigenous control through government ownership. While the direct conflict between the goals of the companies and the government is obvious, at times there have appeared to be contradictions between the government's aims, at least in the short run. In these cases the government has basically been split between a Left emphasizing government control and a Right stressing foreign exchange needs.

A critical point in the history of the Indian oil industry came in 1960 with the offer of the Soviet Union to provide crude oil and refined products at discounted prices on a barter basis. The subsequent refusal of the established majors to handle the

crude oil or refined products through their refineries or marketing facilities, coupled with the unwillingness and/or inability of
the Indian government to force them to do so, set the stage for
virtually all subsequent developments in the Indian oil industry.

While the established majors won their basic fight to maintain sovereign control over their refining and marketing facilities,
they generated considerable ill will in all segments of the Indian
government; as a result these companies have been blocked from
new refineries in the growing Indian oil market of the 1960s.
Instead, the Indian government has pursued various policies
aimed at reducing the country's dependence on the established
majors, including: (a) trying to force reductions in the imported
crude oil and refined products prices of the established majors;
(b) working with other foreign oil companies through joint ventures; (c) expediting the development of government exploration, refining and marketing efforts, largely with Soviet aid. The
Indian government's efforts to dominate the oil industry in India
have brought strong negative reactions from Western governments, particularly the United States, and the major international
financial organizations, especially the World Bank. This Western pressure has been particularly obvious in recent years in the
case of fertilizer projects, a frequent adjunct of oil refineries.

Looking back it is clear that the "public sector" has made
considerable strides vis-à-vis the private sector, not only in petroleum but in other energy areas. In 1955–56, the last year of
the First Five Year Plan, there was virtually no public sector
participation in oil, only about 3 percent of total coal production
came from the public sector, and the largely public hydroelectric capacity was extremely small. Ten years later, at the end
of the Third Five Year Plan, the picture had changed considerably. The public sector produced almost one-half of the crude
oil moving through Indian refineries, controlled one-fourth of
the refining capacity and distributed about one-third of all refined products consumed. In addition, government production
accounted for one-fourth of India's coal and almost all of a
greatly expanded hydroelectric sector. In sum, aggregating the
various energy sources (on a fuel oil equivalent basis) reveals

that the public sector share of total energy production had increased from 10 percent to about 30 percent in this ten-year period.[1]

What about the future? According to the draft outline of the Fourth Five Year Plan, published in August 1966, the Indian government hoped to continue this trend, aiming for public sector control of over one-half of the total energy production targeted for 1970–71. The specific goals for the public sector were 65 percent of oil refinery capacity, 40 percent of coal production and all hydroelectric and nuclear power capacity. Since these goals lie within the framework of a projected two-thirds increase in energy consumption during this five-year period, the size of the public energy sector would have to triple between 1965–66 and 1970–71. It should be noted that this extremely rapid growth target implicitly reflects a government policy decision to increase oil's share of India's total energy production. Thus, by 1970–71 refinery capacity was projected to more than double over the 1965–66 level, compared to an 85 percent increase in hydroelectricity and a 50 percent increase in coal production.

In turn, the emphasis on oil in India's future energy plans undoubtedly stems from a complex interaction of the previously discussed factors which ultimately determine energy policy. First, in terms of the economy as a whole, the Fourth Plan Draft continued to emphasize industrial growth over that of agriculture.[2] While this does not necessarily favor petroleum over coal, it tends to favor both of these fuels over hydropower which is more closely connected with irrigation and rural flood control. More important, the decision implies a very rapid rate of growth of total energy consumption since industrial output requires far more energy input than does agricultural production. In addition, certain industries closely related to oil are planned to increase extremely rapidly: fertilizer production is targeted to increase sevenfold during the Fourth Plan period, compared to a doubling in steel production, a major consumer of coal.

Second, it seems obvious that the recent successes in finding and producing indigenous crude oil must have swayed the

Indian government in favor of oil. Nevertheless, this could not have been decisive since targeted domestic crude oil production will still be less than half of the projected refining capacity, implying the need to import even greater quantities of crude oil in the future. Moreover, in terms of foreign exchange costs, both initial and recurring, a decision to favor oil relative to coal and hydro is likely to further worsen the foreign exchange shortage in India. An indigenous mining equipment industry for expanding coal production is well developed in India, while expansion of oil refinery capacity is almost totally dependent upon foreign equipment.[3] At the current cost level it is even more obvious that favoring oil over other fuels will worsen the foreign exchange situation, given that both coal and hydropower are available within India while crude oil, at least, will still have to be imported.

The high foreign exchange cost, both current and capital, of favoring oil consumption growth has put Indian government planners in an apparently contradictory position. On the one hand, they claim to seek to maximize public sector control and ownership of the energy sector.[4] On the other hand, they seek to promote the most rapid development in that energy subsector, oil, which requires the largest amount of foreign exchange. This need, in the current Indian setting, can only be seen as a need for foreign resources, which in oil are mainly private.

In retrospect India, like other underdeveloped countries, faced three basic policy choices with regard to the crucial question of ownership in the oil industry: (1) having total government ownership, which in India would have meant nationalization of the existing companies; (2) leaving it all to private enterprise, which in practice means private foreign oil companies; (3) allowing both private and government activity, either in separate operations or in joint ventures. While the Indian government has been theoretically committed to the first of these alternatives, government monopoly, in practice it has chosen the third alternative, involving "peaceful coexistence," albeit with continual conflict. As we have seen, in the years since passage of the 1956 Industrial Policy Resolution reserving energy to the government, not only has the government failed to take complete control of the energy sector, but in fact the private sector in both coal and oil has continued to grow. Moreover, the "public sector" growth in oil (and fertilizers) has partly been illusory. Joint-venture projects are now defined to be in the public sector if the government owns at least half of the equity even if operational control resides with the private sector.

The specific reasons for the failure of the Indian government to achieve its goal of complete dominance in the petroleum field have been explored in detail in the preceding chapters. At a more general level, two sets of interrelated factors may be said to have been the underlying causes of the government's failure. First, economically, while the Indian government has made considerable progress in the industrial sector, it has failed to generate significant growth in the dominant agricultural sector of the economy. This situation has made India increasingly dependent upon foreign aid, both for food to meet recurring agricultural crises and for industrial raw materials and equipment to attempt to maintain industrial growth, which has slowed in recent years.[5] As should be abundantly clear by now, the power of the Western "aid-givers" is usually aligned in favor of the private sector and against the public sector in India.

Second and related, politically the Indian government seems to be fundamentally controlled by the conservative wing of the ruling Congress Party. In the words of the *Far Eastern Economic Review*: "Theoretically a party of the left, the Indian National Congress has generally been dominated by men of very conservative temper."[6] The political influence of the Indian upper class has been highlighted by an Indian legislative committee:

As the Election Commission and others bemoan, the statutory limit on candidates' expenses has remained a legal fiction. Worse still, the rate of spending has been going up in each successive election to the detriment, as one Legislative Committee observed this year, of the legislators' ethical standards. The need for large election chests opens the door to the influence of rich patrons. The committee quoted Mr. V. K. Krishna

Menon's estimate that a single business house has leverage of one kind or another over 100 MP's (out of a total of 748 for both Houses of Parliament).[7]

Indirectly, this conservative control of government has contributed to slowing the growth of a public energy sector by failing to generate growth in Indian agriculture; the latter, of course, would require the widespread agrarian reform which would threaten the position of the Indian landlords. More directly, conservatives have effectively retarded public sector control of energy and petroleum because of their fear of endangering internal political stability and/or external economic relations with the Western aid-giving institutions. Insofar as India could build a public energy sector "on the cheap," through Soviet assistance, this might be acceptable and perhaps desirable to India's conservative forces; however, where the growth of government control in the energy sector threatened a confrontation with the West, as in the Soviet barter oil offer, these same forces would have no part of it.

A fundamental question for India's future (as in many other underdeveloped countries) is whether the basic issue of the public versus private sector is even seen as relevant. Perhaps because of the failure of India's Second and Third Five Year Plans, one senses an increasing feeling in India that the only thing that matters is economic growth, no matter in which sector. Thus, for example, Geoffrey Tyson in a study of Nehru states:

> But the debate (*now quiescent*) on the respective merits of state and private enterprise was largely unreal; India requires them both and is seeking to evolve a system which gives ample opportunity for each to be developed.[8]

Nevertheless, even the pragmatists grant that the public-private debate is theoretically meaningful, even though they believe it of little practical significance. Thus, Tyson states in a footnote:

> . . . though I agree with the statement that "the country entering currently on nationhood is faced, at least in principle, with the interesting problem of selecting an economic system." [9]

Far from being of academic interest, the public versus private question is of the utmost practical significance for the development not only of the energy sector but the whole economy of an underdeveloped country. The basic reason is that in most underdeveloped countries the growth of the private sector in the crucial areas of mining and manufacturing for all practical purposes means the growth of *foreign* private investment. The difficulty with foreign private investment for the underdeveloped country is that it generally remains concentrated in highly profitable, narrow segments of the economy, with the profits being exported to the foreign country rather than circulating throughout the indigenous economy, where it could stimulate widespread economic development. Foreign investment thereby tends to help build and perpetuate a dual economy, of which India today is a good example.

Moreover, insofar as the high profits on foreign investment are in fact reinvested, this can relatively quickly lead to the even more serious problems of loss of indigenous control of the economy and major balance of payments crises. To illustrate this, consider the following hypothetical model of a newly independent underdeveloped country with a virgin economy untouched by foreign investment. The country's planners face a situation in which the total capital stock of the country is indigenously owned, the average and marginal capital-output ratios equal 3:1, and the savings-income ratio equals 0.10. (The capital-output ratio is defined as the amount of initial capital investment required per unit of yearly output that will be derived from the investment; the savings-income ratio is the proportion of yearly income [or output] which is available for new investment.)

Under these assumed conditions, utilizing a simple mathematical economic growth model of the classic Harrod-Domar variety, it could be expected that total national income would grow at about 3.3 percent per year indefinitely (savings-income ratio divided by the capital-output ratio).[10] With population perhaps expected to rise at close to 3% per year, thereby reducing the rate of growth of per capita income to a snail's pace of less than 1% per year, it is extremely tempting to the planners to try to attract foreign investment in order to accelerate the rate of growth of national income.

Now, under what conditions would the foreign investment be willing to come in? As we have indicated earlier, the most basic requirement would be a minimum rate of return of about 20% per year after taxes. Note that this means that foreign investment would only enter those sectors of the indigenous economy which have a very high profit-output ratio and/or a very low capital-output ratio or some combination of the two. This is because, assuming a tax rate of 50% on gross profits, in order to earn 20% after taxes the company must earn 40% before taxes. If the capital-output ratio on a proposed project were the average for the economy, 3:1, then the company could not possibly make 40% before taxes on its investment: even if the total output was profit (i.e., no costs) the before-tax rate of return on investment would only be 33%. Let us assume then that foreign investment enters projects which have an average capital-output ratio of 2:1, which in turn requires that the gross profit-output ratio be 0.8, in order to yield an after-tax rate of profit on investment of 20%. (Incidentally, this requirement of a high profit-output ratio explains why foreign investment tends to be channeled either into resource extraction for external markets, e.g., oil, or production for internal markets characterized by inelastic demand and monopolistic control over supply, e.g., fertilizers, drugs.)

Let us further assume that the high profits on the foreign investment are not simply sent back to the home country but rather are all plowed back into further investment within the indigenous economy. Under these assumptions we can then speak of a "foreign sector" of the underdeveloped country's economy with a capital-output ratio of 2:1 and a "savings-ratio" (after-tax profits divided by total output in the foreign sector) of 0.4. Under these conditions, while output in the indigenous sector of the economy would tend to grow at only 3.3% per year, output in the foreign sector would tend to grow at 20% per year, thereby lifting the growth rate of the total economy from what it would have been absent the foreign investment.

The exact amount of the "lift" depends of course upon the amount of foreign investment which comes in. Assume that in the base year foreign investors put $10 into the economy and that the stock of native capital equaled $285. Under the assumed conditions it can easily be shown that total output in the economy as a whole would increase by an average of 5.0% over a ten-year period, or a 50% faster growth rate than would have existed absent the foreign investment. (See Table 20-1.) Note that all this simply assumes initial investment by foreigners equal to less than 1/25th of the native capital stock. Further, it presupposes no additional investment of external funds by the foreigners, but simply a plowback of their indigenous profits.

If we look deeper into this apparently idyllic picture, there are some disquieting implications for the underdeveloped country. For one thing, if we examine the rate of growth not of total output, but of total output less profits belonging to the foreign investors, this rises not by 5.0% per year but by only 4.4% per year. Moreover, by the end of the ten-year period, while the stock of native capital will have grown somewhat, the rapid growth of the stock of foreign capital will have increased the foreign investor's share of total invested capital from nothing to 14% and the proportion of total output in the economy controlled by the foreign sector will have risen from zero to 19% (Table 20-1).

Taking a somewhat longer view, the benefits and drawbacks become even more dramatic. Over 20 years, the average rate of growth of the economy as a whole increases to 6.8% per year (owing to the increasing relative size of the rapidly growing foreign sector). Even total output less profits of the foreign sector increases by an average of 5.6% per year. However, by the end of the period more than 40% of the capital stock of the country will be owned by foreigners and more than half of the country's total output will be controlled by the foreign sector. In addition, the value of the capital stock owned

TABLE 20-1

IMPACT OF FOREIGN INVESTMENT ON INCOME, GROWTH, CONTROL OF ECONOMY, AND BALANCE OF PAYMENTS: ILLUSTRATIVE NUMBERS

	Years after initial foreign investment				Average per annum growth rates		
	0	10	20	30	0–10	0–20	0–30
(a) Capital stock of foreigners	$ 10	$ 62	$384	$2,376			
(b) Income produced in foreign sector	5	31	192	1,188			
(c) Native capital stock	285	396	546	762			
(d) Income produced in native sector	95	132	183	254	3.3%	3.3%	3.3%
(e) Total capital stock (a + c)	295	458	930	3,138			
(f) Foreign share of total capital stock (a/e)	3%	14%	41%	76%			
(g) Total income (b + d)	$100	163	375	1,442	5.0%	6.8%	9.3%
(h) Profits of foreigners (0.2a)	2	12	77	475	20.0%	20.0%	20.0%
(i) Total income less foreign profits (g − h)	98	151	298	967	4.4%	5.6%	7.9%
(j) Foreign sector share of total income (b/g)	5%	19%	51%	82%			
(k) Exports at 10% of total income (0.1g)	$ 10	$ 16	$ 38	$ 144			
(l) Exports rising from 10% to 20% of total income (0.1g to 0.2g)	$ 10	$ 24	$ 76	$ 288			
(m) Balance of payments margin: exports minus potential profit remittances							
(k − h)	$ 8	$ 4	−$39	−$331			
(l − h)	$ 8	$ 10	−$1	−$187			

BASES: Foreign investment assumed to have a capital-output ratio of 2:1 and an after-tax profit rate of 20% per year; all profits in foreign sector are assumed to be reinvested, so that sector grows at 20% per year. Native investment assumed to have a capital-output ratio of 3:1 and a savings-output ratio of 0.10; latter figure is consistent with a 10% per year profit margin on sales (output) and complete reinvestment, so that sector grows at 3.3% per year.

266

by foreigners will be greater than the total national income, and the annual profits of the foreign sector will amount to one-fifth of the total national income; if at this point foreigners ceased to plow back their profits, one-fifth of each year's production would have to be shipped abroad just to meet profit repatriation.

Finally, if one carried the assumptions out for an additional ten years, the whole economy would be totally swamped by the foreign sector. Long before that time, however, the economy would probably either have broken down completely or ground to a halt owing to the balance of payments crises engendered by the foreign investment.

For example, assume that in the initial period exports equaled about 10% of national income, i.e., $10 per year. Further assume that there was no foreign exchange surplus. If exports over the years remain at 10% of total output, after ten years the potential profit remittances would be 75% of the foreign exchange earned by exports; after 20 years the potential drain of profits on foreign exchange would be twice as great as earnings from exports. Even if exports increased to 20% of total output, after 20 years the potential profit remittances would be as great as export earnings.[11] (See Table 20-1.)

Of course, in the real world such extreme conditions as are depicted in this model would normally never come to pass. This is primarily because there are not enough high-profit investment opportunities in underdeveloped countries to induce foreign investors to continually plow back their profits.

Nevertheless, that the related problems of foreign investment leading to increasing foreign control of a country's capital and an increasing burden on exports are not merely mathematical constructs can be seen from a brief examination of India's situation. In 1962–63 (latest year for which data were available) the total assets of all nonfinancial corporations in India (including some government corporations) amounted to about $14 billion. Of this total, outright branches of foreign companies accounted for an estimated $1.3 billion and foreign-controlled Indian companies an additional $1.9 billion; thus, foreign investors controlled almost one-quarter of the assets of all nonfinancial corporations in India.[12]

Moreover, it would appear from other data that in the crucial manufacturing sector foreign investment has been growing more rapidly than indigenous investment (private plus government). According to Reserve Bank of India data, between June 1948 and December 1962 foreign investment in manufacturing (including petroleum) increased from $200 million to $1 billion, or an average compounded growth rate of 12% per year. (During this same period, foreign investment outside of manufacturing increased by only 3% per year, so that manufacturing's share of foreign investment increased from one-third to two-thirds.) While there are no comparable data on total investment by indigenous groups in manufacturing during the same period, we do have Reserve Bank estimates of "tangible wealth" in 1949–50 and 1960–61, for large-scale factories, both private and public, and small enterprises.[13] The estimates show total tangible wealth for these groups of $4.4 billion in 1949–50 and $10.9 billion in 1960–61. While these figures cannot be used for directly estimating foreigners' share of total manufacturing investment in the country, they can be used as an indicator of the growth rate for total manufacturing investment in the country. For the period 1949–50 to 1960–61, this growth rate was about 9% per year. This suggests then that foreign investors' share of total investment in the manufacturing sector has been rising. If we assume that total investment in manufacturing since June 1948 has been rising at 9% per year, compared to 12% per year for foreign investment, it would appear that foreign investment's share of total investment in manufacturing has increased substantially, from about one-sixth in 1948 to one quarter in 1962.

Furthermore, it seems clear that, in line with our hypothetical model, the bulk of foreign investment has been built up by simply plowing back profits earned in India. Thus, data collected in one study show that between June 1948 and December 1960 reinvested earnings of foreign private capital amounted to $386 million, or more than two-thirds of the increase in that period in total private foreign investment.[14]

With respect to the impact of foreign investment on India's export and balance of payments problems, we quote at length from a more recent study entitled "Social and Economic Costs of Private Foreign Investments" (which appeared in the economic journal of the All India Congress Committee):

It is evident from Table 1 that the burden of repatriation of foreign business investments on India's balance of payments has been continuously increasing. . . . The payments on account of royalties and fees have also gone up. . . . In case of profits too the burden has been increasing. The overall burden of private foreign investments on our balance of payments has more than doubled between 1953 and 1961. . . . As a matter of fact it is one of the important factors contributing to foreign exchange crisis. A major portion of the export earnings will have to be diverted towards the payment of dividends on foreign capital and its repatriation. Not only that, we will be required to pay dividends at higher rates even after the foreign exchange component of the original capital contribution is fully repatriated. This is because a large part of the dividends has to be paid on the reinvested profits earned by the foreign companies in India. This is nothing but a return on the use of our own resources which leaks out of the economy every year. Thus, the long-term economic costs of private foreign investments in terms of mounting and persistent pressure on our balance of payments tends to be exhorbitant.

Another disquietening [sic] feature about private foreign investments in India is that the outflow of investment income has been greater than the net inflow of private foreign capital. This trend is evident from the following figures:

Table 2 reveals that in practically every year the payments on account of repatriation of capital are greater than the gross inward flow of private foreign capital, thus exercising a net burden on our balance of payments. Another significant fact about private foreign capital is that reinvestment of earnings constitutes an important part of private foreign investments in India, which means that it is the earnings of these investments within our own country which is largely responsible for increasing our balance of payments commitments. Thus, private foreign investments cannot be looked upon as a source of foreign exchange resources. In this direction their contribution is negative for the last so many years. The growth of private foreign investments is mainly based on reinvestment of profits and investment in kind which means that private foreign investments have not been instrumental in easing the foreign exchange situation in the country while they contributed to a considerable extent towards its deterioration.[15]

Finally, there is evidence that in recent years the pace of foreign penetration in the crucial modern sector has stepped up. For example, the data on foreign participation in new capital issues approved by the Indian government shows that between 1956 and 1964, of the new capital for manufacturing, 30% was to be provided by foreigners; for 1963 and 1964 the figure was 35%.[16]

Several conclusions emerge from this analysis which, while not logically inconsistent with conventional development theory, point up a vital aspect of foreign private investment which is rarely discussed in the economic development literature. The traditional view is that private foreign investment is difficult to attract or keep within the underdeveloped country, but if it can be attracted it is basically desirable. In the words of a prominent book in the economic development field:

> The central problem now is for the recipient country to devise policies that will succeed in both encouraging a greater inflow of private foreign capital and ensuring that it makes the maximum contribution feasible toward the achievement of the country's development objectives. . . .
> From the standpoint of national economic benefit, the essence of the case for encouraging an inflow of capital is that the increase in real income resulting from the act of investment is greater than the resultant increase in the income of the investor. If the value added to output by the foreign capital is greater than the amount appropriated by the investor, social returns exceed private returns.[17]

Our hypothetical model assumes that the underdeveloped country is completely successful both in attracting the initial foreign investment and in getting it to expand indigenously by reinvesting its profits. Moreover, our model meets the test that the increase in real income resulting from the investment is greater than the profits of the investor. Nevertheless, as our model shows, complete success in attracting foreign investment comes at the price of increasing foreign domination of the under-developed country's economy. Among other things this implies an economy which is increasingly dependent upon external decisions; as expressed by Europeans who fear increasing United States takeover of their countries' businesses, "decisions will be made in New York, Washington or Detroit, rather than in Paris, London or Hamburg."

Moreover, this decline in independence is furthered by the inherent instability in the foreign exchange situation caused by the accelerating potential profit and capital repatriation outflows. Even if the foreign investment was not export-oriented (as is most foreign investment in underdeveloped countries), the need to generate foreign exchange earnings to meet these potential outflows ultimately makes the country increasingly dependent on exports and world markets. After all, no rational foreign investor is going to continue to plow back money into a country which shows signs of not being able to cover profit and capital outflows; at the first signs of such a situation there is usually a "run on the bank" which quickly brings the ultimate danger to immediate realization. At this point the underdeveloped country typically has to turn to the International Monetary Fund for assistance, and the process of loss of independence is virtually complete.

A less tangible but nonetheless real danger implicit in allowing wide-scale foreign investment is that the foreign investor comes to feel that he has a stake in the country and a right to a voice in its operations, insofar as that is necessary to protect or expand his investment. Particularly today, when the modern corporation is so tremendously growth oriented, the pressures on all divisions, both domestic and foreign, to expand sales and profits are especially great. Thus, from an initial attitude of simply seeking an investment climate which "safeguards" the foreigner's investment, his mental set ultimately changes to a belief that he has a "right to participate" in the growth of the

271

indigenous economy. It is obvious that such an attitude must lead the foreign investor to an increasing concern with many phases of the underdeveloped country's political, social, and economic life. Given the powerful governments and institutions which the foreign investor can call on for assistance, the dangers to the independence of the underdeveloped country should be obvious.

Finally, for development planning the question should not be foreign investment versus no investment, but foreign investment versus other alternatives open to the country. One possibility of course is to seek foreign loans. While this might appear to be a simple solution, it should be noted that the available supply of these loans for underdeveloped countries is limited. Further, the main sources, the Western governments and in particular the United States, and the World Bank, basically provide loans only for the infrastructure; their aim is to lay a foundation for private investment to invest in the more profitable sectors of the economy. Moreover, governmental loans usually contain at least implicitly some degree of obligation on the part of the recipient country.[18]

The really fundamental alternative open to an underdeveloped country is to gather and harness effectively its own indigenous resources. This can be done particularly by using unemployed and underemployed labor and natural resources and by increasing the savings rate so that a greater proportion of output can be invested.[19] Such a policy does not necessarily imply greater autarchy, since part of the increased output can be channeled into exports to pay for importing foreign technology and goods.

Moreover, if the government of a country is seriously intent on economic development, existing savings rates can usually be raised significantly. In the words of a leading development economist, W. A. Lewis, "What is lacking in most of these countries is not the means but the will." [20] Conversely, since raising savings rates and harnessing unused resources generally require major economic, political, and social changes, the fact that a government is unwilling to undertake these changes is perhaps the best indicator of a lack of serious will to promote rapid economic development.

To sum up, a fundamental difficulty with relying on private foreign investment is inherent in the basic requirement for attracting such foreign investment: a high profit rate. As discussed above, such a high profit rate can usually be found only in those segments of the underdeveloped country's economy in which the profit-output ratio is also very high, e.g., oil production. But, it is precisely in these segments that the greatest savings rates can also be achieved; if the original investment in these sectors were undertaken by indigenous groups, then the savings could relatively easily be plowed back into further economic development rather than ultimately leaking abroad. In short, opening the door to private foreign investment normally ensures that the potentially most important and productive sectors of the economy will pass into foreign hands, leaving the more intractable sectors such as agriculture and light industry as the preserve of the indigenous investor (government or private).

Moreover, in those situations where the real capital-output ratio may be so high that ordinarily foreign investment could not reap a sufficiently high rate of profit to induce it to enter, foreign investment may be attracted only by artificially creating scarcity; such scarcity in effect raises the monetary value of the output, while not changing its real value. This may be the case where foreign investors receive, for example, promises of high prices in the fertilizer field. On a wider scale, as we have seen, the monetary value of crude oil has been internationally maintained at an artificially high price through the control of supply; such control, however, calls for a less-than-active seeking of supply within the underdeveloped country.

What all this suggests is that relying on private foreign investment for developing key areas of the economy may effectively be selling out a nation's birthright for a mess of pottage; while it may appear attractive in the short run, the long-run implications must be carefully analyzed. With respect to this problem in India, we quote the conclusions of the previously cited study on the costs of foreign investment in India:

II. India: A Case Study

The policy of encouraging private foreign investments in Indian economy is not in the long term interests of the country and is fraught with grave consequences, particularly when ours is a planned economy with specific objectives and priorities. It may speed up the rate of growth in the country to some extent but it is likely to damage the social content of our process of planned development and will also involve us in balance of payment difficulties. Under its most acute form it may amount to a loss of our economic liberty. Even under its modest impact it will lead to an abandonment of serious planning in the country. The trend in this direction is already visible. Before granting any concessions to private foreign capitalists, we must very closely bear in mind the economic and social costs involved in such a process. The excessive inflow of private foreign capital is likely to create conditions of monopoly capitalism in India. It will distort the pattern of our development by weakening the priorities laid down in our plans. It will also tend to make the distribution of wealth more inequitable, to breed the conditions suitable for encouraging the exploitation of Indian consumers and in the long run to impose a burden on our balance of payments which may be beyond our capacity to bear.[21]

If there is any validity in these warnings, they would hold particularly for a crucial sector like oil, which has so many additional political and social ramifications. In the chapters that follow we attempt to analyze the range of possible public versus private oil and energy resource development policies within different socioeconomic settings. These brief studies, along with the Indian case, will serve better than theorizing to indicate the potential benefits and drawbacks of various public-private policy decisions.

PART III

The Range of Petroleum Policies
in Underdeveloped Countries

CHAPTER TWENTY-ONE

Public Energy in a Public Economy: China

COMMUNIST CHINA is an obvious candidate for serving as a case study of the role of public energy in a public underdeveloped economy. Not only is China far and away the most important underdeveloped socialist country, but also comparison of her performance with that of neighboring India is natural; since both countries started two decades ago from a similar low level of economic development and have moved along very different roads, such a comparison is not only intellectually appealing but also operationally significant. Hence in the analysis of China that follows her comparative performance will be highlighted. However, because in a totally planned public economy petroleum policy is only one part of a larger energy and economic policy, the analysis must also necessarily deal with the development of nonpetroleum energy resources and the totality of the economy. This will serve also to highlight again the key relationships between energy and economic growth.

Before proceeding, a word should be said about the sources and reliability of the data presented here for Communist China. This is necessary because, in the words of one Western student of the Chinese economy, Professor Dwight Perkins of Harvard:

A widespread feeling that Chinese statistics are not reliable is based on the fact that China's economy is both underdeveloped and Communist. This feeling has been reinforced in recent years by Chinese Communist admission that agricultural production figures for 1958 were exaggerated and the obvious fact that 1959 data bear little relation to reality. This belief in the unreliability of all Chinese statistics, including those prior to 1958, has an element of truth in it. This is particularly true if one compares these data with comparable figures for the United States or Western Europe. But the real issue is whether Chinese statistics are sufficiently reliable to be used to answer various important questions about China's economy.[1]

Professor Perkins goes on to conclude, as have most Western specialists on the Chinese economy, that there is no evidence of deliberate falsification of data and that China's data, carefully used, can help answer these important questions.[2]

It is in this spirit of using rough "order of magnitude" figures to answer crucial questions and highlight crucial relationships that the present chapter is undertaken. Such an approach is particularly necessary for the years following the Great Leap of 1958, when the principal problem is the paucity of data from Communist China. As a precaution, therefore, we have limited ourselves to the use of estimates generated by Western "China watchers" in Hong Kong and elsewhere; these estimates may be presupposed to have a conservative bias. For energy developments through the 1950s we have relied heavily on a study prepared for the Hoover Institute on War, Revolution and Peace (Stanford University) by Professor Yuan-li Wu, *Economic Development and the Use of Energy Resources in Communist China*.[3]

The war-torn economy which the Communists inherited in China in 1949 was one overwhelmingly dominated by coal. The decline in the coal sector prior to 1949 had been even greater than that in the economy as a whole, owing to:

> . . . the destruction of productive capacity in Manchuria as a result of the deliberate industrial stripping by Soviet occupation troops; the shutting down of small mines in Free China after the return of government and industry to the coastal areas; and the general economic dislocation which occurred in Japanese-occupied areas. . . .
> The continued decline of coal output in 1946–48 was . . . in part the result of the civil war, the economic dislocation, and the inflation that characterized those years.[4]

Thus, coal production, which had reached a peak of 66 million tons in 1942, slid precipitously to 14 million tons in 1948, and

278

did not return to the wartime peak until 1952. Hence, 1952 is a good point of departure for evaluating the performance of the Chinese Communists in the energy sector (as well as the economy as a whole).

The other primary energy sources were virtually totally undeveloped in China at that point. Hydroelectric capacity equaled only 340,000 kilowatts, or less than one-fifth of the small total electric capacity; hydropower accounted for less than 3 percent of the total primary energy supply. The petroleum sector was even more underdeveloped with total indigenous crude oil production in 1952 of only 400,000 tons, or 1 percent of the total energy supply.[5] That pre-Communist China, like India, was essentially a marketing outlet for the refined products of the international oil companies can be seen from the fact that in 1947 over 2 million tons, or 14 million barrels, of petroleum products were imported into China.

There is a striking similarity, both in level and in composition, in the energy bases of China and India in 1952. Thus, converting Chinese coal to Indian coal on the basis of their average caloric content, in 1952 total energy consumption in China was about 56 million tons, of which more than 95 percent was supplied by coal; India's total energy consumption was about 41 million tons, with coal accounting for 90 percent. Since China's population was about 50 percent greater, on a per capita basis consumption of total commercial energy was almost identical in the two countries. Moreover, in both countries hydroelectric capacity was virtually undeveloped, indigenous crude oil and refined product production was minimal, and the petroleum industry was controlled by the international majors who simply marketed imported products.

Since 1952, even by the most conservative estimates, energy production in China has increased tremendously. Here, one must distinguish between the period before and after the Great Leap of 1958. Between 1952 and 1957 production of coal in mainland China doubled from 66 million tons to 131 million tons. Coal production was then officially reported to have leaped to 425 million tons by 1960. However, it seems clear that the real value of this increase in coal production is greatly reduced by the fact that much of it came from small "native mines" (with inadequate facilities for removing impurities) rather than modern mines, and hence, much of it was of poor quality. Even after allowing for the decline in quality, Wu estimates that in terms of millions of tons of "standard coal equivalent," Chinese production rose from 93 million tons in 1957 to 130 million tons in 1958 and 205 million tons in 1960, or a doubling in the 1957–60 period.[6]

Since 1960, while no reliable data are available, it would appear that Chinese coal production at first fell sharply, then leveled off, and in the last few years has been on the rise. According to the most authoritative Western "guesstimate," prepared by K. P. Wang of the United States Bureau of Mines, production of raw coal in China was 250 million tons in 1961–62, but reached 300 million tons by 1965.[7] Since the evidence suggests that the cutback in coal production from the 425 million ton peak of 1960 came in the poorest quality coal, it seems reasonable to assume that the average quality of Chinese coal has returned to the 1957 level. On this basis it would then appear that as a rough estimate coal production and consumption in China increased about fivefold in the 1952–65 period, or an average growth rate of about 12 percent per year.

The rate of growth of other primary energy sources in China has been even more rapid. Hydroelectric capacity has increased ninefold, from 340,000 kw in 1952 to an estimated 3,000,000 kw in 1965.[8] The petroleum sector has made even more rapid strides. Production of crude oil was estimated by Wang to have increased over twentyfold, from 0.4 million tons in 1952 to 10.0 million tons in 1965. Indigenous production of refined products has increased correspondingly, to an estimated 9.0 million tons, so that, according to Wang:

> Additional production altered the overall supply position, as the country truly became virtually self-sufficient in oil. Imports, primarily consisting of refined products and from the Soviet Union, were much smaller than a few years back . . . On the other hand, Chinese offers to sell crude oil to Japan appeared to be serious, and there was talk about extending a pipeline from

Manchuria to North Korea. More oil was distributed to the civilian economy, and the price of gasoline was cut 18.6 percent.[9]

Summing up, China's total commercial energy production and consumption appears to have more than quadrupled in the 1952–65 period. It has reached a level about three times as great as that of India on an absolute basis and twice as much on a per capita basis. In other words, over the comparable period the rate of growth of energy production and consumption in China has been about twice as fast as that in India.

What are the reasons for China's faster energy growth? On the supply side, one cause of China's ability to expand more rapidly has been its continued primary reliance upon indigenous coal. While Wang notes that "In early Communist Chinese planning, natural resources were given a prominent position and the policy has not wavered . . . ,"[10] with specific regard to coal Wu comments:

An interesting development in the Chinese coal industry is the continual increase in volume of deposit reported in spite of the remarkable rate of coal output during the last decade and the correspondingly rapid depletion of existing reserves. This is obviously a result of the extension of geological surveys, especially in areas hitherto inaccessible to survey teams. . . .

Between 1949 and 1958 the number of coal prospecting workers increased from 2,500 to 60,600 and included 5,600 technicians. They were organized into 97 teams. Drilling machines increased from 80 to 13,000 sets during the same period, and annual drilling was raised from an average of 800,000 meters during the First Five-Year Plan to over 2,000,000 meters in 1958. As a result of these efforts, about 200 new major coal fields with a combined reserve of 230 billion tons were reportedly discovered.[11]

This is not to say that a similar exploration effort has not been made in the petroleum industry. In Wu's words:

One might not be far wrong in maintaining that the principal effort exerted by Communist China in the petroleum industry since 1949 has been concentrated in the search for oil deposits

. . . exploratory drilling for oil between 1907 (when the first drilling was done at Yen-ch'ang in Shensi province) and 1948 amounted to a cumulative total of only 34,000 meters. By contrast, a total of more than 18,000 meters was drilled in 1953 alone, and the amount increased to 408,000 meters in 1956. Through the introduction of new techniques the speed of drilling was increased from 30 meters a day in 1952 to 318 meters in 1957 while the depth explored increased from less than 1,000 meters to over 2,000 meters. More than 6,000 persons were involved in prospecting for oil in 1956.[12]

Communist China's estimated oil production in 1965 was only about twice that of India compared to coal production almost four times as great. Part of the reason would seem to be that China has devoted relatively much more of its resources to seeking coal than oil, according to Wu's figures employing about ten times as many workers in coal prospecting as oil exploration. This in turn appears related to China's general demand pattern for energy fuels.

On the energy demand side, the fundamental reason for the more rapid growth of consumption in China than in India would appear to be the more rapid rate of industrialization of China than of India. Today, when Communist China is already an important nuclear power it is perhaps hard to realize that in 1952 the Chinese industrial sector was relatively more backward than that of India. The energy sector itself mirrored the backwardness of China's industry: coal, which accounted for 98 percent of China's energy, was mostly consumed in households, while industrial consumption of coal stood at about the same level as in India. In the vital rail transport sector, Indian railways in 1952 consumed about 11 million tons of coal compared to only 6 million tons for Chinese railways; while Indian railways carried only three-fourths the amount of freight that China's did, they transported more than three times as many passengers. Finally, taking a crucial industrial commodity, pig iron, 1952 output in China was 1.9 million tons or barely more than the 1.7 million tons produced in India.

Pig iron production is also indicative of the much more rapid strides made by heavy industry in China than in India. By 1965, according to Wang's estimate, Chinese pig iron production had increased tenfold to 19 million tons. During the same time period, India's pig iron production had increased fourfold to 7.2 million tons. Thus, Chinese pig iron production had grown more than twice as fast as India's.[13] Another index of China's more rapid industrialization was the growth of the electric power sectors in both countries. While electric power generation increased sixfold in both countries between 1952 and 1965, in China it had recorded a tenfold increase by 1960; India, on the other hand, had relatively constant growth over the period. The implication is that China has much greater electric generation capacity right now, but is not fully utilizing it because of the cutbacks in industrial production, particularly in the related areas of iron and steel and coal. In the words of one study:

> The growth in consumption of electric energy in Communist China in the years 1949–59, and the decline since, is mainly a reflection of the growth and decline in industrial production. . . .
>
> Consumption of electric energy in agriculture is the only category of consumption that has shown consistent gains. Much of the increase is going to networks of irrigation and pumping stations being built for purposes of draining farmland in time of flood and irrigating it in time of drought.[14]

Thus, when industrial production gets back on the track it is likely that the rate of growth of China's electrical generation will be much higher than that of India, e.g., as measured over the 1952–70 period.[15]

Finally, it is instructive to analyze China's performance in another critical field related to energy, namely fertilizers, and compare it with that of India. In 1952, consumption of chemical fertilizers in China amounted to only 67 thousand tons of plant nutrient equivalent, or virtually the same level as India.[16] Between 1952 and 1958 consumption increased tenfold, to 696 thousand tons, but then declined somewhat to a low of 561 thousand tons in 1961. However, this was still more than 8 times the 1952 figure, and 50 percent greater than consumption in India in 1961.

Although definitive data are not available, it is clear that since the 1961 low point production and consumption have again risen rapidly. Thus, we have the following 1966 report from Hong Kong:

> As to production, the most one can say is that it is probably going up fast since it is receiving much prominence. Last year's chemical fertilizer output has been variously estimated at figures between 4.5 million tons and some 9 million tons [presumably gross tonnage rather than plant nutrient equivalent], the latter figure being directly based on the percentages and so on issued by the Chinese themselves. The figure seems very high for what is known of Chinese production capacity, but on the other hand the Chinese do not usually publish misleading information.[17]

Assuming that the average ratio of plant nutrients to gross tonnage was the same in China in 1965 as estimated for 1963,[18] 22 percent, these figures imply production in 1965 of between 1 and 2 million tons of chemical fertilizer, on a plant nutrient basis. Furthermore, since China's 1965 fertilizer imports were 2.5 million gross tons,[19] or about 500 thousand tons on a plant nutrient basis, China's 1965 fertilizer consumption on a plant nutrient basis would be between 1.5 and 2.5 million tons, or more than 2–3 times as much as the 1958 peak.

By way of comparison, in 1965 China's chemical fertilizer production was three to five times as great as India's, while its consumption was 67 to 200 percent higher. Moreover, while China's consumption has been increasing faster than India's, in contrast to India China has become increasingly less dependent upon foreign supplies: in 1956–58 the import share of fertilizer consumption was two-thirds, in 1962 two-fifths, and in 1965 between one-third and one-fifth. Perhaps even more significant, at the same time that India was becoming increasingly dependent upon Western, and particularly United States, foreign aid for its fertilizer supplies, the Chinese became independent of Soviet assistance in this area.[20]

What accounts for the fact that in China, according to fertilizer expert Jung-chao Liu of McGill University, "Annual [fertilizer] production was extremely low in comparison with that of most advanced countries, but the rate of expansion was higher than in any other country in the world during the same period of time"? [21] Hong Kong observers explain the rapid growth, particularly in recent years, as follows:

> This has been done partly by the Chinese policy of "walking on two legs"—i.e. using both modern, large-scale techniques, and simple, small-scale local plants. . . . A recent news item from *Hsinhua* pointed out (overlooking the bad years of 1960–62) that the "big leap" which had begun in 1958 was still going on.
>
> This is perhaps more true of the fertiliser industry than any other, as the Chinese still follow the "two legs" policy introduced in 1958, but today's small local factories are much more efficient than anything which operated in that chaotic year. Synthetic ammonia is not something which can be manufactured by enthusiastic peasants; as the Chinese themselves have wisely pointed out, "The equipment . . . require(s) strict supervision because of the high temperature, high pressure, inflammability, explosiveness and the poisonous character of some of the elements involved". The advantages of the small, local plants is that the investment can more readily be found (by the local authorities) and they take only a short time to build.
>
> It is the proliferation of these small plants, plus the operation of five or six new large plants, in addition to the old-established ones which have been enlarged and re-equipped, which is today given as accounting for the increase.[22]

(An underlying factor of course in the success of the Chinese fertilizer industry has been the rapid development of the Chinese coal industry, since in China coal is thus far the most important fertilizer feedstock.)[23]

At a more general level the most crucial factor in the tremendous expansion of the chemical fertilizer industry in China has been the efforts of the state. In Jung-chao Liu's words:

> Obviously technology and raw materials alone cannot insure the rapid growth of any industry. The simultaneous develop-

ment of the machine-building industry, of industrial organization, and of the transportation network are equally vital. However, the development of the chemical-fertilizer industry in Communist China provides a good example of how a government decision can divert the necessary resources to a single industry of strategic importance. If one looks at the overall development of the chemical fertilizer industry in Europe, Japan, and the United States, it appears that rapid growth has usually been associated with new inventions, wars, depressed raw-material prices, and/or an increase in farmers' purchasing power. In China, none of these factors seems to be mainly responsible for the expansion, which has resulted rather from intensive government stimulation and intervention.[24]

In summing up, the performance of Communist China in the energy and fertilizer fields has been very closely correlated with its overall industrial performance. This overall performance has on the whole been impressive, and even more so when contrasted with that of India. China has managed to industrialize much faster than India, while at the same time building up major military (including nuclear) power. India has received far greater quantities of foreign aid, and now has a huge debt burden, while China, which received significant amounts of Soviet aid in the 1950s, repaid all of this by 1965; moreover, the abrupt halt of Soviet assistance to China in 1960 undoubtedly contributed to a loss of growth momentum for a number of years.[25]

The key to the differential energy performances would seem to lie in the immediate and intensive efforts of the Chinese to discover and harness indigenous energy resources. India, on the other hand, relied for too many years on external sources; for example, it will be recalled that the Indian government did not launch its own oil exploration effort until eight years after the government had come to power. China pushed ahead immediately with its own indigenous exploration efforts, not only in coal but in oil, despite the fact that in the early years of the regime Soviet imports were readily available.

The importance of relying upon one's own efforts rather than external sources and/or foreign aid may now be belatedly being recognized in India too. Thus, we have the following report about the new head of the Indian Planning Commission:

> In a lecture last year he sounded the theme of self-reliance and criticized the earlier commission for having placed too great a dependence on foreign aid.
>
> In an allusion to the United States, he said this was "self-defeating" in view of "the known bias of the outsiders" who give aid "under conditions which deflect us even further from our avowed objectives." [26]

In the meantime, however, India has lost many years in its race for economic growth. More important, the critical question is whether the Indian government and social system can yet prove that India can successfully harness and mobilize its indigenous resources to fulfill the proclaimed goal of growth through self-reliance. The evidence to date offers little encouragement for optimism on this score.

CHAPTER TWENTY-TWO

Public Energy in a Private Economy: Mexico

PETRÓLEOS MEXICANOS, commonly known as Pemex, the government oil monopoly of Mexico, has probably been the most controversial oil company in the underdeveloped world. Its creation following nationalization of various international oil companies in 1938 ensured Western hostility from the very beginning. (See Chapter 24.) The purpose of this chapter, however, is not to analyze or evaluate the heated controversy which surrounded Pemex's birth as a nationalized oil company. Rather, the aim here is to analyze and evaluate, as quantitatively as possible, the role and impact of a public oil sector in a fundamentally private enterprise economy. This will be done in two stages: (1) examining Pemex's performance as an investor of public funds, in terms of the criterion of social rate of return on investment; (2) examining Pemex's wider impact upon the Mexican economy as a whole.

Central to both of these issues is a correct conceptual framework for analyzing the impact of any investment, be it public or private, on economic development in an underdeveloped country. As we shall show, much of the criticism of Pemex's performance has been largely irrelevant to these central issues.

Thus, for example, the previously-cited Levy report prepared for the World Bank discussed Pemex's performance between 1938 and 1959, and concluded that in the past Pemex had failed in a vital area since it had been unable to earn an adequate return on its capital investment; as a result Pemex also was unable to accumulate funds for steady future expansion. This essentially negative conclusion appears to be based on viewing a public corporation's contribution to the economy in terms of the criterion used for evaluating the success of private corporations:

namely, reported net profits (divided by capital invested). In fact, however, this is hardly a correct criteria in the case of Pemex in the light of the fact (as recognized by the same Levy report) that the company's earnings have been intentionally kept low in order to subsidize the rest of the Mexican economy (through such means as low prices and high social expenditures). With a virtual monopoly in such an essential commodity as petroleum it would be quite easy for Pemex to keep prices high and thereby have an enormous rate of return on its investment. Yet this would not necessarily make Pemex any more or less a valuable contributor to economic development. In order to make such a judgment one needs a wider and more dynamic notion of "returns" to the economy as a whole.

In evaluating the contributions of a public enterprise in a private economy ideally one should know two things. First, what has been the change in the situation as the result of the actions of the public enterprise? Second, what would have happened in the absence of the public enterprise, i.e., what if anything would private enterprise have done in the same area? Since the second question is far more speculative, we have largely left our answer for Mexico to an argument by analogy—by examining in the next chapter the experience of Iraq, an underdeveloped economy in which the private sector has always controlled petroleum. In answering the first question, since no two countries would ever show exactly similar results from the same oil policy, no claim is made that the specific quantitative conclusions we have made about the Mexican experience are automatically transferable to other underdeveloped countries.

The first question we address ourselves to is: "What has been the rate of return to the Mexican economy of Pemex's investment?" Fortunately, sufficient data are available to roughly estimate the rate of return on Pemex's investment in the most crucial level of the oil industry—exploration. One conceptual problem here is how to measure the value of oil which is found in a country like Mexico which already had sizable reserves. A good conservative method is to assume that all previously discovered oil would be consumed first and then estimate when in the future

TABLE 22-1

EXPENDITURE AND RETURN FROM PEMEX EXPLORATION INVESTMENT, 1938–1966

Year	Investment in millions of dollars (a)	Change in reserves of oil and gas in million barrels (b)	Oil and gas production in million barrels (c)	Oil and gas found annually[a] (b + c) (d)	Finding costs (a/d) (e)
1938–1945	$ 1	275	322	597	$0.002/bbl
1946	5	−78	54	−24	—
1947	11	−49	63	14	0.79
1948	13	−21	66	45	0.29
1949	12	283	70	353	0.03
1950	14	−43	85	42	0.33
1951	19	301	94	395	0.05
1952	19	323	96	419	0.05
1953	12	−8	91	83	0.14
1954	11	316	102	418	0.03
1955	11	201	113	314	0.03
1956	11	209	116	325	0.03
1957	23	414	121	535	0.04
1958	29	703	148	851	0.03
1959	42	278	162	440	0.10
1960	41	439	167	606	0.07

1961	53[b]	203	179	382	0.14
1962	38[b]	17	196	213	0.18
1963	34[b]	143	206	349	0.10
1964	50[b]	77	226	303	0.17
1965	66[b]	−149	231	82	0.80
1966	91[b]	278	240	518	0.18
Totals	$606	4,112	3,148	7,260	$0.08 av

[a] Production plus change in year-end reserves; figures indicate millions of barrels, with gas converted to oil equivalent.
[b] Estimated.

SOURCES: Data for 1938–1961 from: Antonio J. Bermúdez (Director General, Pemex, 1947–1958), *The Mexican National Petroleum Industry: A Case Study in Nationalization* (Stanford University: Institute of Hispanic American and Luso-Brazilian Studies, 1963). Data for 1962–1966 from: *World Petroleum*, April 1967. Oil exploration investment data for 1961–1966 not published and estimated by author on basis of published data on total drilling costs (exploration plus development) and number of exploration and development wells, respectively, drilled. Based on 1958 data each exploration well assumed to be twice as expensive as each development well; nondrilling (survey) exploration costs assumed to be 25 percent of exploration drilling costs.

the newly discovered oil would be used. Then, for each year multiply the estimated consumption of the newly discovered oil by the estimated price for imported crude oil which otherwise would have had to be imported in those future years, absent the new discoveries (after deducting from the import price the relatively minor real costs necessary to bring the indigenous oil to the surface, once discovered). Finally, compute that discount rate which would equate this hypothetical stream of revenues from the newly discovered oil with the amount of money invested in exploring to find that oil. This discount rate would then be the social rate of return on the exploration investment.[1]

Between 1938 and 1966 Pemex invested an estimated $600 million in exploration efforts (see Table 22-1). These efforts resulted in finding 7 billion barrels of oil reserves, including the crude oil equivalent of natural gas (see Table 22-1). Putting a value on this crude oil of a minimum of $1.00 and a maximum of $2.00 per barrel [2] would indicate that this nearly thirty years of exploration effort resulted in finding $7 to $14 billion worth of crude oil and gas, or a return of $12 to $24 for every $1 invested.

However, because this oil will not be used until many years later, one cannot simply say that the rate of return on the $600 million investment was 1200 to 2400 percent. Assuming that all of the previously existing crude oil reserves had been used first, the newly discovered crude oil would have not been required until 1956 and it would probably be used through 1981. (See Table 22-2, for data and bases of calculations.) Even with a conservative forecast of oil production growth, the social return to the exploration investment in the 1938–66 period would be extremely high—between 28 and 35 percent, compounded through 1981. That is to say, the return on this exploration investment is approximately equal to what would have been earned if each year the Mexican government had deposited in a bank what Pemex invested in oil exploration that year, had received a 28 to 35 percent compound interest rate, and had allowed all the money to accumulate through 1981.

One further word should be said about the historical profita-

TABLE 22-2

DCF RETURN ON PEMEX EXPLORATION INVESTMENT,
1938–1966

Year	Net returns (in millions of dollars) on oil valued at		
	$2.00/bbl	*$1.50/bbl*	*$1.00/bbl*
1938–45	—$1	—$1	—$1
1946	—5	—5	—5
1947	—11	—11	—11
1948	—13	—13	—13
1949	—12	—12	—12
1950	—14	—14	—14
1951	—19	—19	—19
1952	—19	—19	—19
1953	—12	—12	—12
1954	—11	—11	—11
1955	—11	—11	—11
1956	55	37	23
1957	219	159	98
1958	267	193	119
1959	282	201	120
1960	293	210	126
1961	303	216	126
1962	354	256	158
1963	378	275	172
1964	402	289	176
1965	396	281	165
1966	389	269	149
1967	508	381	254
1968	538	404	269
1969	570	428	285
1970	604	453	302
1971	640	480	320
1972	678	509	339
1973	718	539	359
1974	762	572	381
1975	808	606	404
1976	856	642	428
1977	908	681	454

	Net returns (*in millions of dollars*) on oil valued at		
Year	*$2.00/bbl*	*$1.50/bbl*	*$1.00/bbl*
1978	962	722	481
1979	1020	765	510
1980	1082	812	541
1981	58	44	29
DCF Rate of return (compounded annually) =	35%	31%	28%

SOURCES: Net returns for 1938–66 based on oil exploration investment data of Table 22-1 and estimated timing of production from oil and gas discovered in 1938–66. The basic assumption that all reserves existing in 1938—1,240 million barrels—are used up first implies that first year of production from new reserves is 1956—32 million barrels from new reserves required. Annual production after 1966 is based on assuming a conservative 6 percent per year growth rate, so that all reserves discovered in 1938–66 will be used up in the 1956–81 period.

bility of Pemex's oil exploration effort, as measured by the social rate of return on investment. As noted above, the Levy report indicated that through 1959 Pemex had not been very successful. A similar view of the historical lack of success of Pemex was expressed in 1967 by *The Wall Street Journal:* "Until recently, Pemex seemed to be in the rut of many government oil monopolies—suffering from production declines, inefficiency, red tape and corruption." [3] In fact, however, analysis of Pemex's oil exploration investment between 1938 and 1959, when $266 million was invested, shows about the same social rate of return as in later years. (See Table 22-3.) During this time Pemex discovered almost 5 billion barrels of oil, worth between $5 and $10 billion, and the estimated rate of return was 28 to 36 percent per year. Thus, it is clear that Pemex's exploration efforts in the earlier period, correctly evaluated, were in fact extremely successful in terms of a social return to the public investment. [4]

Finally, there are some data available [5] which allow very crude order of magnitude estimates of the return to Pemex for its "non-oil-exploration" investment (in refining and distribution) in the 1938–58 period. (See Table 22-4.) [6] Depending on the specific assumptions the nonexploration investment of $615 million would yield gross returns ranging between $1.3 billion and

TABLE 22-3

DCF RETURN ON PEMEX EXPLORATION INVESTMENT,
1938–1959

| Year | Net returns (in millions of dollars) on oil valued at | | |
	$2.00/bbl	$1.50/bbl	$1.00/bbl
1938–45	−$1	−$1	−$1
1946	−5	−5	−5
1947	−11	−11	−11
1948	−13	−13	−13
1949	−12	−12	−12
1950	−14	−14	−14
1951	−19	−19	−19
1952	−19	−19	−19
1953	−12	−12	−12
1954	−11	−11	−11
1955	−11	−11	−11
1956	55	37	23
1957	219	159	98
1958	267	193	119
1959	282	201	120
1960	334	251	167
1961	356	269	179
1962	392	294	196
1963	412	309	206
1964	452	339	226
1965	462	347	231
1966	480	360	240
1967	508	381	254
1968	538	404	269
1969	570	428	285
1970	604	453	302
1971	640	480	320
1972	678	509	339
1973	718	539	359
1974	762	572	381
1975	780	585	390
DCF Rate of return (compounded annually) =	36%	33%	28%

SOURCES: See note to Table 22-2 for basic methodology. Only difference here is that oil discovered in 1938–59 period becomes exhausted by 1975.

TABLE 22-4

DCF RATE OF RETURN ON PEMEX
REFINING INVESTMENT, 1938–1958

Year	Refining investment in millions of dollars	Quantity of refined products from refining investment (in million barrels)	Net returns (in millions of dollars) with oil refining valued at:	
			$2.00/bbl	$1.00/bbl
1938	$ 2	0	$−2	$−2
1939	6	0	−6	−6
1940	16	0	−16	−16
1941	4	0	−4	−4
1942	4	0	−4	−4
1943	5	0	−5	−5
1944	10	2	−6	−8
1945	24	5	−14	−19
1946	18	9	0	−9
1947	7	17	27	10
1948	17	15	13	−2
1949	16	20	24	4
1950	28	21	14	−7
1951	22	26	30	4
1952	26	30	34	4
1953	44	36	28	−8
1954	65	42	21	−23
1955	68	46	24	−22
1956	60	47	34	−13
1957	84	52	20	−32
1958	91	61	31	−30
1959		70	140	70
1960		72	144	72
1961		75	150	75
1962		75	150	75
1963		75	150	75
1964		75	150	75
1965		75	150	75
1966		75	150	75
1967		75	150	75

Year	Refining investment in millions of dollars	Quantity of refined products from re- fining investment (in million barrels)	Net returns (in millions of dollars) with oil refining valued at:	
			$2.00/bbl	$1.00/bbl
1968	75		150	75
1969	75		150	75
1970	75		150	75
DCF Rate of return on investment (compounded annually)			23%	11%

SOURCES: Data on refining investment estimated from Bermúdez, by subtracting exploration investment from Pemex's total capital investments for all purposes; the estimate thus actually equals all nonexploration investment and hence overstates refining investment. 1938–61 quantity data also based on Bermúdez.

$2.6 billion. On a DCF basis the compounded annual rate of return would be between 11 and 23 percent. As would be expected, this is considerably lower than the return to oil exploration. Needless to say, however, without the refining (and distribution) investment, the crude oil would have been of much less value to Pemex or Mexico.

Finally, the speculative question may be raised as to whether in the absence of nationalization the foreign oil companies might have proceeded to develop Mexico's oil in a way which would have left the country as well off, if not better, than under Pemex. There would seem to be two basic answers to this, both of which point to a negative conclusion. First, in the prenationalization era of the international oil companies, Mexican oil was primarily for export. And, the companies were losing interest in Mexico since the most lucrative fields had been (wastefully) exploited already. As *The Wall Street Journal* noted:

> Mexico's production had been falling sharply even before ex-
> propriation. As some of Mexico's oil fields became exhausted,
> foreign companies turned their attention to Venezuela. Mexico's
> output was only 46.9 million barrels in 1937. Few oil men gave
> Petroleos Mexicanos much chance of survival . . .[7]

In the light of the vast crude oil reserves of Venezuela and the Middle East available after World War II to the international oil companies, it is highly unlikely that they would have been greatly interested in exploring in Mexico. Moreover, since internal demand in Mexico appears to have been stimulated by subsidization pricing which the oil companies would never have voluntarily undertaken, it is not likely that under private auspices Mexico would have been a rapidly growing market for oil.

Second, even if the companies had been willing to undertake the oil exploration, assuming the traditional 50–50 profit split, close to half of the $14 billion worth of oil discovered in the 1938–66 period would have flowed out of the country in the form of profits. Giving up $7 billion worth of oil in order to save $600 million in oil exploration investment would seem to be a poor bargain for any underdeveloped country. Morcover, the cost to the Mexican economy from having shared with the international oil companies might well prove to have been even greater. In the words of an early director general of Pemex:

> In the calculation of crude reserves, Petroleos Mexicanos has always followed a conservative policy. Indeed, for diverse reasons, petroleum technical experts not connected with Pemex have invariably made calculations of Mexico's proved crude oil reserves much higher than those of Pemex.[8]

Turning now to the question of Pemex's overall role in the economic development of the Mexican economy as a whole, there seems little doubt that petroleum has been a (if not *the*) leading sector in the Mexican development process. While there is no way of definitively proving this, evidence for it can be derived from an examination of the growth of Pemex in relation to the growth of the economy as a whole, and also the manner in which the "locked up capital" available to Pemex from crude oil discoveries has been utilized.

The Mexican economy has shown a remarkably steady growth rate in real output of about 6 percent per year over the last two decades. During the period in which Pemex was assertedly "unsuccessful," the annual growth rate of 6 percent in total production was associated with a 7 percent increase in output in the manufacturing sector. Since output in the oil sector increased by 8 percent per year, and petroleum accounts for over nine-tenths of all energy used in Mexico, it is clear that total energy consumption was rising at a faster rate than either the economy's growth or that of manufacturing.

Undoubtedly an important factor leading to this rapid growth in petroleum demand was the government policy of keeping petroleum prices very low (a cause of much of Pemex's financial problems). While it may be argued whether this method of subsidization of the overall economy is desirable, there can be no doubt of the direction in which it operates. When joined with the government policy of also keeping electricity prices relatively low, and considering that petroleum is the fuel in most Mexican thermal-electric plants, it can be seen that Pemex's subsidization of petroleum prices played a dual role in reducing the cost of energy. Energy's specific relationship to manufacturing in Mexico has been described as follows:

> Between 1945 and 1960, electric energy consumption increased at nearly 11 per cent a year, a rate almost double that of the increase in gross national product . . . At the war's end, industrial consumption accounted for 60 per cent of the total, and residential and commercial consumption another 20 per cent. By 1960 industry's share declined to 45 per cent while that of the two other principal end-users increased to 30 per cent. (Industry's declining share was due not only to the faster growth of demand in some other sectors, but also to the increasing substitution by industry of cheap oil products for electricity.) At the same time, changes in the pattern of industrial power consumption prove that Mexican industrialization was progressing on all fronts. In 1945, the manufacture of capital goods had absorbed only 30 per cent of the power bought by the industrial sector as against 70 per cent used in the manufacture of consumer goods. Fifteen years later, these proportions were reversed, partly because a new and important customer group emerged in the manufacturing sector—the chemical producers. By 1960, the chemical industry alone accounted for one fifth of the energy consumed by Mexico's industrial activities.[9]

In addition to charging low prices, Pemex has also subsidized the rest of the economy through such various means as paying higher than necessary wages to its workers, and providing a variety of social services such as "paving roads, building schools, providing water systems and other public works that contribute to the economy of rural regions." [10] Again it may be argued whether this is the best method of utilizing the locked up capital which Pemex taps. But the decisive point is that the capital is only available because of Pemex's activities in oil exploration; while its activities may not be sufficient to ensure efficient economic development, they are a way of providing the necessary capital foundations.

Pemex also indirectly subsidizes the economy in ways which may be recognized as more directly and conventionally fitting the role of a dynamic leading sector which moves ahead of the rest of the economy, dragging it along:

> As Mr. Reyes Heroles [Director General of Pemex] sees it, a national oil company should be a useful social tool. "There are some areas of the country where there is no justification, from a profitability point of view, for a gas line; however, its laying breaks the vicious cycle hampering the progress of those areas," he explains.[11]

Another important way in which Pemex has played the role of leading sector has been described by Director General Bermúdez:

> Petróleos Mexicanos has invariably followed the policy of favoring the purchase of nationally-manufactured goods, and of actuating investments that produce, within the country, articles used by the petroleum industry. However, this policy has not consisted in preferring national goods indiscriminately by sacrificing quality and price. In initial operations, certain unfavorable specifications in domestic materials which did not appear in foreign purchases were tolerated. As a result of this policy, however, national industry has been able to supply in increasingly greater proportions, the needs of Pemex while improving quality until such time as competition could be made on even bases.[12]

A dramatic example of this starts with large diameter pipe, which through Pemex's instigation was manufactured by a local businessman; this in turn stimulated production of sheet steel by Altos Hornos de Mexico, the leading steel producer in Mexico.[13] (Between 1953 and 1962 the petroleum sector's steel consumption equaled 20 percent of Mexico's steel production.[14]) As a result of this general preference policy, in the 1946–57 period Pemex's purchases from all domestic manufacturers rose from virtually nil to $56 million, or about half of its total purchases;[15] by 1966 Pemex's total purchases of goods and services, excluding payroll, was up to $200 million.[16]

Finally, Pemex is consciously providing a major stimulus to overall economic development through its activities in the crucial petrochemical sector. Thus, in a paper prepared by Pemex, appropriately entitled "Natural Gas Reserves in Mexico as a Factor of the Social and Economic Development of the Country by Means of Nitrogenous Compounds," Pemex indicated its view of its role as follows:

> Petróleos Mexicanos (Pemex), by being an integrated petroleum institution, and now also a petrochemical institution, plays a very important part in the economic development of Mexico. The development activities of Petróleos Mexicanos have been oriented to benefit all the levels and sectors of the population. The great achievements in industrial development in Mexico have caused an unbalanced situation between the two most important sectors of the country, the industrial one in full prosperity, and the largest sector, which is the agricultural population, in a depressed situation. In order to correct this imbalance, Petróleos Mexicanos has developed this programme, by means of which it is expected that the agricultural and rural sectors will receive the benefits of integrated development. . . .
>
> Petróleos Mexicanos is proposing to contribute to the correction of this situation by applying a more intensive technology and promoting a wider use of fertilizers. These proposals can be implemented in a short period of time, and, in consideration, of the low income of the agricultural sector Petróleos Mexicanos proposes to make ammonia and nitrogenous fertilizers accessible to the agricultural and rural sector, at prices considerably lower than at present. Reduction in the price of ammonia by 30 per cent is being suggested. . . .

Several means are being proposed by means of which Petróleos Mexicanos could reduce the price of ammonia and nitrogenous compounds. However, the one which is bound to produce the greater impact is the construction by Pemex of ammonia and urea plants with the greatest capacity that can be built at the present and by the utilization of the most advanced technology in the manufacture of fertilizers.[17]

Pemex's ability to undertake the role of providing modern technology for the development of the overall economy is unique in Latin America, as a brief survey prepared by a United States engineering firm indicates:

> Mexico is fortunate and far-sighted in having a large group of professional chemical engineers and skilled technicians, both of which are indispensable for the application of modern technology. These engineers and technicians made special note of the full use and understanding of automatic instrumentation, whose operation must be proper and reliable for the use of continuous processes.[18]

This "fortunate" situation is a far cry from that at Pemex's birth in 1938, when the firm could barely operate the existing small petroleum industry. However, the result does not appear to be accidental, but instead seems to reflect the natural development of a state oil industry oriented toward indigenous growth both for the industry and the economy as a whole. It does not seem any more fortuitous than the fact that a 1964 UN survey of the petrochemical industry in Latin America revealed that 40 percent of the petrochemical capacity in operation or under construction was owned by Pemex.[19]

Today, Pemex is busy on all fronts, building pipelines, refineries, petrochemical facilities, and buying the nucleus of an oil tanker fleet. In all of this, however, it has not lost sight of the fundamental role of crude oil exploration:

> Mr. Reyes Heroles has given top priority to efforts to increase oil reserves. Pemex has launched in the Gulf of Mexico the most ambitious offshore drilling venture of any national oil company

anywhere, with expected ultimate reserves of 10 billion barrels of oil-and-gas-bearing hydrocarbons.[20]

And, as Mr. Reyes Heroles reported vividly in March 1967:

> At sea, with an exploration and drilling investment of 203 million pesos [$16 million], probable reserves of over 150 million barrels were obtained. Could this be considered a bad operation for Mexico? Only the blind and the deaf, who also deserve to be mute, have dared insinuate it.[21]

Thus the evidence, both quantitative and qualitative, suggests that Pemex as a public oil sector has played and will continue to play a vital leading role in the development of the Mexican economy. It is little wonder that the previously cited *Wall Street Journal* story on Pemex was headlined: "Model monopoly —nationalized oil agency in Mexico so successful it worries the industry—firms fear other lands may follow example of Pemex," and continued as follows:

> Private oil company executives are worried about the impact an increasingly efficient Pemex might have on private oil operations in the rest of Latin America and in the Mideast. "As a successful government venture, Pemex is the model for other countries wanting to nationalize their oil," says an apprehensive vice president of a U.S.-based international petroleum concern.[22]

In order to fully appreciate the reasons for these oil company "apprehensions," it is useful to compare the role that "private oil" has played in the economic development of countries which are, as Mexico once was, using their petroleum resources primarily for export. We turn to this task in the following chapter on Iraq.

CHAPTER TWENTY-THREE

Private Petroleum in a Private Economy: Iraq

IRAQ AFFORDS A GOOD TEST CASE of the effect on economic development of private foreign ownership of oil in an oil-rich private enterprise economy. There are two basic reasons for the choice of Iraq as the "model" of an oil-exporting country. First, we have the benefit of a recent comprehensive study relating to this problem, *Financing Economic Development in Iraq: the Role of Oil in a Middle Eastern Economy* by Abbas Alnasrawi; this work explicitly includes "a study of the impact on and the response of the non-oil sector of the economy to the stimulus which the oil sector provided. . . ." [1] Second, Iraq represents a most fair test of the role played by privately owned oil in generating economic development because, unlike some oil-exporting underdeveloped countries which are almost total desert, Iraq has always had other important resources. As Alnasrawi notes:

> Iraq possesses most of the salient features of an underdeveloped country. These include a very low per capita income, a high rate of illiteracy, a high rate of unemployment, a high degree of resource immobility, an unequal distribution of income and wealth, a stagnant agricultural sector, and a low productivity of labor. . . .
>
> Iraq, however, can be distinguished from other underdeveloped countries in her potential for growth. Iraq, for instance, is spared the pressing problem of population surplus. With an area of 168,000 square miles, Iraq should be able to support more than her present total population of about 7 million. What supports this contention is that Iraq has large areas of cultivable land. While the per capita potentially cultivable land in Iraq is 3 to 3.5 acres, it is half an acre in Asia and less than one-

third of an acre in Egypt. Together with the cultivable land, proper management of Iraq water resources, the Tigris and Euphrates rivers, should give rise to a developing and prosperous agriculture.[2]

A consortium of international oil companies has always controlled oil production in Iraq, ever since the first concession was granted to the consortium, now known as the Iraq Petroleum Company (IPC) in 1925; the parents are British Petroleum, Shell, Jersey, Mobil, and CFP.[3] Various concessions have been negotiated by IPC, running through the year 2000 and (up through 1960 at least) blanketing the entire country. While the first commercial quantities of oil were found in 1927, until 1950 oil production was relatively low, never rising above five million tons per year (or about 35 million barrels). Since then production has increased dramatically, reaching 47 million tons in 1960 and 67 million tons in 1966. The growth of Iraqi oil production paralleled that of the rest of the Middle East in response to the rapidly rising demand after World War II, but initially it lagged by about five years, owing to the need for construction of new pipelines.

The growing crude oil production in Iraq has been completely oriented to the export market, with less than 5 percent of the crude produced being used at home. Iraq has only one oil refinery (government-owned) with an annual capacity under 3 million tons. Thus, despite its enormous petroleum resources, consumption of total energy in Iraq, even on a per capita basis, is amazingly low: less than one-half the world average, and, interestingly, less than two-thirds that of Mexico. Hence, any significant impact of the oil industry on Iraq's economic development must perforce have come through the oil revenues received by the Iraqi government.

In this respect the period analyzed by Alnasrawi, 1950–64, is particularly favorable to Iraq because early in 1952 the government's revenue was increased from the former royalty payment of a flat 33 cents per barrel to a 50–50 share of profits, or closer to 80 cents per barrel. Thus, over the 1950–64 period the Iraqi government has received relatively huge oil revenues,

amounting in total to $3.0 billion. This figure is equal to over
$400 for every man, woman, and child in the country. The criti-
cal question is what has been the impact of this enormous government oil revenue upon the overall development of the Iraqi
economy.

Superficially, progress over the 15-year period under study
might be considered moderately impressive. Thus, real national
income (in 1956 prices) has been estimated to have risen by
about 7 percent per year. Moreover, even excluding the income
generated in the oil sector,[4] total nonoil national income grew at
6 percent per year, about in line with that of Mexico. Even after
subtracting population growth of 2.3 percent per year, the per
capita increase in real income has averaged about 3.7 percent per
year over this fairly long period.

However, considering the enormous government oil revenues relative to the size of the economy, the Iraqi performance
appears disappointing indeed. Alnasrawi has made the following
estimate of the effect of Iraqi government development expenditures from oil revenues:

> In order to know the impact of development expenditures on
> the rate of growth . . . we have to know the capital/output
> ratio. . . . The estimates of the former vary from 3:1 to 6:1.
> . . . With development expenditure averaging . . . 20 per cent
> of the 1951 national income, then national income can be said
> to have increased by 5 per cent per annum ($20 \times \frac{1}{4}$) as a result
> of development expenditures.[5]

Even this 5 percent figure may be an incorrect overestimate.
Since development expenditures over the whole 1950–64 period
averaged only 14 percent of nonoil national income, use of Alnasrawi's 4:1 capital output ratio would imply that nonoil national income increased by only about 3.5 percent per year ($14
\times \frac{1}{4}$) as a result of the government's development expenditures.
Whatever the correct figure, 3.5 percent or 5 percent, it seems
clear that the impact of these enormous oil revenues has been
relatively marginal. In the words of Alnasrawi:

> . . . Iraq's experiment in planning and development has failed
> to evoke a strong response from the non-oil sector to the de-

velopment stimulus and has failed to achieve major structural changes in the economy. This failure has increased rather than lessened the dependence of the development effort and the entire economy on the oil revenues.[6]

In order to understand the reasons for this failure we must first examine Iraqi government policy in utilizing its oil revenues.

One cause of failure lay in the inability and/or unwillingness of the Iraqi government to fully utilize its oil revenues for capital investment in economic development; of the $3 billion total government oil revenues only half were used for development expenditures. Part of this "leakage" was the result of deliberate government policy. Thus, when the Iraqi Development Board was originally created in 1950 by law it was to receive all government oil revenues. After the 50–50 company-government profit sharing agreement of 1952, the law was changed to reduce the Development Board's share of government oil revenues to 70 percent, with the remainder allotted to the ordinary budget for current expenses. After the revolution of 1958, the Development Board's share was further reduced to 50 percent of the government's oil revenues.

Moreover, the Development Board's actual expenditures lagged behind both planned expenditures and actual revenues. Over the whole 1950–64 period various plans envisioned development expenditures of $3.2 billion, while oil revenues of the Board were $1.9 billion and actual expenditures only $1.5 billion. Thus, not only did the Development Board fail to tap significant nonoil revenue sources, but in fact it piled up a "surplus" from oil revenues of over $300 million.

A second cause of failure was the sectoral allocation of the government investments. The overwhelming emphasis of the development programs was on "social overhead capital," particularly irrigation projects, along with roads and bridges, and buildings. In the first four Development Plans, covering the prerevolution period (1951–58), such social overhead capital as a percentage of the total allocated investment funds was: 89 percent, 71 percent, 75 percent, and 80 percent, respectively. Moreover, actual (as opposed to planned) investment in social overhead capital turned out to be an even higher proportion of actual total investment.[7]

The virtual ignoring of direct investment in industry or agriculture in this period appears strange in light of Iraq's structural situation.[8] The explanation for the Iraqi government policies in the period before 1958 would appear to lie in the problem common to many underdeveloped countries: namely, control of the government by a relatively small group of landlords who wish to preserve the status quo. Great concentration of land ownership can be seen from the fact that while there were 4.4 million landless agriculturists in Iraq, 3,600 landlords owned over 80 percent of the cultivated land.[9] This concentration of land ownership and its associated political power can be seen as, simultaneously, a cause of agricultural problems, a deterrent to their solution, and an explanation of the development program undertaken. In Alnasrawi's words:

> Such a system of land tenure gave rise to two interrelated problems of income distribution. Within the rural sector a minority of landlords received the greater share of the yield and at the same time paid little attention to the methods of production or the quality of the produce. But since the majority of the population is rural peasantry . . . an impoverished rural majority cannot be expected to generate savings for investment on the land to improve yields.
>
> The other feature of agriculture is that crop rotations are rare and the general system of cultivation is fallow farming and usually about 40 per cent to 50 per cent of the land is left uncultivated during the year.
>
> The third problem of agriculture is the inadequacy of the drainage system . . .
>
> With these fundamental problems, one would expect the development programs to include an agricultural development policy which would lead to a better distribution of land ownership, better methods of cultivation, and a different system of irrigation and drainage.
>
> The introduction of agrarian reform, to solve the land problem, was not possible because of the very nature of the political

structure of the country where the landed class dominated the legislative apparatus. . . .

Instead, a considerable part of the Board's expenditures went to irrigation, which actually included the projects for flood control, water storage, and irrigation. . . .

Given the land tenure system at the time, the benefits from the stored waters had to accrue to the already privileged minority. . . .[10]

The lack of industrial development in the pre-1958 period would also seem to be due to conservative forces. Interestingly, the World Bank sent a mission to Iraq in 1952 which found that:

Conditions . . . are generally favorable to further industrial development. An expanding agriculture should provide more materials . . . for processing and when the standard of living among the rural population is raised there should be a growing domestic market for industrial products. In oil and natural gas the country possesses a cheap source of power and fuel as well as an important source of raw materials.[11]

On the other hand, the Bank mission went on to recommend that government ownership of industry should take place only where absolutely necessary. It then offers the following policy guidance:

Particular care should be taken that government assistance does not foster inefficient industries at the expense of the country's standard of living. This danger is by no means unreal. . . . Iraqi-owned industrial enterprises can qualify for these benefits as long as they primarily use raw materials available in Iraq or produce goods imported in considerable quantities. Neither of these conditions is necessarily relevant to efficient production; and it is therefore suggested that the law be amended to stipulate that in general, such benefits should be accorded only by enterprises which have a reasonable prospect of becoming efficient enough to withstand foreign competition.[12]

The Development Board received similar strategy advice from other Western consultants.[13] Alnasrawi's analysis of this strategy, which emphasizes agricultural development at the expense of industrial development, correctly indicates its inapplicability to Iraq (and to many other underdeveloped countries):

> One of the reasons for this argument is that since Iraq is an agricultural country with a relative abundance of water and cultivable land, she has a comparative advantage in developing agriculture. . . .
> The other reason underlying these recommendations is that Iraq is not equipped with the technical skill needed to carry out a comprehensive program of industrialization. . . .
> The third justification for this policy recommendation is that the protection which the industrial sector receives will tend to further the imbalance in the distribution of income between the rural and the urban sectors of the economy. . . .
> . . . the concentration on the development of agriculture for the world markets would have exposed the economy to the instabilities associated with the inevitable fluctuations in the proceeds from the export of primary commodities. This would have led to the defeat of the very goal of the development policy, that of loosening the dependence of the economy on oil.
> Furthermore, Iraq is in the unique position in that the supply of capital is elastic. The waste which would result from industrialization could be easily absorbed. . . .
> The second line of reasoning behind the experts' advice against forced industrialization, i.e., the lack of skilled workers, can be criticized . . . Importation of foreign talents can be accompanied by a thorough program of training. It goes without saying that the supply of skilled workers will increase as the process of industrialization picks up momentum. To put it differently, the factories, though not efficiently run, will serve as training centers. Hence the recommendation that Iraq should give special attention to handicraft industries is of only limited advantage. With an abundant supply of capital, Iraq could afford to follow a policy of rapid economic development achieved by means of rapid industrialization without restricting consumption. . . .
> The third line of reasoning, that the solution must be sought not in protecting industry but in agrarian reform, can be said to be unrealistic. . . .[14]

Put more bluntly, the advice which Iraq received from these consultants, if followed in practice, would have left the country essentially as a "hewer of wood and a hauler of water," where handicrafts equal wood and water equals oil. The advice of the World Bank mission undoubtedly reflects its strong bias against government enterprise along with its intense interest in promoting world trade.[15] It is important to note that in the prerevolution period, "The Board seemed to have subscribed to the advice of the experts. This is evident not only in the allocation of a small share of its funds to industrial development, but also in the failure of the Board to achieve its own modest targets." [16] It is hard to believe that the willingness of the Board to go along with this advice was not at least partly due to the political power of status quo forces. In particular, the indigenous business community in very underdeveloped countries like Iraq typically is dominated by commercial rather than industrial capitalists. The importers, along with their foreign supporters including major foreign trading groups, clearly benefited most from the "nonindustrialization" policy. Between 1950 and 1957 imports rose from $37 million to $313 million, or almost twice as fast as national income, and by 1957 amounted to almost one-third of national income.

The revolution of 1958 brought some attempts at changing the previous development policies, including steps in the direction of limited land reform and greater emphasis on industrialization. However, little progress was made in either area.[17] One cause of the continuing failure of development policy after the revolution of 1958 would appear to be the political instability which ensued.[18] At the same time, it should be remembered that ever since the 1958 revolution the various Iraqi regimes have been involved in major struggles with the international oil companies and the Western governments:

The Iraq Revolution of July 14, 1958, led by Abdul Karim al Kassim, touched off American occupation of Lebanon to prevent the revolt from spreading to Jordan, which had entered a kind of union with monarchial Iraq. When Kassim announced that the IPC concession would not be touched, it was reported in Washington that "intervention will not be extended to Iraq as long as the revolutionary government in Iraq respects Western oil interests." [19]

After three years of verbal conflict between the new government and IPC over such issues as changing the 50–50 profit sharing formula, increasing Iraqi oil production, giving the Iraqi government an equity share in IPC, and having the company relinquish some of its concession (which then covered the whole country), negotiations collapsed in October 1961. At this point, as discussed in Chapter 6, the Iraqi government seized the nonproducing part of IPC's concession; the situation is still unsettled at the time of this writing.

Whether in the period of "political stability" under an oppressive monarchy, or in the later period of "political instability" under various military regimes, it is difficult to disagree with Alnasrawi's previously quoted dismal conclusion: "Iraq's experiment in planning and development has failed to evoke a strong response from the nonoil sector to the development stimulus and has failed to achieve major structural changes in the economy." [20] The potential role that oil could have played as the leading sector in Iraq's economic development during this period can perhaps be illustrated through presenting some hypothetical quantitative growth patterns that the Iraqi economy might theoretically have taken under rational development policies.

Because several arbitrary and mechanistic assumptions have to be made in this kind of aggregate quantitative model-building, it cannot be concluded without much deeper disaggregate analysis that such growth patterns would have been attainable in practice; perhaps they can best be viewed as indicative of the magnitude of potential growth tendencies.

The fundamental assumption underlying the following analysis is that since the oil sector was relatively insignificant for Iraq in 1950 (with government oil revenues accounting for less than 2 percent of national income), the subsequent enormous growth in government oil revenues was essentially a "windfall" which could have been used in toto for investment in industry and agriculture without in any way reducing current consumption. (Some idea of the real magnitude of the government's oil revenues in 1950–64 of over $400 per person can be seen by analogy; for India this would be equivalent to $200 billion, or ten times as much as that country received in foreign aid in the same period.)

To quantify the potential impact on Iraqi development of this huge capital windfall (all of which was in foreign exchange), assume that the government in 1950 had divided the economy into two sectors: the existing nonoil sector, which we shall call the traditional economy, and a new (nonoil) modern sector which was to be created on the basis of the government oil revenues. In 1950 the traditional sector generated over 90 percent of national income, or about $400 million. Further assume, conservatively, that the government decided to leave this traditional sector completely on its own and that production remained stagnant throughout the period. Given the actual inflation in Iraq's nonoil sector of over 4 percent per year, income in 1964 prices from this sector would have been $700 million.[21] Assume Alnasrawi's incremental capital-output ratio of 4:1, and that all of the government's actual oil revenues were invested in capital projects with such a ratio. Then, using a classic Harrod-Domar econometric model (see Chapter 20), to estimate the oil revenues' effect on the growth in national income it only remains to make an assumption about the proportion of income generated by the new capital projects which could have been saved and invested as opposed to consumed. (See Table 23-1.)

One assumption might be that 80 percent of the newly generated income could be reinvested and 20 percent could be passed on to the population in the form of current consumption. Under this assumption which implies that real "modern sector" income would equal one-fourth of each year's government oil revenues, compounded annually at 20 percent per year, by 1964 national income (in 1964 prices) generated by the newly created modern sector would have been $4.1 billion per year, or seven times that from the traditional sector.[22] (See Table 23-2.) Total per capita income from both sectors (excluding oil) in 1964 would have been $700, compared to the actual figure of well

TABLE 23-1

ACTUAL AND HYPOTHETICAL INCOME GROWTH IN IRAQ,
1950–1964

Real Income Stream Derivable from
Investing All Government Oil Revenues

Year	Government oil revenues in millions of dollars	Real annual income (in millions of dollars, 1950 prices) derivable by 1964 from investment of all government oil revenues in modern sector, assuming capital/output ratio of 4:1 and various savings/income ratios in modern sector:		
		80%	*50%*	*20%*
(*a*)	(*b*)	(*c*)	(*d*)	(*e*)
1950	$ 15	$ 48	$ 20	$ 7
1951	37	99	63	18
1952	91	203	94	41
1953	140	260	128	60
1954	183	283	149	75
1955	236	304	171	92
1956	193	208	124	72
1957	136	122	78	48
1958	224	168	114	75
1959	243	151	109	78
1960	266	138	107	81
1961	265	115	94	77
1962	266	96	85	73
1963	308	92	87	81
1964	353	88	88	88
Total	$2,956			
Total annual[a]		$2,357	$1,511	$966

[a] Modern sector income generated in 1964 from investing all government oil revenues (million $, 1950 prices).

under $200. At the other extreme, if one makes a much more conservative assumption—that only 20 percent of the additional output created by the government oil revenue investment would be reinvested—the modern sector by 1964 would still have generated national income in 1964 prices of $1.7 billion per year,

TABLE 23-2

ACTUAL AND HYPOTHETICAL ANNUAL INCOME IN IRAQ, 1950–64

		National income (in millions of dollars)			Per capita income[a]	
	Gov. oil revenues	Nonoil "traditional" sector	Modern sector (hypothetical)	Total	Nonoil	Total
Actual (current prices)						
1950	$ 15	$ 408	$ 0	$ 423	$ 77	$ 78
1964	353	1,148	0	1,501	164	214
Hypothetical (1964 prices)[b]						
Modern sector savings ratio (0.80)	353	705	4,108	5,166	688	738
Modern sector savings ratio (0.50)	353	705	2,614	3,672	474	525
Modern sector savings ratio (0.20)	353	705	1,671	2,729	339	390

[a] Population estimates of 5.3 million in 1950 and 7.0 million in 1964 derived from 1964 figure and Alnasrawi's estimate of population growth at 2.3 percent per year—Alnasrawi,[1] p. 87.

[b] Hypothetical 1964 income figures based on multiplying total income figures in 1950 prices—columns (c), (d), and (e) of Table 23-1 by 1.73 to allow for 4 percent per year inflation.

or more than twice the traditional sector. Under this assumption nonoil per capita income in 1964 would have equaled over $300. For an intermediate savings ratio, 50 percent, 1964 nonoil per capita income would have equaled $500 (see Table 23-2).

In other words, under these various assumptions Iraq could have used its "windfall" oil revenues to achieve in 15 years a nonoil per capita income far above most underdeveloped countries,[23] and one in the neighborhood of countries like Mexico which appear on the brink of sustained economic growth. Note that almost from the beginning per capita consumption would also have been rising steadily. Note further that at the end of this "takeoff" period government oil revenues would amount to 10 to 15 percent of total national income, so that the economy's dependence on oil revenues would have been less than it is now. Moreover, with the new higher level of per capita income, and close to three-fourths of it being generated in the modern sector, it would probably be relatively easy to maintain an overall savings ratio of at least 20 percent. Thus the economy could have a relatively assured long-run growth of at least 5 percent per year *even if all oil revenues were eliminated.*

Regardless of these numbers, choice of appropriate industrial investment projects, such as fertilizer and petrochemical plants, heavily utilizing petroleum and natural gas resources, could have speeded the development process by allowing the government to capture an even greater proportion of the increased output than might otherwise be the case. The potential here was clearly enormous: "Every day 150 billion cubic feet of gas are flared in Iraq; fortunately the government has had better luck than Venezuela in gaining the right to use this gas free of production charges." [24] The potential in agriculture also was great. As late as 1959 the Food and Agricultural Organization estimated that Iraqi agriculture could contribute about $1 billion in 1965. By 1963 the actual contribution of agriculture was less than one-third of that, while investment in agriculture had fallen far short of that projected by the FAO.[25]

Finally, what is the lesson of Iraq's experience in terms of the potential effect of oil upon the economic development of underdeveloped countries? The fundamental answer would seem to be that even abundant supplies of oil represent only a *potential* for development, and are no guarantee that development will take place.[26] A basic prerequisite to economic development today is the determination to develop combined with the willingness to use state power to overcome the forces which have a vested interest in underdevelopment. When governments are fundamentally controlled by conservative landed classes, allied with indigenous commercial tradesmen, military groups, and foreign companies and their home governments, all of which have a stake in the preservation of the status quo, there appears little hope for major economic development regardless of the amounts of capital potentially available.

In Iraq, like many underdeveloped countries, these groups form a natural alliance whose common interest was in maintaining the country in an underdeveloped condition. Above all, the landed classes fear genuine agrarian reform which would both reduce their own agricultural holdings and make hired labor more expensive;[27] on the other hand, the distorted type of development investment undertaken in Iraq in the irrigation area serves to enrich this small minority of wealthy landowners. Indigenous commercial interests, along with foreign traders, profit enormously from the tremendous import flow which is encouraged and abetted by wasteful development policies, and which is threatened by genuine industrial development that might initially displace large amounts of imports. Similarly, the professional officer class benefits from the kind of "development planning" utilized in Iraq:

> The DEP [Detailed Economic Plan of 1962] gave the impression of emphasizing housing by allocating 25 per cent of the funds for housing and buildings. However, an inspection of the figures will reveal that out of a total of ID 140.1 million [$392 million], ID 40 million was allocated to the Ministry of Defense, ID 10 million to army officers' housing . . . and only ID 14 million for popular housing.[28]

Iraq allocated 14 percent of its national income for defense expenditures in 1964 compared to 6 percent for Egypt, and 3 to 4 percent for 22 underdeveloped and 15 developed countries analyzed in a recent study.[29]

Finally, the foreign oil companies and their home governments clearly benefit from the maintenance of existing conditions. The more that Iraqi governmental revenues are dependent upon oil revenues, the more are the Iraqi government and indigenous ruling groups lined up against the popular forces favoring nationalization of the oil industry. (See Chapter 6.) Further, to the extent that the threat of nationalization is effectively ruled out, Iraqi governmental pressures for greater profit sharing and participation in the oil industry are correspondingly lessened. From the companies' viewpoint "wasteful" use of government oil revenues for luxury imports, military spending, etc., are certainly preferable to government expenditure for building up the state oil sector; the latter might even seek to purchase marketing facilities abroad or enter into barter deals with governments of oil-importing underdeveloped countries, thereby competing with the companies and threatening their profits. Finally, at the end of the line of a process of true economic development might lie the prospect for the country of largely ending its dependence upon oil revenues. Such a process, which of necessity brings into power popular forces devoted to agrarian reform, etc., might also give these forces the keys for opening the Pandora's box of nationalization.

CHAPTER TWENTY-FOUR

The Naked Politics of Oil: Oil Boycotts

OF ALL THE COMMODITIES moving in international trade, oil is undoubtedly the supremely political one. Because most oil moves internationally, because the trade is of enormous monetary value, because huge profits are to be made, because it is a vital necessity to most oil-importing countries, because it is of crucial importance to the economies of the oil-exporting countries and the balance of payments of the developed countries—for all these reasons, the ordinary day-to-day flows of international oil trade are the resultants of enormous and conflicting political pressures among the companies, governments, and international organizations.

Like an iceberg, the struggles among the parties in the international oil trade are usually subsurface, emerging only slightly in a case like India. It is only in those extreme situations, involving oil boycotts, that the enormous latent political and economic forces always involved in international oil surface fully. Because the oil boycott strikingly reveals both the vital significance of oil and the true nature of the political and economic pressures centering around it at all times, it is important to have some understanding of the mechanics and nature of oil boycotts. No attempt will be made in this chapter to examine in detail the numerous oil boycotts which have taken place in the twentieth century. Instead, we will focus on an examination of the nature and impact of these oil boycotts, and the implications for energy planning.[1]

The boycotted countries naturally divide into two groups: oil-exporting countries and oil-importing countries. Three countries have the dubious honor of being oil exporters who have been the targets of international oil boycotts: the Soviet Union, Mexico, and Iran. Many more oil-importing countries have been the targets or near targets of oil boycotts, including Italy, Israel, Finland, China, Rhodesia, South Africa, Ceylon, and the most significant one of all, Cuba. Because the focus of this book is on the oil-importing underdeveloped countries, only a relatively brief analysis will be made of the boycotts of oil-exporting countries.

The first important boycott of an oil exporter, originated by the international oil companies, came after the 1917 Russian revolution during which oil properties owned by the international companies were among the first to be nationalized:

> . . . on July 21, 1922, Teagle [Jersey], Gustav Nobel, and Deterding [Royal Dutch Shell] met in London to form a united front against the Bolsheviks. Deterding admitted that "the boycott is never an agreeable nor even a clean weapon." To which the British oil publicists, Davenport and Cooke retorted: "Eyewash! Royal Dutch does not usually shrink from any weapon to attain its own aggressive ends." Teagle's "London memorandum" pledged the three to demand, in any negotiations with Moscow, complete indemnification for seized properties, or the restitution of the properties with indemnification for losses. None was to approach the Russians without informing the others. In September of 1922 Deterding presided over a meeting in Paris in which all the major companies with Russian claims—most of them smaller fry French and Belgian firms—pledged themselves to the united front of the big three. Jersey prudently refrained from formally signing the agreement so that it could not be charged with being "at the bottom of the opposition against Russia." [2]

Ultimately this boycott was unsuccessful, largely because there was a shortage of oil on world markets, which undermined any tendency of the boycotters to stick together; in addition, as internal oil demand grew the Russians became less interested in exporting oil.[3]

The next major boycott of an oil-exporting country, Mexico, came in 1938. In this case the immediate cause was also nationalization, but the original issue was the international oil companies' challenge to the Mexican government's sovereignty: the refusal of the oil companies to comply with a wage increase awarded by the Mexican Federal Labor Board, which in turn led to a strike closing down the crucial petroleum industry.[4] After President Cárdenas seized the companies' facilities, the companies (Jersey, Royal Dutch Shell, Standard of California, and Sinclair) launched a boycott on Mexico's oil exports and also on her imports of goods and services to run the oil industry.

The immediate impact of the boycott was to cut sharply Mexico's oil exports, primarily refined products; ultimately, however, this boycott failed also. One reason for Mexico's ability to withstand the boycott was that it was able to overcome the departure of key foreign personnel and restore production. Another factor was the willingness of the Mexican government to pay compensation, along with the inability and/or unwillingness of the oil companies to stick together. Finally, a critical factor in Mexico's ability to withstand the boycott was that it essentially abandoned the export area in favor of promoting internal consumption of oil.[5]

The most dramatic and significant boycott of an oil-exporting country took place in the early 1950s, following the nationalization of the oil industry in Iran in mid-1951. This case became a particularly naked political struggle, partly because the British government itself owned the majority of the Iranian oil industry, through its 51 percent share in Anglo-Iranian (later BP), the sole oil company operating in Iran. In addition, the United States government played a key role in helping both to overthrow the Iranian government and to gain a place for United States oil companies within Iran.

The long-run factors leading up to the nationalization consisted essentially of a widespread belief among Iranians that they got little benefit from the oil industry,[6] and that without control of it they could not rule their own destinies. Related to this were longstanding grievances over the way Iran, historically a political football between Great Britain and Russia, had been forced or "tricked" into unfair oil agreements.[7]

III. The Range of Policies in Underdeveloped Countries

The immediate backdrop to the 1951 crisis was the government's drive to gain greater oil revenues for financing its Seven-Year Plan, inaugurated in 1949. In addition, by 1949 Venezuela and Saudi Arabia had already won 50–50 profit sharing agreements with the international oil companies, while Iran was still receiving a much smaller share. It was accepted by both the company and the government that some changes would have to be made in the Iranian concession agreement and unsuccessful negotiations were carried on from 1947. In December 1950 a special oil committee of the Iranian parliament, chaired by Dr. Mossadegh, began a study of nationalization. Too late the company

> . . . expressed its willingness to discuss a new agreement on the fifty-fifty profit-sharing basis and also make an immediate payment of £5,000,000 against future royalty payments and monthly advance of £2,000,000 for the remainder of 1951. This offer came too late because shortly afterwards Dr. Mossadeq proposed to the Parliamentary Oil Committee that the oil industry be nationalized.[8]

The British government found itself in a quandary as to how to deal with the nationalization. On the one hand, the use of naked force was ruled out by the changing world power structure and particularly by the danger of Soviet intervention. On the other hand, Great Britain was unwilling to simply accept Iran's offers for compensation, both because it might set a precedent for nationalization in other countries, and also because any reasonable settlement for the physical properties of the company would be inadequate to compensate the British economy. Thus, while the company had total assets of about $1.5 billion, part of which lay outside Iran, Winston Churchill noted the drastic negative effects of the nationalization on Britain's balance of payments:

> Now that Abadan refinery has passed out of our hands we have to buy oil in dollars instead of Sterling. This means that at least 300 million dollars have to be found every year by other forms of exports and services. The working people of this country

must make and export at a rate of 1 million dollars more for every working day in a year. This is a dead loss which will affect our purchasing power abroad and the cost of living at home.[9]

Great Britain thus first took its case both to the International Court, which held that it did not have any jurisdiction in the case, and then to the Security Council, which also took no effective action. While pursuing these channels of appeal, Great Britain apparently placed its greatest hope on the cumulative impact of various economic retaliations, particularly the boycott of the export of Iranian oil. According to the United States Ambassador to Iran at that time:

> The concept that financial pressures would bring the Iranians into line and solve the oil problem in Iran was from the beginning the key to the British blunders which proved so costly. This notion springs from a colonial state of mind which was fashionable and perhaps even supportable in Queen Victoria's time, but is not only wrong and impractical today but positively disastrous. It is an attitude which seems to persist in spite of the British experiences in India, Burma, Ceylon and Egypt. In Iran, it was expressed in variations of this theme: 'Just wait until the beggars need the money badly enough—that will bring them to their knees.' I heard that vapid statement so often that it began to sound like a phonograph record.[10]

Some of the steps in the oil boycott have been described as follows:

> The British oil tanker fleet was withdrawn from Abadan so that Iranian oil could not be transported to foreign ports. Public warnings were issued that legal proceedings would be instituted against any and all purchasers of Iranian crude or refined products. In fact British chargé d'Affairs in Iran reportedly averred that the British government would pursue the matter all the way to the North Pole if necessary in search of a court that would adjudicate such a dispute. Diplomatic pressures were also brought to bear upon foreign governments to discourage or prohibit sales of Iranian petroleum products. As a result of these and other embargo measures only an independent Japanese oil distributor, and two independent Italian companies exported any

significant quantities of oil from Iran during the more than three-year span between nationalization and Mossadeq's fall.[11]

The boycott was extremely effective in cutting off Iran's oil income. At the same time, neither Anglo-Iranian nor Great Britain suffered the losses that might have been feared, owing to the company's ability to draw upon its fabulous oil reserves in Kuwait. On the other hand, while the oil boycott caused financial trouble for Iran, it was not directly decisive in achieving the end of overthrowing the Mossadegh government and/or restoring the nationalized properties to the company. The basic reason was that the oil sector was a relatively autonomous island in the Iranian economy, affecting only a small number of Iranians; moreover, owing to the government's small share of oil profits, oil revenues accounted for only about 12 percent of total government revenues. However, in the long run the boycott helped to undermine the government insofar as it forced it to take measures which contributed to internal political discontent, particularly on the part of the Iranian upper class:

> Needing money desperately, on August 21, 1952, the Premier decreed the creation of a commission to collect the immense arrears of unpaid taxes of the rich for the last ten years. The commissions were empowered to throw the wealthy Iranians, so called, "thousand families" in jail and confiscate their property if they did not pay. An additional blow to them was Mossadeq's decree that they cut feudal dues received from sharecroppers and return 10 per cent of the profits derived from the land they worked. . . .
>
> About this time [early 1953] Mossadeq was disputing bitterly with the monarchy over the question of the control of the army. In an effort to minimize the power of the king and to ease the country's financial woes, Dr. Mossadeq decided to cut into the Shah's $720,000 a year government allotment and his $2,000,000 a year income from other sources. The tension between the Shah and Mossadeq was intensified . . .
>
> A chain of ominous events followed one after another. . . .[12]

The decisive cause of the overthrow of the Mossadegh government was the actions of the United States government. In the words of one Western legal scholar:

> Officially, of course, the United States assumed the role of "honest broker" in an endeavor to mediate a dispute between two cherished friends. Notwithstanding its outwardly professed regard for Iran, however, it is clear from a careful scrutiny of American behavior that from the very beginning of the controversy this country, with few exceptions, consistently supported the position of the Anglo-Iranian Oil Company. In so doing, it participated actively in a course of conduct which ultimately brought about the deposition of Mossadegh and stifled the small Iranian nation into an abject submission.[13]

The study of Wise and Ross on *The Invisible Government*, while noting that "the British and American governments had together decided to mount an operation to overthrow Mossadegh,"[14] describes graphically the crucial role of the United States Central Intelligence Agency:

> There is no doubt at all that the CIA organized and directed the 1953 coup that overthrew Premier Mohammed Mossadegh and kept Shah Mohammed Reza Pahlevi on his throne. But few Americans know that the coup that toppled the government of Iran was led by a CIA agent who was the grandson of President Theodore Roosevelt.
>
> Kermit "Kim" Roosevelt, also a seventh cousin of President Franklin D. Roosevelt, is still known as "Mr. Iran" around the CIA for his spectacular operation in Teheran more than a decade ago. . . .
>
> One legend that grew up inside the CIA had it that Roosevelt, in the grand Rough Rider tradition, led the revolt against the weeping Mossadegh with a gun at the head of an Iranian tank commander as the column rolled into Teheran.
>
> A CIA man familiar with the Iran story characterized this as "a bit romantic" but said: "Kim did run the operation from a basement in Teheran—not from our embassy." He added admiringly: "It was a real James Bond operation."[15]

One of the significant aspects of the Iranian boycott was the demonstrated solidarity of the international oil companies with their British "competitor," in marked contrast to the oil companies' behavior in the Russian and Mexican boycotts.[16] The rewards to the oil companies for their solidarity were substantial. Following the overthrow of the Iranian government, rather than the oil properties being restored to Anglo-Iranian, a new corporation was given control of the properties. Only 40 percent of the shares of this corporation went to British Petroleum, with 14 percent going to Royal Dutch Shell, 6 percent to CFP, and 8 percent to each of the five United States majors, who in turn gave one-eighth of their share to eleven other United States companies, including Cities Service. As a footnote to the whole Iranian affair Kim Roosevelt "later left the CIA and joined the Gulf Oil Corporation as 'government relations' director in its Washington office. Gulf named him a vice-president in 1960." [17]

Boycotts of oil destined for oil-importing countries also have a long history. Prior to World War II the League of Nations launched an abortive oil boycott aimed at Italy, following its 1935 invasion of Ethiopia:[18]

> The experts at Geneva reached the conclusion that the one economic sanction most certain to check Italy would be to refuse to supply it with *petroleum and petroleum products.* . . .
> [They] estimated that a *total* embargo by all countries would result in an exhaustion of Italian stocks in something like three to three and one-half months.[19]

Superficially the boycott failed because Italy was able to obtain the oil that it needed, particularly from smaller Western companies.[20] The basic underlying factor in the failure of the League to implement an oil boycott of Italy was the reluctance of the big oil powers, the United States, Great Britain, and France, to take decisive steps against Italy. One lesson from the affair is that in such a boycott originated by governments, the united front of the oil country superpowers is a vital ingredient for success. Thus, even the willingness of the major international oil companies to go along with such a boycott is not necessarily sufficient to ensure the success of the boycott. This is because the amounts of oil required by any one country are usually small enough to be obtainable from a relatively large number of smaller companies; the temptations for a smaller company or even a group of individuals to provide oil for a boycotted country are great, owing to the potential enormous profitability for a small operator. As a result, the willingness of the major governments enforcing the boycott to use force as the ultimate deterrent to the oil trade must be unquestioned in order to at least deter attempts to make a quick profit. As one student of the Italian boycott bluntly concluded: "The record proves that profitable trade will cease only under command." [21]

Probably the most famous and significant attempted boycott of an oil-importing underdeveloped country was the one aimed at Cuba in 1960. Actually, several kinds of boycotts were involved, and oil company boycotts and governmental boycotts became closely entangled during this confrontation. Nevertheless, it is worthwhile to attempt to disentangle these by a careful examination of the sequence of events involved in the various boycotts. It is particularly important here not only because there are numerous lessons to be drawn about the nature of oil boycotts from this case, but also because the Cuban events help shed additional light upon the subsurface pressures involved in the struggle between the oil companies and the government in India (and other underdeveloped countries).

In Cuba as in India the immediate cause of the open struggle was the Soviet offer to supply barter oil at "cut-rate" prices, followed by the refusal of the international oil companies to allow the use of their facilities for handling the Soviet oil; the difference was that in Cuba the companies' refusal was met by seizure of their facilities. Events in Cuba, then, in one sense reflect "what might have been" in India (or elsewhere) had the government pressed its demands to the ultimate limit as in Cuba. Moreover, it seems quite clear that the object lessons of the Cuban experience were not lost on many governments, including that of India.

The key initial event leading to the oil boycott[22] was the signing in February 1960 of a trade agreement between the

Soviet Union and Cuba in which the Russians agreed to buy 5 million tons of sugar from Cuba in five years at approximately the world market price. As was often the case in Soviet trade agreements with underdeveloped countries, crude oil was offered as one of the Soviet products which could be bartered for the sugar. Since virtually all of Cuba's crude petroleum was being imported at a cost of about $80 million per year (10 percent of Cuba's total imports), this barter trade would clearly help ease the pressure on her scarce and declining foreign exchange reserves. Thus, in mid-1960 Cuba asked the Jersey, Shell, and Texaco refineries to each handle 11,000 barrels per day of Russian crude, which on an annual basis would total 12 million barrels, or about one-half of Cuba's oil imports.[23] The three companies refused and in July 1960 were "intervened" or taken over (to the extent of managerial control, although not formal ownership) by the Cuban government.

There are two interesting questions in this sequence of events: (1) Did the Cuban government expect the international oil companies to comply with its request for processing Soviet crude oil? (2) Did the international oil companies, in refusing to process the Soviet crude oil, expect that they would escape seizure by the government?

The published evidence suggests the Cuban government expected that in a showdown the international oil companies would accede to its demands. Such an expectation was probably furthered by reports appearing in the U.S. press. In the words of one study:

> The Revolutionary Government had given the companies advance notice of the importation of Soviet crude oil when the trade treaty was signed in February 1960. It was generally understood, as reported in *The Wall Street Journal* (May 24, 1960) that the companies *would* refine the Soviet oil. "If we didn't take it [the Soviet crude oil], we would be taken over, so what would we gain?" one oil company executive was quoted as saying. . . .
>
> It was reported on June 7 that the oil companies informed the Cubans they would refuse to refine Soviet oil, but company

officials declined comment. Then, on June 11, Castro warned the companies publicly that they would have to refine the oil or accept the "consequences." "Don't let them say afterward that we attacked them, confiscated or occupied them. The Government accepts the challenge and the companies must decide their own fate." The companies—Standard Oil Company (New Jersey), Texaco, Inc., and a British-operated unit of the Royal Shell group—had "*unexpectedly* defied the Cuban Government," *The Wall Street Journal* (June 13, 1960) pointed out. The Soviet oil had not yet been delivered to the refineries, however, and the showdown between the companies and the Revolutionary Government had yet to occur. Fidel Castro, speaking as late as June 24, apparently did not expect the companies to refuse to refine Soviet oil when the deliveries would actually be made. Despite his obvious pessimism in the same speech (he felt the sugar quota would be cut very soon), he referred to the oil question with optimism: "They will quietly start refining Russian oil as soon as we deliver it to them," he said.[24]

If the Cuban government really expected the international oil companies to refine Soviet crude oil in Cuba, it was either naïve and/or badly misled by the reports emanating from the United States press. Any knowledgeable observer of the international oil industry would have predicted with a high degree of confidence that the companies would not take such action. In our analysis of the rejection of the Soviet crude oil offer in India we discussed in detail the reasons why acceptance of Soviet crude was complete anathema to them. The essence of our argument was that since most of the profits in international oil come from the crude oil, the primary reason for the international oil companies to have affiliates in oil-importing underdeveloped countries is as outlets for sale of their own surplus crude oil. Hence, it would make little sense for the companies to accept Soviet crude oil out of fear that if they did not they would lose their outlets for their own crude oil.

The other major reason for the companies' unwillingness to handle Soviet crude in any one country was their fear of the "demonstration effect" on other countries. How could the international oil companies persuade the NATO nations to limit import of Soviet crude oil, let alone dissuade other underdeveloped countries from forcing them to refine Soviet crude oil, once the companies accepted this crude oil anywhere? Specifically, in mid-1960 the same three companies were being asked by the Indian government to process about 1½ times as much Soviet crude oil as the Cuban government was requesting. It seems obvious that if the companies acceded to the demands of the Cuban government, they would find it almost impossible to resist pressures in India and other countries to handle large amounts of Soviet crude oil (and even refined products). This would be particularly true because the Cuban government and the United States government were already at loggerheads, so that the companies could expect maximum political support in this case.

Some rough calculations of the type that would be made by the companies in arriving at their final decision will serve to clarify the argument. The companies in Cuba had been importing at least 20 million barrels of crude oil per year, which with a rough profitability of $1 per barrel (ignoring refining, transporting, and marketing profits) would mean that the value of their Cuban facilities solely as an outlet for crude oil would be $20 million per year. On the other side of the ledger, Jersey estimated its investment in Cuba at $62 million, while that of Texaco and Shell together was about $70 million;[25] if these investments yielded an average profit of 10 percent per year, this would be only $13 million per year, or considerably less than the profits derived from the sale of crude oil. Moreover, once the principle of accepting Soviet crude oil was established, the companies could further lose some $30 million per year on their imports into India, to say nothing of the cost in other countries.

In addition, the oil companies must have realized that the trend in Cuba after the Revolution was running against their future prospects for making money in Cuba from anything but importing crude oil. Thus, for example, in November 1959 the Cuban government nationalized the subsoil, and hence all oil which might be found in Cuba:

> The reasoning behind this and other such moves is to be found in Jimenez's "Geography of Cuba." The author advances the

thesis that in the matter of ore and petroleum deposits, the U.S., represented by big business, has purposely stalled the exploitation of these resources in order to provide America with a reserve in times of war.

The majority of Cubans, including many educated ones, strongly suspect that American oil companies have discovered enormous reserves of oil in Cuba, but have hidden all strikes in order to protect their Cuban market.

On this premise INRA [the Agrarian Reform Institute headed by Jimenez] has seized the records of the oil companies and assumed the task of developing a national petroleum industry.[26]

Again, in May 1960, the government began building its own marketing network for handling Soviet refined products:

On May 23 it cancelled exclusive marketing contracts between the three majors and Sinclair on the one hand and their 2,500 outlets on the other so as to provide markets for Soviet imports.[27]

The significance of this last step should not be underestimated, as one astute observer noted in *The London Times*:

The Cuban government's cancellation of all restrictive selling agreements between the oil companies and their customers is another less publicized feature of the new policy, which may be hardly less irksome than the refining requirements.[28]

Once the Cuban government put itself into marketing, and broke down the exclusive agreements between the majors and their marketers, there was grave danger of intensive price competition at the marketing level which would endanger profits made within Cuba.

All in all, then, to the oil companies at that time it would have appeared foolish to accede to Soviet crude oil in a country like Cuba which was already clearly laying the groundwork for strengthening the position of the state at the expense of the companies. Given the fact that yielding to the Cuban government would probably set off a worldwide chain reaction with major deleterious effects on profits made it seem doubly or triply un-

331

wise. Finally, if a stand had to be taken by the oil companies against Soviet oil, since United States–Cuban relations were already at a low point, this was as good a time and place as any.

Even if at worst the companies' refusal to handle Soviet crude would lead to expropriation, since it was clear that this would lead in turn to strong United States retaliation, such a sequence of events might have been viewed as a good warning to India and other countries. In any event, as one study concluded, both for the United States government and the companies the refusal to refine the Soviet crude was the logical step:

> The refusal to refine the oil was intended to precipitate a crisis. It was a reflection of the tough line of the United States. The companies believed that the Cubans would not be able to operate the refineries and that the economy would come to a halt. The Revolutionary Government would be forced to relent and change its program. Its alternative was, of course, to ask the Soviets for technicians and equipment and to rely on them for help in general. The United States Government and the private oil companies were willing to take this gamble, for they apparently believed that a Cuba in which U.S. interests were expropriated or made unprofitable was little better than a Communist country anyway.[29]

The companies, led by Jersey, which had the largest stake in Cuba, probably also felt that they had enough aces in the hole to win the gamble:

> The companies prepared to shut down their plants rather than refine Soviet crude, even small amounts. They let their crude stocks fall and began to send wives, children and some American employees back to the United States. The deadlock threatened to paralyze the transportation system and leave Cuba without electric power. The companies counted on the lack of Cuban technicians to balk seizure; also on an ability to keep spare parts from going to Cuba if this became necessary; also on Standard Oil's published threat to blacklist any shipping company that carried Soviet crude to Cuba. The State Department supported the companies in their principled position of self-sacrifice to defend the free world from infiltration by Soviet oil.[30]

Given the refusal by the oil companies to process Soviet crude oil, three possible courses of action were realistically open to the Cuban government. First, seize the refineries and thereby force the processing of Soviet crude. Second, simply rescind its demand that the oil companies process the Soviet crude. Third, as was done in India, accept the companies' refusal but seek over the long run to increase the government's role in the oil industry so as to reduce the country's dependence on the companies.

It is certain that a fourth theoretical alternative, namely, punishing the three established majors by refusing to allow them to import their own crude oil, was not a practical option. Ninety-eight percent of Cuba's total energy was derived from petroleum fuels (as compared to less than 20 percent for India). Moreover, Cuba's economy was relatively energy intensive, with per capita consumption of all energy amounting to the equivalent of 850 kilograms of coal, compared for example to 140 for India, 350 for Brazil, and 1,160 for the rapidly industrializing Japan.[31] Not only were transportation and electrical output totally dependent upon oil, but also industry, including the sugar mills (and through the latter, agriculture).[32] Since without oil the Cuban economy would rapidly grind to a halt, a deadlock could not be allowed to continue very long.

That the Cuban government would choose the first course, seizure of the refineries, rather than the second course, complete capitulation to the companies, or the third course chosen by India, "capitulation plus harassment," also should have been predictable. After all, the Castro government had come to power through a revolution which had completely swept aside the existing government with its close ties to the United States. Moreover, from the agrarian reform of early 1959 onwards the revolution was increasingly moving in the direction of socializing the internal economy. This in turn meant coming into conflict with United States property interests in Cuba, leading to a loosening of the ties between the two countries. For Cuba, embarked on a course of socialization and economic independence, or at least reducing her economic dependence on one country, both

333

goals were challenged by the companies' action, and at stake in the government's decision. In the words of one study:

> That Cuban seizure of the refineries would follow the companies' refusal to refine the Soviet crude oil could have been expected for at least two reasons. First, the Soviet oil was necessary for Cuban industry, it was cheaper, and it was obtained in exchange for sugar rather than the dollars the Cubans could not spare. It was, in short, essential to the Cuban program for economic independence, as the revolutionaries saw it. Second, the Cubans regarded this to be a test of sovereignty: was the Cuban Government to determine its own economic program or were U.S. (and British) oil companies to continue to possess a veto power over it? Could Cuba be economically independent? Would the United States permit it? [33]

The Cuban oil boycott actually consisted of two separate but interrelated boycotts. The first, involving the refusal of the oil companies to refine Soviet crude oil, was just one part of the worldwide attempt of the international oil companies to boycott exports of Soviet crude oil and refined products. The leader in this attempted worldwide boycott of Soviet oil was Standard Oil of New Jersey, which as the world's largest international oil company stood to gain the most from blocking Soviet oil. Moreover, Jersey and the other United States oil companies were fortunately domiciled in the world's most powerful country, which at that time also felt itself engaged in a life and death struggle with the Russians; hence the whole Soviet oil boycott became closely tied up with the Cold War.

Following the seizure of the oil refineries by the Cuban government, a second boycott which was a logical derivative of the first, was launched by the oil companies, again led by Jersey. The aim of this second boycott, the Cuban boycott proper, was to prevent Cuba from utilizing the seized refineries, thereby presumably forcing the government to return to the *status quo ante*, including the elimination of Soviet crude oil. The companies hoped to achieve their ends through three separate forms of "subboycott," any one of which if successful would bring the Cuban economy to a halt. The first of these was to be a boycott of skilled labor for running the refineries; the second, a boycott of essential spare parts for the refineries; the third, a boycott on tankers, to cut off the potential flow of crude oil from Russia (or anywhere else) to Cuba. We shall now analyze the reasons why each of these three subboycotts failed.

The failure of the "labor boycott" created by the withdrawal of foreign employees was due to several factors. First, in the confrontation between foreign companies and their own government, the Cuban refinery workers largely sided with their own country and government.[34] Second, as is increasingly true in most underdeveloped countries, foreign employees were only a small percentage of the total number of refinery workers.[35] Finally, the few highly skilled operating personnel who were foreigners could be and were replaced by Soviet technicians. And, if for some reason these had not been available, it is likely that the Cubans could have obtained assistance from individuals in the government oil companies of Latin America; thus, for example, the Cuban Petroleum Institute, established in November 1959, was originally headed by a Mexican who recruited thirteen Argentinians, three Mexicans, and two Peruvians to lead the exploration and drilling efforts.[36]

The fact is that in today's world of spreading technology, no company, group of companies, or single country can realistically hope to maintain a stranglehold on technical skills, particularly one as widespread and long established as oil refinery operation.[37] Because of this diffusion of technological skills, the boycott on spare parts was also doomed to failure.[38] This effort had even less chance of success; such a boycott takes a considerable time to work since there is usually some stockpile of spare parts on hand and/or improvisations can be made. At its most effective, such a boycott may cause inefficiencies but it cannot usually be decisive.

The final attempted boycott, aimed at blocking transportation of crude oil to Cuba in oil tankers, was the most ambitious of all. The heart of the companies' deterrent was a threatened blacklist of any independent tanker owners chartering to the Soviet Union. Thus, Standard Oil of New Jersey stated in a manifesto:

Standard and its affiliates, in making future commitments to charter tankers, will take into consideration which owners are now chartering or selling tankers or hereafter do charter or sell tankers to the Russians for any trade, and also when dealing with brokers will take into consideration whether the brokers are now acting or hereafter do act as intermediaries in Russian business. Russian business as used in this context is intended to cover tankers done by Sovfracht, Deutschfracht, and Polfracht [the Soviet, East German, and Polish companies] and other charters or shippers who handle Black Sea oil.[39]

This Herculean effort was labeled by *The New York Times* "a unique venture into private foreign policy."[40] Nevertheless, it would appear that the effort had the support of the United States government, after the event if not before. Thus, for example, we have the following *New York Times* report of the loss by an independent tanker, which had chartered to the Russians, of the valuable right of registration in Liberia:

Albert J. Rudick, Deputy Commissioner of Maritime Affairs for the Republic of Liberia, confirmed that the four-year-old vessel was deprived Monday of its "citizenship" and the right to fly the Liberian flag. . . .

Reports circulated in marine circles that the Liberian action and [sic] stemmed from State Department suggestions.

The action also was partly attributed to a policy position adopted by the Standard Oil Company (New Jersey).[41]

Again, strong United States governmental support of the oil companies' boycott is indicated by the following reports:

The U.S. Ambassador paid a surprise visit to Gutiérrez Roldan, head of Pemex; a Cuban delegation then in Mexico, it was explained to the Ambassador, consisted only of petroleum students inspecting Pemex properties. There was speculation about another United States-British boycott on Mexico, as in 1938, if she sold oil to Cuba.[42]

* * *

Well-informed quarters say Cuba has unofficially expressed a desire for a sizable amount of Mexico's crude oil as an adjunct to shipments from the Soviet Union. . . .

It is considered almost certain that the United States is prepared to go to unusual lengths to discourage the sale of Mexican crude for refining in the American and British-owned plants seized by the Fidel Castro regime.[43]

The international majors of course would also go to great lengths to block oil from reaching Cuba. In the words of *The Wall Street Journal*:

> The big international oil companies that control all but a trickle of the Western world's petroleum almost certainly wouldn't come back to Cuba while Mr. Castro remains in power. They already are black-listing tankers that haul Soviet crude oil to the seized Esso, Shell and Texaco refineries. They might well put even greater pressure on any Western company that decided to sell oil to Mr. Castro.[44]

The international majors were almost totally successful in getting other Western oil companies to agree to the boycott. In those few cases where they were not, the tanker boycott apparently did the trick: "In need of crude, Havana made a deal with Superior Oil to take some of its Maracaibo crude, but no tankers could be found and the deal fell through." [45]

Blocking the Soviet Union from obtaining tankers for shipping its oil to Cuba was a horse of a different color, however. Some rough numbers may help to give the reader both an idea of the magnitude of the task facing the oil companies and the reason for their failure. In order to ship a year's oil supply to Cuba from the Soviet Union, about 500 thousand tons of oil tanker capacity would be required.[46] In mid-1960 the world oil tanker fleet consisted of some 3,300 ships with a capacity of 63 million tons. While the great bulk of these ships are under the control of the international oil companies, at any moment in time between 8 to 20 percent of world capacity is operating in the spot market.[47] Since 1960 was a period of depressed tanker rates, it is likely that close to 12 million tons of tanker capacity were available for the spot market. Hence, potentially there was available to the Russians more than twenty times the amount of tanker tonnage needed for the Cuban trade (to say nothing of their own tanker fleet, which was over a million tons itself).

Put another way, for the boycott to be successful the oil companies would have to induce the owners of more than 95 percent of the theoretically available tanker tonnage to refrain from dealing with the Russians.

The main "inducement" would, of course, be the fact that the companies are the principal charterers in the world tanker market, and could be expected to have a "long memory" for any independent tank owner who offended them by chartering to the Russians. Working against the companies, however, is the highly competitive structure of the tanker market, in which a large number of operators each owns a relatively small amount of total capacity.[48] Moreover, while the total amount of tonnage needed by the Soviets for the Cuban trade was relatively small, it was sufficiently large to be extremely attractive to any individual tanker owner, particularly if the Soviets were willing to pay higher than market rates for a long-time charter. Furthermore, the very knowledge that small tanker operators would be attracted by Soviet offers might serve as a further inducement to the larger operators; while they might want to continue doing business with the international oil companies, they could also jump on the Soviet bandwagon and argue that they were not the first.

In any event, it is clear that after initial difficulties, the Russians had little trouble in supplementing their own tanker fleet. For one thing, China provided assistance:

> Russia is also striving desperately to make good her scarcity of tanker tonnage through Red China, which has been buying up and chartering tankers, most of them vintage tonnage, for several months.
>
> Now, Red China is reassigning to the Russians many of the tanker charters it has picked up.[49]

Moreover, the ranks of the tanker operators split wide open. On the one hand:

> In London, Aristotle Onassis, Greek supertanker tycoon, said he had been approached for tankers by both Cubans and Russians. "But I turned them down flat," Onassis reported. . . .[50]

338

But, on the other:

> Stavros Niarchos, the Greek magnate, had just signed a 2-million-ton deal with the Soviets.[51]

Most analyses of the failure of the tanker boycott stress the fact that in mid-1960 there was an "abnormally" large amount of idle tanker capacity available:

> Independent shipowners were reluctant to run counter to Standard Oil's policy, but the effects of the depression in the shipping industry soon became apparent.[52]

* * *

> Idle tankers were a drug on the market; soon ships carrying Soviet, Norwegian, Liberian, Greek, and other maritime flags were heading for Havana and Santiago, heavy with crude. . . .
> . . . in an otherwise bleak tanker market, the Norwegians said it was a choice between accepting Russian gold or starving to death—and they preferred to live.[53]

The fact is, however, that basically it was the ownership structure of the international tanker market, with its numerous small independent owners (for many of whom a major deal with the Soviets could bring an irresistible profit) that ultimately determined the fate of the boycott. Even if the period had been one of relatively strong demand for tankers, with high tanker rates, there would undoubtedly have been sufficient capacity available to the Soviets, at the right price, to overcome the boycott. For, as we have seen, even in peak demand periods some eight percent of the world tonnage is available in the spot market, or close to ten times what the Soviets needed for the Cuban trade. (Since the Cubans were crucially dependent on oil they would undoubtedly have been willing to pay virtually any price.) Thus, the United States National Petroleum Council, albeit reluctantly, drew the right conclusion:

> In July, 1960, a major oil company initiated its "Black Sea" policy which denied shipping contracts to any tanker owner moving Soviet oil. This policy apparently caused an increase in the cost of Soviet charters, but it did not reduce the volume of

oil shipments. The results of the "Black Sea" policy indicates that individual action by one company has little effect on the oil movements by the Soviet Bloc.[54]

At this point we can turn to an evaluation of the various effects of the Cuban oil boycotts. Viewed strictly from the economics of the energy sector, it is apparent that the switchover from an oil industry dominated by private foreign companies to one owned by the state, and from dependence on Western oil supplies to dependence on Soviet oil supplies, caused no long-run damage. While indigenous production of refined oil fuels remained constant at 3.5 million tons between 1959 and 1965, rising product imports enabled total consumption to increase from 3.5 million tons to 4.7 millions tons, or 36 percent.[55]

It is true that problems were initially created by the fact that Soviet crude oil is much lighter than Venezuelan oil, and hence tends to yield relatively more gasoline and less industrial fuel. Since all fuel oils (including distillates) accounted for three-quarters of Cuban demand, while gasoline's share was only about 20 percent, this could have been a serious problem. At first Cuba apparently had hoped to sell the surplus gasoline in Great Britain or Canada.[56] However, given the influence of Royal Dutch Shell in Great Britain, and that of the United States in Canada, it is highly unlikely that gasoline would ever have been allowed to be a part of a Cuban trade deal with those two countries; in any event, Cuba did not export any gasoline. Instead, the problem appears to have been solved by a combination of rising demand for gasoline, along with changes in refinery design which allowed a reduction in the percentage of gasoline derived from a barrel of Soviet crude oil.[57] On the other side of the ledger, the Cubans claimed that costs in the oil industry were sharply reduced as a result of the major switchovers:

> Director Gutiérrez estimated that in the second half of 1960 the [Cuban Petroleum] Institute saved $35 million in its operations. Black Sea crude for example cost 20 percent less than cartel crude from nearby Maracaibo, and in addition could be paid for in sugar and other Cuban products rather than in dollars, hard to come by. . . . Most clearly visible were the econ-

omies in distribution. The three refining companies along with Sinclair had maintained their own competing wholesale and retail outlets throughout the island. Consolidation in distribution cut costs.[58]

As to the political effects of the Cuban oil boycott, there is one obvious and important question. Did the method of defeating the oil boycott, even if the only feasible one, simply lead to Cuba's shifting from dependence on one country to dependence on another, thereby negating the expressed purposes ultimately leading up to the boycott?

A typical affirmative answer to this question was given by *The Wall Street Journal*, soon after the refinery seizures:

> Most of all, the Soviets aim at becoming sole suppliers of a vital commodity. In Egypt it was arms; here it is oil, and this makes Mr. Castro's dependence on Moscow almost complete.[59]

The negative view was expressed by Che Guevara soon after. In an interview reported in *Look* magazine Guevara was asked, "Aren't you just exchanging so-called U.S. dominance for Soviet?" His answer was:

> It is naive to think that men who carried through a revolution of liberation such as ours, independently, did so only to kneel before any master. If the Soviet Union had just once demanded political dependence as a condition for its aid, our relations would have ceased at that moment. If we maintain increasingly cordial relations with all the Socialist bloc, it is because the word "submission" has never arisen.[60]

The truth of the matter is somewhat complex, and probably lies between the two extreme views. In one direction Cuba has complete independence from the Soviet Union: if Cuba were to decide to revert back to a close relationship with the United States, or even to take a "Titoist" position, it could undoubtedly obtain United States support, including oil. The real issue then of dependency hinges on the extent of Cuba's ability to maneuver in a "Left" direction vis-à-vis the Soviets.

There is some evidence that at least in the early stages of Soviet-Cuban relations the Russians were not aiming at using

341

their crude oil supply to dominate Cuba. As O'Connor has pointed out in 1962:

> Oddly enough, for those who regarded Cuba as a Communist satellite, the Petroleum Institute devoted its main attention to exploration in an effort to make Cuba self-sufficient in oil, and thus no longer dependent on Russia for its energy supply. Oddly, too, the Russians seemed disposed to aid this effort. The Institute had ten rigs which were set to work in spots regarded as propitious. The Soviets in 1961 were sending two rigs able to penetrate 16,000 feet, and a crew and equipment to survey the promising Camagüey Keys.[61]

In later years Cuba's relationship with the Soviet Union has been in a state of flux. On the one hand, Cuba allowed the Soviets to station missiles on Cuban soil, which according to Castro was for the benefit of the Soviets rather than the Cubans. On the other hand, Cuba rejected that part of the Soviet–United States deal to end the 1962 Cuban missile crisis which would have allowed the United Nations to inspect Cuban soil to verify the removal of the missiles. Probably only time will reveal the true extent of Cuban dependence on the Soviet Union. However, it is interesting to note the following early 1967 analysis and prediction of the future course of Cuba's relationship to Russia:

> Internationally, too, Cuba has been moving consistently to the Left. . . . The strident denunciations of China have disappeared from Fidel's speeches and in their place we find a running criticism, no less clear for being expressed in indirect and guarded terms, of the policies of the Soviet Union and its followers abroad. . . . All of this adds up to a declaration of political independence within the socialist camp, all the more courageous and praiseworthy because of Cuba's continuing economic reliance on the socialist countries as a market for sugar and a source of supply for industrial goods.[62]

The view quoted above would seem to gain support from Cuba's sponsorship of the "Organization of Latin American Solidarity" conference in August 1967; among other things this meeting of revolutionaries, under Cuban leadership "approved a resolution today condemning Soviet economic and technical policy in the Western Hemisphere." [63] Thus far Cuba's increasingly critical position toward the Soviet Union has caused no obvious drastic retaliation, such as cutting off her oil supply. In part this may be because even if the Soviet Union were so inclined, it could not do so because of the major negative reactions within the Communist world; again, with countries such as China and Algeria possessing sizable supplies of crude oil relative to Cuba's needs, it is not obvious that such a Soviet oil cutoff, or even the threat of it, would restrain Cuba. Thus, it seems fair to conclude that the evidence to date does not indicate that Cuba has simply "changed masters" as a result of the events surrounding the classic Cuban oil boycott of 1960.

Finally, at least one general lesson about oil boycotts can be learned from the Cuban experience. That is, where an oil-importing country has the strong backing of a major "oil power," i.e., a country controlling both oil and important military power, it cannot be successfully boycotted in the sense of having its economy brought to a halt through cutting off energy supplies. The labor and capital controlled by the international oil companies can be replaced by that of a major oil power, and transportation for oil can be provided, either through the major power's own tankers or chartered ones. The "strong backing" of a major oil power implies, of course, the willingness to ultimately use force if necessary to break through a physical blockade which could prevent the importation of oil. This was the case with regard to Cuba, as O'Connor has noted:

> Standard's venture into an economic boycott seemed as futile as its previous ventures. Only the threat of armed force could stop the tankers, and Moscow had already warned that it was willing to reply with force.[64]

In the last analysis then, it may truly be said that the failure of the Cuban oil boycott resulted from United States' unwillingness to use force to blockade the island in 1960. This, of course, was just the reverse of the situation in 1962, when the United

States used force to blockade the island in order to obtain the removal of Soviet missiles. However, especially since that near-nuclear confrontation in 1962, the world's major powers have been reluctant to force a similar life-or-death showdown through blockade; an example is the reluctance of the United States to blockade North Vietnamese ports. Since it is increasingly clear that countries will only resort to a blockade where they feel their most vital interests are at stake, particularly where there is danger of a confrontation with another major power, blockades are unlikely to be used in any future oil boycotts. In that sense, the theoretical possibilities of use of force do not negate the basic lesson of the Cuban oil boycott experience. Nor does the fact that at some point an oil-importing country, e.g., the Union of South Africa, might be subject to a physical oil blockade negate this lesson. For, if that time ever comes, presumably such a country will not have the support of any major oil power.

A Footnote on History: Cuba and India Compared

It is instructive to compare the different reactions of the Indian and Cuban governments to the refusal by the international oil companies to handle Soviet crude oil in their refineries. The formal parallelism of the two situations is striking. Both countries received the offer of Soviet crude oil at about the same time. Both countries had serious foreign exchange problems. Both countries were offered not only a lower price for their oil but also oil on a barter basis, which would especially help their foreign exchange situation. Both countries faced the same three international oil companies: Jersey and Texaco (allied with Mobil and Standard of California, respectively, in India) and Shell. Both countries' economies were heavily dependent on the West. India was receiving large-scale aid from a consortium of Western governments headed by the United States, while Cuba was vitally dependent upon the United States for its sugar market. Finally, both countries faced a united front of the established oil companies in their refusal to handle Soviet crude oil.

Why then was the outcome in India so different from that in Cuba? Certainly, the formal similarities of the two situations were not lost on observers. In the words of one Indian newspaper:

> If British and American oil companies continue their reluctance
> to refine—and as in some cases to distribute—Soviet oil, it looks
> as if there will be a multiplication of "Cuba" situations.[65]

Moreover, India was certainly in a stronger position than Cuba in
the event of a showdown with the oil companies. For one thing
the Soviets would probably have been even more willing to go
out on the limb in supporting India than the lengths they demonstrated in helping Cuba. At the extreme, if events in India led,
as in Cuba, to the Soviets "replacing" the West, while the economic cost to the Russians would be high, it would not be unmanageable. In terms of the enormous political victory it would
give them in the then raging Cold War, the cost would be relatively cheap. And, considering that the Soviets were already
having quite strained relations with China, they might well have
viewed it as a cheap price indeed for an important ally on China's
borders.

Precisely because India was viewed as so important to the
United States and Western political interests, particularly as a
potential offset to China, it is questionable whether the United
States and Western governments would have given the oil companies in India the same strong backing that they received in
Cuba. However, this last consideration is more speculative, since
in a final showdown it is conceivable that the Indians would have
faced the same strong Western response, particularly following
the Cuban events.[66]

The basic reason for the different paths taken by the two
governments at this mid-1960 watershed would appear to be the
fundamentally different character of the respective governmental
leaderships. The irony is that at this point in time the words and
deeds of the two countries were inverted. The Cuban government, even in the eyes of the United States Central Intelligence
Agency, started out as a noncommunist one:

> Our conclusion, therefore, is that Fidel Castro is not a Communist; however, he certainly is not anti-Communist.[67]

Without question, however, the government and the revolution were nationalistic:

> The most significant political fact about Cuba today is not Communism, but the wave of nationalism which has swept up almost every Cuban man, woman and child.[68]

A further irony is that the confrontation between a nationalist government and the oil companies helped move the Cuban revolution in a Communist direction:

> Castro in a television program on June 10th said that the oil companies must refine the Soviet crude or take the consequences. Their refusal he saw as a part of a conspiracy between them and the United States government to destroy the Revolution. For the first time since coming to power he applied the Leninist word, "imperialism," to this combination of economic and political sway. The revolutionary government, he said, accepted the challenge.[69]

The Indian government, on the other hand, had always professed to be staunchly socialist. In fact, however, its dominant wing was quite far from that, as pointed out in the *Far Eastern Economic Review:*

> Theoretically a party of the left, the Indian National Congress has generally been dominated by men of very conservative temper. . . .
> . . . Successive Congress governments, for all their socialist ideals, always saw their main domestic enemies on the left, among the Godless Communists.[70]

This conservative position of the Indian Congress Party, as opposed to the radical approach of the Cuban government, stems from their fundamentally different relationships to the key groups in any underdeveloped country. In Cuba, the revolution basically owed its victory to the support of the poor peasants and farmers, particularly in the impoverished areas surrounding the Sierra Maestre. Thus, the first major step of the Cuban revolutionary government was to launch a wide-scale agrarian reform beneficial to its mass base. In India, on the other hand, the Congress Party and the government are basically controlled by the urban and rural upper classes.

The position of the industrialists in India, on a Cuban-type approach to the oil companies, is clearly indicated by the following editorial in *Commerce*, a leading Indian business periodical:

> There is an unfortunate tendency among some commentators in this country as well as abroad to compare the Government of India's endeavours to get more petroleum products at lower prices with the Cuban Government's actions in the same field. Even the Cuban Premier, Dr. Fidel Castro, has sought to quote the Indian Government's request to the three foreign oil companies—Burmah-Shell, Stanvac and Caltex—to refine Russian crude in their refineries as a justification for his violent action against these companies. This is not fair to India. While it is true that the Government of India would have been glad to see its proposal for refining Russian crude accepted by the oil companies, it has not, unlike the Cuban Government, retaliated against them when they gave a firm negative reply. Dr. Castro would have done a distinct service to all the capital-hungry and technologically-backward underdeveloped countries had he retracted his step and followed India's example of respecting its agreements with the oil companies and getting them modified only where possible through mutual negotiations.[71]

The rural political situation in India has been described as follows:

> In an adult suffrage democracy, the Congress Party leadership finds itself torn between its desire to win over the rural masses and its fear of alienating the rural rich whose political power rests on their ability to influence the rural vote.[72]

But, an acid test of the true nature of a government in any underdeveloped country is its real position on land reform.[73] By this criterion it seems clear that the Indian government has been fundamentally aligned with its powerful supporters, the wealthy landlords. Thus, the *Far Eastern Economic Review* stated in 1967:

> In the course of the discussions on the Fourth Plan, there was a great deal of talk this year about land reforms. A Planning Com-

mission Committee turned in a 400-page report urging the Central and State Governments to make with the utmost speed the legal changes necessary to give tenants security of tenure. It warned that unless this was done the benefits of the large investments on agriculture proposed to be made during the Fourth Plan would be appropriated by the rural rich, aggravating the existing income disparities and social tensions in the village.

None of this was new; committee after committee reported in the same strain in the past without making any impact on Indian legislatures dominated by the rural rich. The latest data on land distribution, collected by the National Sample Survey in 1961, shows that top 10% of the cultivators hold 56% of the total cultivated land, while the bottom 50% hold less than 3%. This confirms that the enormous volume of land legislation undertaken since Independence by Congress Governments professing socialism has had singularly little effect. . . . Likewise, the ceiling on rents is a piece of legal fiction. Share-cropping continues unchecked; the prevailing rent is half of the gross produce almost everywhere but in highly fertile areas it can be as much as two-thirds.[74]

Given this evidence, it should be abundantly clear why the Indian government would not (while the Cuban government did) force a confrontation with the oil companies. If the Congress government were unwilling to even carry out substantial land reforms, then it is surely clear they would not take such a radical step as seizure of foreign-owned oil refineries. This is particularly true because of the logic of this type of dramatic event. It can rebound internally, leading to a demand for greater radicalization within the country, thereby threatening the tenure of industrialists, landlords, and governments alike.

CHAPTER TWENTY-FIVE

From Public to Private Petroleum:
Latin America in the 1960s

IN RETROSPECT, 1960–61 was probably the high-water mark for the tide of increasing substitution of public petroleum entities for private oil companies. This turning of the tide has been closely associated with the worldwide swing in the 1960s toward replacement of moderate or radical Left governments by conservative military dictatorships—from Indonesia to Ghana to the Dominican Republic. In Latin America, the turning point undoubtedly came after the Castro revolution of 1959–60. Since that time, in the view of a conservative analyst of coups d'état in Latin America:

> . . . it appears to have been his [Castro's] pronouncements and attempts to expand his influence in the Americas that finally tipped the balance in favor of a reversal of the trends in lands to the south. Between November of 1961 and the same month in 1964 ten rather significant *golpes de estado* [coups d'état] occurred in almost as many Latin American countries . . .
>
> All of the coups of this second cycle were motivated partly and often largely by fear of Communist and/or other radical groups. Few of them, however, were extremely reactionary. They were rather middle-way movements composed mainly of mildly conservative and moderately progressive groups opposed to drastic and rapid political, economic, and social change.[1]

Because oil has been so deeply involved in the events surrounding these coups d'état, the following analysis of oil policy changes in several crucial countries should not only be of intrinsic interest but also serve to throw additional light upon these important political developments.

III. The Range of Policies in Underdeveloped Countries

Historically, Latin America has a long tradition of government oil entities operating at various levels of the industry. In addition to Mexico's Pemex, Argentina has its Yacimientos Petrolíferos Fiscales (YPF), which dates back to 1922 and Brazil its Petroleo Brasileiro (Petrobras), founded in 1952; seven other Latin American countries also have state oil companies.[2] However, not all of these Latin American state oil entities are particularly significant even within their own countries; some appear to serve merely to assuage popular desires for a state oil company and in practice have little real power.[3] The most important ones which we shall analyze below in detail are YPF in Argentina and Petrobras in Brazil. No attempt will be made to develop the historical background of these two companies, but rather the focus will be on analyzing their experience in the last decade in the light of our previously developed analytical framework.

Argentina's experience is particularly significant of events which were to repeat themselves with regularity in many other countries in the 1960s. Thus, O'Connor's book, published in 1962 could say: "Argentina represents the one outstanding victory of the international petroleum cartel in the face of a series of devastating defeats it has suffered in the past ten years."[4] Now, from a later perspective, events in Argentina can be seen as a turning point model rather than as a unique event.

The first crucial event in recent Argentine oil history came in 1958 when Arturo Frondizi was elected to the Presidency and invited foreign oil companies to explore for oil on a grand scale. Until that year virtually all new petroleum exploration had been reserved to YPF, which also owned about two-thirds of the country's refining capacity. While YPF had increased its refining capacity threefold since the end of World War II, its oil production had only doubled in the same period. Hence, there was an increasing need for crude oil imports, which jumped from about 7 million barrels in 1946 to 42 million barrels in 1957.[5] This led to balance of payments pressures which in turn provided the rationale for allowing foreign capital into oil exploration.

At the time, and since, there was considerable debate as to whether or not YPF could do the job alone, without foreign capital. The case for bringing in foreign capital was put succinctly by Frondizi himself:

> The people must know that it will be impossible for Argentina to be self-sufficient in oil by 1960. We lack $800 million to develop our resources and the oil deficit must be covered urgently.[6]

On the other hand, Argentine supporters of YPF disagreed:

> A number of groups hold the view that YPF was in a position to produce the petroleum needed to achieve oil self-sufficiency in Argentina without the assistance of private companies.
>
> They argue that the new drilling equipment acquired by YPF and the two pipelines transporting petroleum and gas from YPF's Salta oilfields to San Lorenzo and Buenos Aires—already under construction and completely financed in 1958—have allowed YPF to almost double its own production in the last four years. Regarding the lack of foreign exchange to finance the purchases abroad needed to expand even more the activities of YPF so as to achieve oil self-sufficiency in Argentina, they argue that many of the materials—such as oil and gas pipes—and the equipment required by YPF were increasingly being manufactured in Argentina, and thus the need for foreign exchange was less than estimated by the Government.[7]

An ironic aspect of the whole policy shift toward private enterprise was that Frondizi was elected essentially on an "anti-imperialist," pro-YPF policy. Earlier, in fact, he had argued as follows:

> The interests of the country demand the rejection of any proposal by private companies, because the control of our petroleum must be entrusted entirely to YPF. We must say frankly and emphatically to foreign negotiators that our petroleum is not for sale or for negotiation; we need urgently to produce it to develop our economy but we want to accomplish this task ourselves. The petroleum is Argentine because it is in our subsoil, has been discovered by YPF, we have the technicians and skilled workers capable of producing it. We lack only machinery. We can provide YPF with the funds needed for machinery and equipment for exploration and drilling through the Central Bank. With dollars, YPF can buy in the United States and else-

where the equipment it needs. If the United States refuses to sell to us, in order to favor their own oil companies, it will assume a heavy responsibility before world public opinion, and YPF will be able to resort to any other source of supply without waiting for permission from interested parties.[8]

Why the dramatic turnabout? One of the most perceptive long-time students of U.S.–Latin American relations, Simon G. Hanson (editor of a weekly newsletter on trade and investment in Latin America) bluntly lays out the causes and their historical background. In 1960 he wrote:

And when Perón [dictator of Argentina until 1955] whom Milton Eisenhower in apparent adulation had found "a great leader of a great nation," indicated that he would accept a quid-pro-quo deal involving U.S. government financing of a great industrial plant in exchange for the granting of a petroleum concession to a private American company the [State] Department was veritably overwhelmed with the "success" of its Argentine policy. . . .

When President Eisenhower arrived in Argentina, he came fully prepared to intervene as actively for Frondizi as the Department had intervened for Perón. For this time, there was even greater cause for elation. The Frondizi regime in pre-election and post-election commitments secretly arrived at had signed away proven oil reserves to certain foreign interests, some reputedly associated with the White House on terms so prejudicial to Argentina in the light of prevailing industry practice that many veteran oilmen were astounded when the announcements appeared.[9]

And, Mr. Hanson is no radical critic of United States foreign policy, but fundamentally a conservative one who feared that United States' overall interests, including those of business as a whole, were being sacrificed in favor of the oil companies. Thus, Hanson raises and answers his own questions:

Again, as happens so often in any analysis of the good-*partner* policy, the question arises: Is there a difference between the national interest and the interest of petroleum-concession hunters? Does a regime, whatever its other attributes, become worthy of direct intervention by the United States to effect its survival because it has come forth with petroleum concessions? What exactly is the role of the United States in this matter? . . .

What Mr. Rubottom [at the 1957 Senate hearings on his nomination as Assistant Secretary of State for Inter-American Affairs] was saying essentially was that a policy contrary to the interests of an oil company is automatically contrary to the interest of the U.S. government, i.e., contrary to the national interest. Presumably, this implied that what is good for the company is good for the United States.

But there is every reason to believe that this is a completely false basis for the foreign policy of the United States, a basis that is dangerous in the extreme. Yet, the State Department was demonstrating in Argentina that it stood by the enunciation of Mr. Rubottom as to where the national interest lies. It was aware, if not from its own intra-departmental analysis then from the comment of oilmen not involved in the concessions in Argentina, that the terms of the concessions diverged widely from what foreign countries were able to negotiate in this era in their relations with foreign investors. There was no recent precedent for the terms on which proven reserves had been surrendered. The Department was aware that the very idea of the surrender of proven reserves was so repugnant to large elements of the Argentine population that the very least the U.S. could do was to make sure that the terms were not so open to criticism. And above all, at a moment when the Department was applying pressure on Brazil to surrender potential petroleum resources on terms to be dictated from abroad, it could not have failed to appreciate that it was opening up the U.S. to charges of a revival of "dollar diplomacy" which would not only destroy the good-*partner* policy but would make it more difficult for a successor administration to reinstitute the good-*neighbor policy*.[10]

In any event, in November 1963, President-elect Arturo Illia, who had campaigned on a promise to annul these long-term contracts, in fact did so. His grounds were that the contracts were illegal because they were not ratified by the Argentine Congress but rather were put into effect by executive decree of Frondizi. The Illia government had come under strong United States pressure not to take this step, including U.S. government threats of a cutoff of public and private investment.[11] A show-down confrontation was avoided by bringing the question of "compensation" to the oil companies for the canceled contracts before the Argentine courts.[12]

For several years negotiations dragged on between the Argentine government and the companies on the compensation question. At the same time the Argentine Congress was investigating the circumstances surrounding the signing of the contracts. In July 1964 it was reported that:

> The Argentine congressional group investigating the cancellation of oil contracts has been repeatedly requested to show proof of the unsubstantiated allegations that bribes were offered in the awarding of contracts. Most recently, the probers have been given what allegedly is described as "photographic proof"— with at least three companies being named.[13]

Finally, in late 1964 the congressional committee reached its conclusions:

> A congressional investigating committee has issued one majority and two minority opinions on the 1958–59 agreements between the Frondizi Government and 13 private oil companies.
>
> The majority report labeled the contracts fraudulent and said they cost state-owned YPF $167-million in lost profits through 1963 and would have cost $915-million if continued up to their termination dates.
>
> One minority report, on the other hand, said the contracts were sound and promoted national production to such an extent that Argentina attained virtual oil sufficiency, which it has since lost.
>
> The second minority report condemned the way the contracts were signed but recommended reform of YPF and new agreements with private oil companies.[14]

United States government pressure on behalf of the companies appears to have been increased, with aid disbursements the principal coercive weapon.[15] Finally, by early 1966 most of the oil companies reached out-of-court settlements with YPF.[16] (The total sum to be paid to the companies ultimately amounted to well over $200 million.)

Nevertheless, the fundamental policy problems facing Argentina were still to be resolved. Oil production since the annulment of the contracts had remained virtually stagnant while consumption had risen sharply, leading to increased crude oil imports and further pressure on foreign exchange reserves. The Illia government still seemed indecisive in its attitude toward the role of YPF versus the private oil companies in Argentina. In February 1966 it was reported that a large-scale oil exploration program was to be undertaken by YPF, with the oil companies limited to a partial role as subcontractors.[17] However, financing such a program was the critical problem, and the Illia government appeared to be desperately grabbing at straws for a solution:

> The Government hopes to help finance the expansion plans of state-owned YPF and Gas del Estado with loans from the World Bank—PIW [*Petroleum Intelligence Weekly*] learns in Buenos Aires. Negotiations are still continuing.
>
> YPF would use the proceeds of any loan to speed up its ambitious oil exploration and development plans.[18]

The companies appeared to reject their projected role as simply YPF subcontractor. According to *Petroleum Intelligence Weekly*:

> Some oilmen on the scene, however, remain skeptical of the Government's ability to reach its [production] objectives unless it offers broader incentives needed to bring the big foreign international companies back to Argentina. Only these companies, some Argentine oil circles say, possess the financial resources to take the risks associated with exploration. And they won't take the risks until they can expect to make a profit. These sources maintain that local firms don't have the capacity to meet the state's drilling plans, and that, under present conditions, most foreign firms are unlikely to bid.[19]

In any event, all vacillation was swept away when in mid-1966 the military, which had always been the real power in Argentina,[20] established an open dictatorship. Illia was replaced by General Ongania,[21] whose military dictatorship quickly made it clear that its weight was to be thrown on the side of foreign private oil companies and against YPF. In the fall of 1966 the government drew up plans to shift control of oil policy from YPF and attract the international oil companies in on a large scale.[22]

Even within the government formed by the military dictatorship, the drive to put oil development in the hands of the foreign companies met resistance:

> Disagreements over Argentina's oil law led to a split last week between the government's nationalists and liberals and a major shake-up of President Juan Corla [sic] Ongania's cabinet. The new bill, unsigned after 7 weeks in the hands of high government officials, . . .
>
> Odds that the bill will become law soon are improving as a result of the cabinet shakeup. Four ministers strongly opposing the bill have resigned, and their replacements are reported to regard it more favorably.[23]

Not surprisingly, in mid-1967 a new law was passed which opened up Argentina to the foreign oil companies on a much wider scale:

> Bids for concessions—open to foreign ownership for the first time in 32 years—should be taken by fall. The bill, designed to lure foreign investment back into the country, allows the government to grant exploration licenses and concessions for exploration and transportation of hydrocarbons to private enterprisers. Companies owned by foreign governments are ineligible for rights under the new law. . . .
>
> Another incentive for foreign firms is a tax feature providing a 55% maximum tax, including royalties and rentals, on net income. Under the plan, the tax rate could be reduced over a 10-year period for those companies taking quick advantage of the new rights by applying and receiving concessions within 18 months of the bill's passage.
>
> Though YPF will keep most of the producing areas—leaving areas considered expensive and "difficult" for development open for bids—foreign firms will have some promising spots for consideration.
>
> The government firm, hesitant to invest in costly offshore operations, will put the entire Continental Shelf, along with un-

committed areas on shore, up for grabs. And, too, YPF has promised to negotiate with private firms for some onshore areas which are located close to known reserves.[24]

Thus, in ten short years, oil exploration policy in Argentina had moved almost 180 degrees to its opposite. Starting from a government oil monopoly on new exploration, through the stage of foreign companies exploring for oil as government subcontractors, oil policy moved to outright oil exploration concessions for the foreign companies. These political economy developments, so accurately and frankly analyzed by Simon Hanson, raise serious questions about the reassurance given in the Levy paper to the United Nations that:

> As a final note it should be emphasized that private participation, by its very terms, is limited in tenure. Upon completion of contract or concession, a going oil operation reverts to the country —to be continued under state or private auspices, according to the dictates of national policy.[25]

Events in Brazil similarly represented a dramatic swing from public to private oil. When Petrobras, the state oil entity, had been established in 1953 it was given not only the sole right to oil exploration but also all rights to increase refining capacity. Only marketing was controlled by the private oil companies, which handled the oil produced or imported by Petrobras. Brazil is one of the largest oil markets in Latin America, with its consumption topped only by that of Mexico and Argentina; in fact, the Brazilian oil market is 50 percent bigger than that of India. Thus, Brazil has always been particularly attractive to the international oil companies, especially those with large crude oil reserves and refineries in nearby Venezuela. With marketing dominated by Jersey, Shell, Gulf, and Texaco, the vesting of Petrobras with exclusive rights to oil exploration and expansion of refining capacity threatened these companies with ultimate loss of a large and rapidly growing outlet for crude oil and products. Hence, from its very beginning Petrobras had been opposed by the companies, as noted by O'Connor:

In 1954, fought bitterly by the conservatives, Vargas resigned the presidency after his labor minister, "Jango" Goulart, had been ousted from office. Soon after, Vargas committed suicide. In a posthumous message to the nation, he declared that the pressure of the oil companies and their allies had ruined his administration. "Once more, anti-national forces and interests, co-ordinated, have become infuriated with me," he cried. "I wanted to create national freedom in realizing our national wealth through Petrobras; hardly had it begun to function when the wave of agitation grew. They don't want the people to be free." [26]

Nevertheless, Petrobras took giant strides in the 1950s. Its refining capacity increased from 2 million barrels in 1953 to 16 million in 1954 and 38 million barrels in 1959; by then it controlled two-thirds of Brazil's refining capacity.[27] During the same period, crude oil reserves increased from 50 million barrels to 600 million barrels and production from less than 1 million barrels to 24 million barrels (and over 30 million barrels in 1960).[28]

At this point, an "all-out effort" was made by the international oil companies and the United States government to "open up" the Brazilian oil industry. This has been vividly described and analyzed by the knowledgeable Mr. Hanson:

What happened in 1959–60 to cause the State Department suddenly to fear the consequences of the deterioration in relations with Brazil?

Intoxicated with the success of the assistance it had rendered U.S. oil companies, in breaking down resistance to the granting of concessions in Argentina and Bolivia on terms wholly inconsistent with current industry practice and as such prejudicial to the Latin Americans, by a skillful use of promises of financial assistance by the U.S. Government, the Department had made an all-out effort to break down Brazil's resistance. Despite the fact that the marketing situation for its chief export commodity was such that normally special assistance would have been thought in order, the word went out now that Brazil must be broken. It was now or never. There could be no further accepting of Brazil's resistance to introduction of the petroleum companies, particularly those where there was a special association with this administration on a personal basis, into exploration and exploitation of Brazil's petroleum potential.

Reviewing the various avenues of assistance, the U.S. Department of Commerce reported that in 1959 our cooperation with Brazil had been reduced 75%.

So, day by day, the State Department waited for Brazil to collapse, to reach a point where from its knees it might be compelled to accept terms on petroleum. . . .

But Brazil refused to break down. The economy survived the pressure. At least it was still retaining its independence when Eisenhower came to Rio. . . .

By March 28, 1960, the State Department found itself before the Senate Foreign Relations Committee confessing sheepishly that despite the U.S. pressure on Brazil, per-capita gross national product in 1959 had increased by more in Brazil than in any other major Latin American country.

The whole campaign against Brazil had been carried on, for public-relations cover-purposes, in terms of Brazil's unwillingness to accept a stabilization program such as Argentina had consented to.[29]

While the drive to penetrate the Brazilian oil market failed at this time, this was not the end of the matter. The weak link in Petrobras' position was that while it had increased crude oil production substantially, crude oil imports had risen dramatically (from 1 million barrels in 1953 to 31 million barrels in 1959). While Petrobras' expansion of refinery capacity had at the same time sharply cut refined product imports (from 48 million barrels down to 30 million barrels), the combined imports of refined products and crude oil were costing the country over $200 million per year, an amount about equal to its total balance of payments deficit.[30]

Moreover, the increase in oil production which had been achieved by Petrobras came entirely from an old field, Bahia, which while successfully extended by Petrobras, could not be expected to sustain increased production indefinitely. Between 1955 and 1960 Petrobras spent an estimated $300 million for wildcat crude oil exploration without significant success.[31] This failure provided a wedge for the oil companies to continue their campaign to get into the Brazilian market.[32] However, from the companies' viewpoint the situation in Brazil in the early 1960s went from bad to worse. Particularly in the last months of the Goulart regime (the same Goulart whose ouster as Labor Minister had triggered Vargas' resignation in 1954), Petrobras' role had accelerated, at the expense of the foreign companies. Petrobras continued to increase its refinery capacity, which in 1964 was scheduled to reach close to 150 million barrels per year, or 90 percent of the country's petroleum refining capacity. In its drive to obtain crude oil, Petrobras was beginning to lean more heavily on Soviet barter oil, which amounted to almost 15 million barrels in 1964. The final straw, which had been foreseen for several months, came in March 1964, with the nationalization of the existing private refineries.

Less than one month later Goulart's government was overthrown by the military and an open military dictatorship was established. The foreign oil industry clearly foresaw better times, as indicated in the May 1964 editorial in *World Petroleum:*

It now seems likely that the change in administration in Brazil will soon bring a change in the climate in which business is carried on, not only for the local private interests but also for investors from other lands. The economic difficulties resulting from Brazil's continued political problems have affected all types of industry, not only petroleum. . . .

Today, however, investors are facing problems which are all quite familiar to the oil industry—expropriation, taxes, profit-limiting laws, and the like. Brazil took over International Telephone in 1962. Goulart had recently threatened nationalization of the huge iron ore holdings of M. A. Hanna. In January, Goulart had signed a bill limiting profits which foreign companies can take out of Brazil.

The World Bank apparently grew discouraged as long ago as 1959. It has not approved any projects there since that time, because "economic prospects are so uncertain." . . .

Goulart's turn toward the left, toward the Communist line, in the past eight to ten months has not been much help. Not more than a month before he was deposed, he announced the nationalizing of the six remaining privately-owned refining companies, thus giving the government a monopoly over every phase of the petroleum industry except marketing. Here too the freedom of action on the part of the oil companies has been curtailed by government action. Under the new administration it is hoped for the economic well-being of the Brazilian nation, that many of the recent measures will be cancelled or at least postponed until they can be studied carefully.

An early action of the new regime placed former officials of Petrobras under arrest and dismissed those officials who were active in agitating for nationalization of privately owned oil refineries. This suggests that the government's attitude toward private enterprise may be far different in the future. If so the nation may benefit by a return of confidence on the part of foreign investors.[33]

Benefits for the international oil companies from the new regime came quickly. First, as noted, pronationalization officials and also union leaders in Petrobras were immediately fired. Within a year the planned nationalization of private refineries was reversed, and private investment was to be welcome in the petrochemical sector, previously solely reserved for Petrobras.[34] Finally, development of oil shale reserves was similarly opened to private investment for the first time.[35] Ironically, Soviet assistance was also to be used to further promote the private shale oil industry in Brazil.[36]

By late 1966, Phillips Petroleum, backed by the U.S. government and the World Bank, was moving into Brazil's fertilizer industry with a $70 million investment.[37]

While the foreign oil companies were aggressively moving into fertilizers and oil shale, crude oil exploration, which presumably had been the ultimate rationale for bringing in foreign capital, was still restricted to and only actively pursued by Petrobras.[38]

A review of the Brazilian oil industry in 1967 showed that Petrobras still controlled five-sixths of the country's refining capacity; however, of the seven projects being carried out in petrochemicals, Petrobras was involved in only one.[39] The same survey noted the temporary (albeit probably unstable) equilibrium which appeared to have been reached:

In spite of possible ministerial changes, the present government's policy regarding oil and Petrobras should not change.

Only recently Prof. Roberto Campos, Minister of Planning, in reply to an inquiry made by one of the foreign correspondents in Rio stated that the government had already gone as far as it should go as regards encouraging the private initiative by opening to it access to the petrochemical industry and shale oil, adding that "There is no intention of altering the legislation in connection with the monopoly of (oil) production and refining." [40]

Whether this intention is honored in the practice rather than in the breach remains to be seen.

It was not only in the major Latin American countries that the swing from public to private oil took place in the 1960s. Thus, for example, in Colombia in 1965 the government sought a "rapprochement" with the oil companies. Once again, the ostensible reason for encouraging the companies to come in was to step up crude oil exploration.[41] Even in the little Dominican Republic the United States oil companies' drive for markets appeared to be associated with political upheavals. Thus, we have the following April 1963 analysis by I. F. Stone:

Bosch's [the newly-elected president of the Dominican Republic] first steps in power seem designed to disengage himself from Washington. His speech February 19 on returning from his trip to the U.S. and Europe emphasized the latter rather than the former as the source of new development contracts. . . . Soon after the inauguration Bosch announced the signing of a 15-year $150,000,000 agreement with a Swiss consortium (including the Bank of America) to build two dams for irrigation and hydroelectric power, and for an aqueduct to supply fresh water to the capital. Bosch admitted he might have had better terms and lower interest rates if he had financed the projects from international organizations but argued that it would take a year or two for those banks to make preliminary studies and come to decisions and that work was needed urgently to cope with unemployment and hunger. But one wonders whether another motive in dealing with this international consortium was not to lessen his dependence on Washington, which plays the major role in such international institutions as the World Bank and the Inter-American Development Bank. Bosch also touched a tender nerve when he complained that the Council of State which ruled the country before the elections had secretly concluded a contract with Standard Oil [Jersey] for a refinery on terms unfavorable to the Dominican Republic. Bosch said he had received better offers in Europe and had warned Standard Oil that the contract would be reviewed as soon as he took office. Bosch said he did so although he knew "my attitude would be used in spreading throughout the world the report that I am hostile to foreign private investments in this country, that I am a Communist, that I am a Fidelista, or that I am something else still more radical." The President elect's prediction proved correct. Though little has appeared in the U.S. press about the Standard Oil contract, a campaign to picture Bosch as somehow linked to Communists has already begun, though he and his entourage are—like most Socialists elsewhere—passionately anti-Communist and anti-Castro.[42]

Finally, it should be noted that the rollback of state oil entities and the resurgence of foreign oil companies,[43] associated with right-wing coups d'état, was far from restricted to Latin America. Thus, in Ghana, Nkrumah had arranged to have Soviet crude oil imported for the country's sole refinery, jointly owned by the government and ENI; the established marketers (Texaco, Shell, Mobil, BP, and ENI) were forced to market the refined products from this Soviet crude. After the overthrow of the Nkrumah regime the companies were given half of the crude oil supply contract. This led *Petroleum Intelligence Weekly* to comment:

> The thought that half a market is better than none probably won't entirely console the six companies, who will be eager to reopen negotiations with Ghana towards the end of the year. They hope to convince the government either to renew their contract for a longer period than six months, or to reinstate the former processing agreement.[44]

Again, the overthrow of Sukarno in Indonesia in 1966 was followed by these developments, which also illustrate the forces and motives in the petroleum area:

A freshet of activity is also due in the petroleum industry, now that the government has scrapped Sukarno's plans to nationalize foreign companies. Caltex, a joint venture of Texaco Inc. and Standard Oil Co. of California, is increasing its Sumatran oil production, which already accounts for 70% of national output. But offshore is where the action is—in more ways than one.

Four relatively unknown companies—two American and two Japanese—have signed contracts with free-wheeling Maj. Gen. Sutowo, director general of oil and natural gas, for concessions under the shallow seas of Kalimantan (Borneo) and Java. The controversial pacts give Indonesia management control, at least nominally, plus 65% of future profits.

Maj. Gen. Sutowo's superior, the minister of mining, tried to cancel the agreements and award the offshore rights to better-financed oil companies, but acting president Gen. Suharto has backed Gen. Sutowo's decision. For one thing, it's known that the Japanese government, with whom Indonesia is currently attempting to negotiate a loan agreement, exerted pressure on behalf of the two Tokyo-based firms, one of which is partly owned by the Japanese government.

Another reason for the maneuver has also come to light: Maj. Gen. Sutowo concedes he uses state pertoleum [sic] earnings to help finance Gen. Suharto's army, which can't exist on regularly budgeted funds.

The American companies involved are Independent Indonesian-American Petroleum Co., a joint venture of Rocky Mountain independent oil firms, and Refining Associates, Inc. Together with the Japanese concerns, they possess another advantage over major oil companies in the bid for Indonesian concessions: The larger international concerns refuse to meet management and profit-sharing demands of the Indonesian government, largely because they fear a precedent that might be used against them elsewhere.[45]

Finally, there were the "oil events" surrounding the 1967 coup in Greece. Two months before the coup it was reported that:

The profit battle between foreign oil companies and the countries in which they do business is shaping up in still another area: Greece. Oilmen there are saying that they will be forced to cut investment even further this year unless the "extremely restrictive" oil policy is changed. This warning, which represents a wide variety of industry opinions, comes from experts who are closely watching the activities of the non-political Lambroukos committee that is currently studying oil industry profitability in the country. The oil picture is generally described as "bleak." . . .[46]

Immediately after the coup we have the following petroleum trade press "speculation":

Forecasts are difficult when a government suddenly changes, but some western oilmen are already speculating that if Greece now leans further toward the political right, it might review its Communist Bloc trade policies. If so, the country's crude oil supply contracts with Russia could be cut (see PIW-Feb. 20, p. 6), leaving some slack for western oil firms.[47]

Six months later the new regime's general philosophy on foreign investment was spelled out in a report in *The New York Times:*

There's only one kind of Westerner the new army-led Greek Government is pursuing harder than the tourist. He's the foreign investor who may be looking for a place to build a hotel or a factory or perhaps only locate a sales or branch office.

Not the least of the changes in Greece since the army seized the government, in a swift, bloodless coup on April 21 has been in the attitude toward foreign investment and toward the basic philosophy of internal economic development.

With the army-led Government, Greece got a new Secretary-General, the key job, for its economic-coordinating ministry.

He is Costas A. Thanos, who has a master's degree from McGill University, Montreal, and a doctorate from Columbia University in New York City.

In a European age when central planning is still in vogue and economic nationalism is making a comeback, Mr. Thanos has some novel things to say. For example, on foreign investment:

"We've finished with complexes about it. We want it, among other reasons, to provide the competition for Greek industry and investment. If the foreigner gains the initiative over the Greek, well, let him."

On economic philosophy:

"We want to do away with central planning. It's been presented as a sort of panacea. It's high time the developing countries used the old methods that succeeded in the developed countries. Let the individual be free." [48]

Perhaps most startling was the means chosen by the regime to achieve its long-run goals:

The new Government also moved swiftly to conclude years of negotiations with Litton Industries of Beverly Hills, Calif., for an immense development agreement that Litton officials hope will become the example for developing countries.

In Litton's own words, it is "pioneering in a program which brings to bear a combination of free enterprise, the newest technologies and a systems-management approach" to develop the Island of Crete and western Peloponnesus.

Litton and the Greek Government have signed a contract for 12 years, subject to periodic review, in which investment goals call for at least $240-million by 1970 and up to $840-million by 1978. Litton will not provide any of the money but has undertaken to raise it.

Litton will have complete charge of the program, *unique in scope and concept for a private company in a foreign land.*[49]

The "uniqueness" of this Greek approach to development, which was being considered by the previous regime also, can hardly be denied. It is undoubtedly the first time in history that the government of a legally independent underdeveloped country has virtually turned over its most important function, economic development, to a private foreign company. Surely this is the "ultimate" in the swing from public to private enterprise, and changes in this direction in the Greek oil sector can hardly be far behind. What remains to be seen, however, is whether Litton's hopes of spreading such a program to other underdeveloped countries are realized. Perhaps, on the other hand, history will record the Greek developments as the truly unique high-water mark of the private enterprise tide in the 1960s.

Oil and the Underdeveloped Countries: Problems, Prospects, and Possible Strategies

THE BASIC PROBLEMS of the underdeveloped countries in the arena of international oil stem directly from the economic and political structure of that crucial industry. The international oil industry is presently dominated by vast, enormously powerful, private profit-maximizing corporations of developed Western nations, particularly the United States. The strength of these corporations partly derives from their control of various economic and geographic levels of the industry, and in particular of the huge quantities of low-cost crude oil in the oil-exporting underdeveloped countries. In addition, because within their own home countries the companies are powerful institutions as well as vital to the overall economy, they can usually count on the full support—economic, political, diplomatic, and military—of their home governments and the Western-controlled international organizations in any struggle with the underdeveloped countries. This is particularly important because the economics of the international oil industry generally bring the profit-maximizing goals of the companies into direct conflict with the economic development *cum* political independence goals of the underdeveloped countries.

For the oil-exporting underdeveloped countries, the conflict is clearest with regard to the struggle between their governments and the companies over the division of the oil pie: every dollar of profits which the companies get is one less dollar potentially available to the country for investment in economic development. Less obvious, but no less real, is the continuing struggle over the efforts of the oil-exporting countries to gain increased knowledgeability about and participation in the international oil industry. The profit-maximization goals of the companies generally lead them to oppose these efforts on the part of the underdeveloped countries. The fear is that, at a minimum, increased oil activity of the underdeveloped country might interfere with the companies' balancing of their own worldwide interests, and, at a maximum, the countries' increased activity might lead to the elimination of the companies from the oil-exporting countries via nationalization.

The conflicts of interest between the oil-importing underdeveloped countries and the international oil companies take analogous forms. Clearly, consonant with their profit-maximization goals, the companies have a great incentive to charge the oil-importing countries as high a price as possible. In addition, the companies have a general interest in keeping the governments of the oil-importing countries out of the oil business as much as possible. Here again governments can reduce the companies' profitability by interference and/or entry, with the ultimate danger being nationalization of refining and marketing facilities.

The broad aims of the international oil companies vis-à-vis the oil-importing underdeveloped countries are specifically manifested in three areas previously analyzed at length: oil exploration and production, refining, and transportation. With regard to exploration, since the companies generally have large quantities of low-cost crude oil outside the oil-importing countries, they have little incentive to explore for oil within the countries and a great incentive to keep the governments from undertaking exploration. In the area of refining, the companies' main interest is in obtaining outlets for their external crude oil; hence they generally seek to prevent government ownership of refineries, particularly since this might lead logically to greater governmental efforts in the field of oil exploration. As for transportation, the companies' control over the international tanker fleet provides the underlying basis for reaching their goal of getting the highest profit-maximizing price for providing these services; an important mechanism is their use of a historically oriented price-averaging system which inadequately reflects the long-term trend toward lower transport prices resulting from continually declining transport costs.

Finally, to this picture of the international oil industry should be added two increasingly important features which help give rise to additional tensions within the industry as well as possibilities for change. One major feature is the competition among the international oil companies, particularly between the long-established seven majors and the newcomers. The second factor is the re-entry of the Soviet Union as a major world oil power, offering not only to sell or barter oil at relatively low prices but also to help the underdeveloped countries build up an indigenous publicly-owned oil sector.

Given this overall structural setting, the crucial policy issue for the underdeveloped countries would seem to be the extent to which they should rely on the international oil companies to own and control their oil sector, as opposed to developing an indigenous, and normally public, oil sector. For an oil-exporting country the question appears as one of where to position itself along a line which at one extreme has the government solely a collector of taxes on oil company profits and at the other has the government the sole operator of the industry. The specific point on the line chosen by a country is of course normally a derivative of a more basic choice on the strategy of development—in particular, on the extent of a revolutionary versus an evolutionary approach. Naturally, while in theory the range of alternatives is great, in practice the choice is severely limited by a country's existing social structure and the power forces which its government reflects.

Put crudely, the question might be: Is long-run economic development most likely to come about through the efforts of a government collecting oil taxes and plowing part of them back into a "moderate" development program, possibly backed by assistance from Western aid-giving nations and institutions? Or will economic development proceed faster under a government which nationalizes the industry and is strongly opposed by the oil companies and their home governments? Or is there some other "intermediate" position or different combination which would be both possible and more effective?

For the oil-importing underdeveloped countries, government policy must also position itself along a similar line. The main questions which have to be answered here are primarily: Who should undertake oil exploration and the building of refineries within the country? If these functions are to be left to the companies, which ones should undertake them? If the government undertakes them, what will be the likely effect of an adverse reaction from Western governments and organizations? If the government stays out of these functions, particularly oil exploration, and the companies do not fill them, what is the cost to the country, both in purely economic terms and in reduced political independence, that arises from not controlling an oil supply within its own borders?

These are some of the principal underlying oil policy issues which have faced underdeveloped countries in recent years and will continue to do so in the foreseeable future. It is hoped that the historical and theoretical analyses of this book will be of assistance to government planners in coming to grips with these questions. As a further aid we conclude with our summary judgments on the likely major future trends and developments in the international oil industry, along with some suggestions as to how the underdeveloped countries might capitalize on them.

There seems little doubt that, barring a major collapse in the economies of the developed countries, the demand for oil should increase over the foreseeable future. At the same time, the already proven physical supply of oil is so great as to be more than adequate to meet this demand. In fact, the supply of oil available to the market is likely to increase even faster than the demand, thereby tending among other things to drive the price of oil even lower in the future. The basic reason for this tendency is the gap between price and cost in the international oil industry, which is still so great as to encourage cut-price marketing by those who already have large amounts of crude oil. Furthermore, this gap will promote additional exploration for oil by new-comers in highly promising areas like North Africa.

Specifically, in the coming years one should expect much more vigorous overseas exploration programs by French, German, Italian, and Japanese oil companies, possibly subsidized by their home governments. The supply of European and Japanese private capital available for investment has been greatly augmented by the prosperity of the last two decades; moreover, there are signs that the rate of growth of internal industry may be slowing, which could make investment in potentially highly profitable overseas oil exploration increasingly attractive. In addition, since oil has become an ever more important element in these countries' balance of payments, the respective governments may well give further assistance to stimulate overseas exploration by their own nationals.

Implicit in this perspective is a continuation of past trends toward fragmentation of control over the world oil supply. This weakening of the position of the seven international majors has potentially great significance for all the underdeveloped countries. Obviously, insofar as it results in continued price cutting, it raises opportunities for benefit to the oil-importing countries while posing serious problems for the oil-exporting countries. At the same time the dispersal of power within the international oil industry among more companies and countries offers the possibility to each oil-exporting country to greatly strengthen its control over its own oil industry. This is true whether the country's ultimate goal is nationalization, increased participation in operations, or simply a greater share of the profits.

This last conclusion can most clearly be illustrated by reference to the crucial and controversial issue of nationalization. Thus, in the absence of total unity among the important oil-exporting countries, the question is whether any single major oil-exporting country can nationalize its oil industry and still survive. Many seem to believe that, particularly in the light of the Iranian developments of 1951–54, such a step is impossible. It would be a mistake, however, to draw any firm conclusions from

371

that earlier experience given the major changes in the political economy of oil since that time. For one thing, the attainment of political independence by a number of underdeveloped countries since the early 1950s has opened up potential markets for "nationalized oil" which did not exist at that time. In addition, the growth of oil consumption in oil-importing socialist countries, which would presumably be sympathetic to any nationalization, provides additional possible markets. It is true that these markets are generally not sufficiently large to absorb the present exports of a major oil-exporting country. However, considering that generally most of the oil-exporting countries' economies are largely insulated from the oil sector, given a governmental determination to husband and reallocate the country's reduced oil revenues, these oil markets could be quite important. In particular, their existence might help an oil-nationalizing country to get through the initial and most dangerous years of an oil boycott. (As the case of Mexico shows, if one can survive these early years, eventually some accommodation may take place.)

Moreover, given the projected increasing fragmentation and economic rivalry among the developed Western nations, even larger markets for nationalized oil might be found in maverick countries such as France. When the latter's specific oil interests are coupled with its general desire to win influence in the underdeveloped world, mutually profitable deals on a large scale might well be possible. One supporting piece of evidence is France's recent willingness not simply to buy nationalized oil in Iraq but to actually help produce it.

As a matter of strategy, such a community of interest could be deliberately cultivated in the future by the oil-exporting countries. For example, some of the enormous oil revenues of these countries could be used to buy refining and marketing facilities from independent companies in a country such as France. Then, were nationalization to take place, the oil-exporting country might have a continuing market for its oil. Undoubtedly even a country like France would come under strong pressure from the United States and Great Britain to prevent importing nationalized oil. Nevertheless, it might be difficult for the French Government to disallow such oil on the grounds of sanctity of contract if in turn this would abrogate contracts between an oil-exporting country and a refinery legally incorporated in France. Thus, purchases by oil-exporting countries of "downstream" facilities in countries which have some freedom of maneuver vis-à-vis the United States and Great Britain could be good preventive medicine against a possible oil boycott.

Even if an oil-exporting country does not aim at ultimate nationalization, purchase of such overseas facilities would still seem to be a desirable step. Thus, the mere existence of such nationally owned facilities increases the credibility of possible nationalization and could thereby give oil-exporting governments valuable leverage in negotiating for their various aims short of nationalization, e.g., greater participation in the industry or a greater share of the profits. Ownership of such facilities would also give the oil-exporting country invaluable on-the-job experience in the transportation, refining, and marketing ends of the oil business and provide useful tools in dealing with the international oil companies.

A final way that the oil-exporting countries might improve their long-run position would be to invest a large part of their oil revenues in indigenous industries which are relatively large consumers of oil, so as to develop a significant internal market. An example is fertilizer production, which can aid indigenous agricultural production as well as be sold or bartered to other underdeveloped countries and which would be extremely hard to boycott in the event of an oil nationalization. Investment in fertilizer production, therefore, seems to recommend itself to the oil-exporting country immediately. Again, plastics is another industry which might have great potential.

For the oil-importing underdeveloped countries, the crucial strategy decisions lie elsewhere. One need is for government officials to integrate into their planning the fact that in the short run at least, oil is a scarce and expensive resource; recognition of this can influence and interact with other key strategies such as the kind of transportation investment to be made or the types and location of industrial factories. This, however, does not obviate the fact that in the long run, unless these countries develop indigenous crude oil supplies, their economic development is likely to be hampered; this will be true because the country will have to rely on less desirable but locally available fuels or else will be forced to utilize scarce foreign exchange for oil imports. Moreover, the present study indicates clearly that in most oil-importing underdeveloped countries, if a major exploration effort is to be undertaken, the task will perforce fall to the state. This is both a crucial problem and an opportunity, which should be recognized and dealt with at the highest levels of government economic planning.

Until the time when an oil exploration effort is undertaken and proves successful, important opportunities still exist for the oil-importing countries to improve their oil position. One way would be for them to become increasingly knowledgeable about the economics of the international oil industry in general, and pricing in particular, so as to bring all possible power to bear to minimize the oil-import bill. Another possibility which might be pursued concurrently would be to attempt to work out mutually profitable state-to-state deals with the oil-exporting countries. From a long-run viewpoint, one economic basis for such arrangements is generated by the large gap between the price paid for oil by the importing countries and the revenues received by the exporting countries, which gap is far greater than the real costs of production and transport. In other words, the large profits of the international oil companies made in the intra-underdeveloped world oil trade provide an opportunity to squeeze the international oil companies as middle men.

A further potential community of interests exists because the oil-importing underdeveloped countries generally lack capital while the oil-exporting ones have relatively large amounts of it. One specific example of how to realize this potential was shown when the state oil company of Iran provided capital for a refinery in India, thus obtaining an outlet for its crude oil. In return the Indian government was given a favorable opportunity to participate in oil exploration in Iran.

374

A very general area of potential cooperation between the two groups of underdeveloped countries lies in the field of petrochemicals, particularly fertilizers. On the one hand billions of cubic feet of natural gas are being burned off uselessly in the course of oil production in the oil-exporting countries, even though these gases are a major component of fertilizers and the oil-exporting countries have the capital to build fertilizer factories; on the other hand the oil-importing countries are generally desperately short of fertilizer and could profitably use additional amounts to increase agricultural production and thereby pay for these fertilizers. Thus, the capital and natural resources of the oil-exporting countries plus the hirable skills of international construction firms plus the markets of the oil-importing countries could potentially be brought together with profit for all parties.

Finally, the world of the foreseeable future is likely to be one which is increasingly fragmented, both politically and economically. As the present process of the breakup of the capitalist and Communist blocs continues, this fragmentation, particularly in the developed capitalist countries, could provide more maneuverability for the underdeveloped countries to improve their position. Offsetting this, the series of right-wing coups d'état in the 1960s has sharply reduced the number of "independent" major countries in the underdeveloped world and thereby reduced the bargaining power of the underdeveloped world as a whole. Given this situation, it is particularly important for the remaining relatively independent underdeveloped countries to do all in their power to help overcome this loss of bargaining power.

In the oil sphere, one arena for concentrating the power of the smaller but numerous underdeveloped countries would be the United Nations, particularly through the General Assembly where each country has one vote. The independent underdeveloped countries have a large stake in pressuring the United Nations into taking a far more active role in the area of oil, which could include not only providing comprehensive training and education of personnel for the underdeveloped countries but also seeking to harness investment funds for oil projects in underdeveloped countries.

If any major UN effort in this latter area is to be undertaken, the stress needs to be placed on the potential harmony of interests between the oil-exporting underdeveloped countries and the oil-importing ones. This must be done in order to overcome any fears of the oil-exporting countries that the UN's intervention into the oil arena would undermine their position and profits. When the focal point of any program has become the possibilities for mutual assistance between the two blocs of underdeveloped countries, then the UN could potentially play a significant role in indicating and developing this harmony of interests. Such a program will not even be begun, however, unless the underdeveloped countries themselves take the lead in demanding a start.

Notes

CHAPTER ONE

1. See particularly Albert O. Hirschman, *The Strategy of Economic Development* (New Haven: Yale University Press, 1958).

2. The role of petroleum, particularly kerosene, in fulfilling this lighting function may be particularly important in underdeveloped countries seeking to spread literacy into rural villages which lack electrification.

3. United Nations, Economic Commission for Latin America, *Energy Development in Latin America* (Geneva: 1957), p. 4.

4. Data for energy costs in transport are for Latin America. (See note 3 above.) Other data for heavy and light industry are for European Common Market countries, reported in *The Wall Street Journal*, October 28, 1963.

5. Note 3 above.

6. Hirschman, note 1 above, pp. 98–112.

CHAPTER TWO

1. In formal economic terminology, the upper limit to price would be the value of the marginal product derived from using the particular refined oil product.

2. This is modified to some extent by the fact that over time automobiles themselves can be adjusted to reflect, and in turn affect, motor gasoline prices:

> The demand for gasoline is generally considered as quite inelastic, and in the short run it is (though recent Italian experience makes one wonder). But over the long run it is not; for if it were, how could we explain the difference in car design, and in average consumption per mile, between Europe and the United States? [M. A. Adelman, "The World Oil Outlook," in Marion Clawson, ed., *Natural Resources and International Development* (Baltimore: Johns Hopkins Press, 1964), p. 84.]

If motor gasoline prices are high relative to automotive diesel oil, there is a tendency to substitute diesel engines in automobiles, thus putting a downward pressure on the demand for motor gasoline and hence its price.

3. The general principle for "profitable" or "optimum" refinery operation is that the refinery should be designed to produce as much as possible of the highest priced refined product, subject to the constraint that the cost of producing an additional gallon of high priced product—e.g., motor gasoline—plus the cost of foregoing the potential production of a gallon of lower priced product—e.g., fuel oil—be less than the price of the additional gallon of gasoline; this rule would apply to production of any specific refined product.

4. J. E. Hartshorn, *Politics and World Economics* (New York: Frederick A. Praeger, 1962), pp. 143–44.

377

5. *The New York Times,* October 2, 1966.

6. From a paper presented at the Fourth Arab Petroleum Congress, November 1963; cited in Hussein Abdel-Barr, *The Market Structure of International Oil with Special Reference to the Organization of Petroleum Exporting Countries* (University of Wisconsin Ph.D. thesis, 1966), p. 9.

7. Adelman, note 2 above, p. 58.

8. This average is supported by figures revealed to *The Oil and Gas Journal* by a spokesman for the consortium of international companies that control Iran's production. According to this authoritative source, the per barrel "production costs" are $0.14 in Iran, $0.08 to $0.09 in Saudi Arabia, and $0.06 in Kuwait—*The Oil and Gas Journal,* June 26, 1967.

9. See Chapters 11 and 12 on refining and transportation for more precise data.

10. Adelman, note 2 above, p. 81, emphasis in original.

11. P. H. Frankel, *Mattei: Oil and Power Politics* (New York: Frederick A. Praeger, 1966), pp. 123–24.

12. The data also exclude "bunker oil" used as fuel for international shipping, since this is not usually classified as an export of a country.

13. A minor weakness is that some crude oil which is imported by underdeveloped countries is then re-exported in the form of finished or refined oil products. We have minimized this overlap by excluding refined product exports of certain underdeveloped countries which primarily re-export imported oil (Netherlands Antilles, Trinidad and Tobago, Canary Islands).

14. This is true despite the fact that published comparisons of reported company earnings with their payments to governments frequently show the latter to be far greater than the former; for example, the First National City Bank of New York collates the Eastern hemisphere earnings and payments to governments in that area for the seven international majors, and found for 1964 a payments-to-earnings ratio of 1.7. (*Petroleum Intelligence Weekly,* November 7, 1966). It is true that widespread discounting off posted prices had reduced the companies' share of profits below the nominal fifty-fifty level, based on posted prices. However, this is somewhat offset by the fact that reported earnings are reduced by various "non-real" bookkeeping charges such as for depreciation and depletion.

CHAPTER THREE

1. Phrase used by the late Enrico Mattei, head of the Italian state oil company (ENI); quoted in David Hirst, *Oil and Public Opinion in the Middle East* (New York: Frederick A. Praeger, 1966), p. 28.

2. The titles of some recent important works support this view: *The Politics of Oil; Politics and World Oil Economics; Mattei: Oil and Power Politics.*

CHAPTER FOUR

1. None of this is inconsistent with management rewarding itself highly for its skills. In any event, these rewards are usually only a very small share of

total profits, and part of them are in the form of stock options which temporarily at least harness management directly to the aims of the stockholders.

2. See, for example, Andrew Hacker, "The Making of a (Corporation) President," *The New York Times Magazine*, April 1967, p. 27:

> "It's remarkable how many of the young fellows we take in go along for years without realizing that their real job is to make money for the company," one executive remarked. . . .
> . . . The up-and-comer, however, soon learns to think first and foremost in business terms: the specialized skills he was taught at college are only useful if they can be applied to augmenting the firm's earnings. At a certain point he may have to compromise with professional standards in making or promoting a profitable but less-than-quality product. How he reacts to this challenge will be noted by his superiors.

3. For the dating of the spread of the DCF method in the oil industry, as well as a good discussion of its strengths and weaknesses, see: A. S. Mekkawy and H. A. El Banby, "A Modified Technique for Investment Analysis in Petroleum Production Industry," paper presented at the Fifth Arab Petroleum Congress, Cairo, 1965.

4. A survey article on "Financial Planning for International Oil Operations" notes, "The D.C.F. analysis is regarded generally as the most reliable instrument for investment decisions"—*World Petroleum*, June 1965, p. 67.

5. This may be defined as the rate of return required by the investment community in order to induce it to supply capital to the company (both debt and equity).

6. This is consistent with the fact that over the years the average rate of return on investment in manufacturing as a whole, after taxes, has been in this range. In effect, if the average firm over the years can earn 10 percent on its investment, the cost to it of raising new capital by bringing in outside investors is about 10 percent, owing to the need to share the profits among more investors.

7. Assuming there were no "tie-in" effects which would make the real return from this complementary investment considerably higher, as is often the case in the oil industry.

8. A study of Jersey reported: "It's opportunity—not geography—that counts. The return target is usually commensurate with the risk, meaning some ventures must promise 20% or better if the overall average is to be decent"—*The Oil and Gas Journal*, April 15, 1963, p. 89.

9. For the mechanics of oil exploitation decisions, see *World Petroleum*, August 1965, pp. 46–58.

10. *Barron's*, March 18, 1968.

11. P. H. Frankel, *Mattei: Oil and Power Politics* (New York: Frederick A. Praeger, 1966), pp. 78–86.

12. *Ibid.*, p. 87.

13. These newcomers had been propelled into their search for overseas oil partly by the increasing evidence of the enormous profitability of Middle Eastern oil. This can be seen from their reaction to the formation of an international consortium to take over the Iranian oil industry in 1954 (following the overthrow of the Mossadegh government which had nationalized the industry):

. . . Washington had made its approval of American participation in the Iranian consortium contingent upon the offer of a 5% share to U.S. independent firms not yet represented in the Middle East. . . .

For the additional members of the Iranian consortium, their minor share proved an exceedingly profitable investment and the prospectus which had been drawn up by a large firm of chartered accountants showed clearly for all concerned to see that even after the compensation payable to Anglo-Iranian any stake in that venture was like getting "a licence to print money." This experience proved to be of some major historical significance: it put a whole pack of American oil companies, who had never before ventured to operate outside the U.S., on the scent of new bonanzas; this oil rush had a lot to do with the number of newcomers to international oil . . . [*ibid.*, pp. 95–96].

14. This opportunity cost is particularly great owing to the high discount factors used in the international oil industry, especially for surplus crude oil; in Kuwait the discount factors used by the companies appear to be infinite, since there is more crude than they can ever hope to sell before their concession expires.

15. One possible long-run solution might be for newcomers with surplus crude oil to merge with large marketing and refining companies lacking sufficient crude oil; recent examples of this were the proposed mergers of Amerada Petroleum and Ashland Refining in 1967 and of Occidental Petroleum and Signal Oil in 1968.

16. *World Petroleum*, February 1961, p. 40.

17. Frankel, note 11 above, pp. 128–29.

18. *Ibid.*, p. 137.

19. *Ibid.*, p. 93, emphasis in original.

20. *Petroleum Intelligence Weekly*, October 25, 1965.

21. *Ibid.*, November 14, 1966.

CHAPTER FIVE

1. As an indicator of this, in the Eastern hemisphere, whose oil consumption is dominated by that of Western Europe and Japan, in 1964 the seven majors controlled 83 percent of crude oil reserves, 78 percent of crude oil production, 61 percent of refining capacity and probably a similar figure for product sales—source: ENI, *Energy and Petroleum in 1964*, p. 79. In the Common Market the seven majors own 53% of all refining capacity—*Petroleum Intelligence Weekly*, October 24, 1966.

2. Petroleum's share of total energy consumption increased from 18 percent in 1955 to 44 percent in 1965 for Western Europe and from 18 percent in 1955 to 57 percent in 1965 for Japan; in the United States the share remained constant at about 40 percent. (United Nations, statistical papers, series J, *World Energy Supplies*; various years.)

3. *Fortune*, June 15, 1968.

4. *Ibid.*

5. Data for 1962 from United States Senate, Subcommittee on Antitrust and Monopoly, *Economic Concentration: Part 1*, 1964, pp. 390–94.

6. *Fortune*, note 3 above.

7. *Fortune*, September 15, 1967.

8. Source: U.S. Department of Commerce, *Survey of Current Business*. This partly reflects the fact that the book value of U.S. investment in oil producing countries greatly understates the real value, owing to the omission of any value for the 150 billion barrels of oil proven to be in the ground.

9. Charles Issawi and Mohammed Yeganeh, *The Economics of Middle Eastern Oil* (New York: Frederick A. Praeger, 1962), p. 108.

10. *Ibid.*, pp. 88–89.

11. From testimony at 1948 Congressional hearings of James A. Moffett, former executive of Jersey, Texaco, and Socal, quoted in Robert Engler, *The Politics of Oil* (New York: The Macmillan Company, 1961), p. 199.

12. The growing shortage of "international liquidity," stemming both from the failure of the world gold supply to increase as rapidly as international trade and the increasing unwillingness of the developed countries to hold dollars instead of gold, has had widening repercussions on oil. There has been an increasing tendency for all countries, particularly in Western Europe, to assist their indigenous oil companies in order to improve the respective balances of payments. See, for example, *Petroleum Intelligence Weekly*, February 13, 1967.

13. The Chase Manhattan Bank, *Balance of Payments of the Petroleum Industry* (New York: n.d., October 1966), p. 4:

> The Bank's survey covered 29 U.S. petroleum companies, accounting for well over half of the petroleum operations carried on in the free world. The information obtained by this survey, when used in conjunction with published data and adjusted to cover the entire U.S. petroleum industry, permits a more detailed and complete appraisal of the petroleum industry's balance of payments position than would otherwise be possible.

14. *Ibid.*, p. 11.

15. Interestingly, the largest single item in the service account is the $139 million in managerial and other service fees received by U.S. petroleum companies from their affiliates (compared to only $11 million received by U.S. petroleum companies from nonaffiliated foreigners). Clearly affiliates contribute to the profits of the parent company not only directly through income earned on the investments but also indirectly through managerial and other service fees. These various fees paid to parents by affiliates have sometimes been a point of issue between the international oil companies and the governments of underdeveloped countries.

16. Chase Manhattan Bank, note 13 above, p. 14.

17. *The Economist* (London), May 15, 1965.

18. J. E. Hartshorn, *Politics and World Oil Economics* (New York: Frederick A. Praeger, 1962), p. 235.

19. *The Economist* (London), March 26, 1966.

20. That crude oil production profits are the key can be seen from the fact that British Petroleum accounted for only 37% of the two companies' crude oil imports, but 51% of their combined profits. British Petroleum's enormous reserves of crude oil (particularly through its one-half share in Kuwait), along with its relatively limited marketing position, force it to sell close to 40% of its crude oil to other companies.

21. Aside from the direct monetary flows resulting from the existence of international oil affiliates, there are various indirect ways in which the overseas oil investments affect the home country's monetary position. Perhaps most significant for Great Britain is the fact that a large part of the enormous revenues received by the governments of Middle Eastern oil-producing countries has flowed back to Great Britain in the form of sterling investments either in British securities or as deposits in British banks (see Issawi, note 9 above, pp. 148–150). Oil investments also indirectly help the balance of payments of both the United States and Great Britain through the dollar and sterling inflows coming from the oil-exporting countries' purchases of British or American goods. As the Chase study for the U.S., note 13 above, noted:

. . . Host countries tend to spend a substantial portion of U.S. investment funds to purchase U.S. exports and to build up dollar deposits in the U.S. . . . their importance is suggested by the fact that in 1964, Venezuela, The Netherlands Antilles, and Trinidad and Tobago added $143 million to their dollar deposits in U.S. banks, while importing $650 million of American merchandise. Also, the Middle Eastern oil producing nations and Libya added $163 million in U.S. bank deposits, while importing $417 million from the United States. These figures indicate that any cutback in investments could reduce such payments inflows [p. 17].

22. Hartshorn, note 18 above, p. 233.

23. Testimony of U.S. State Department's World War II Petroleum Attaché in the Near East, quoted in Engler, note 11 above, p. 249.

24. Engler, note 11 above, pp. 191–193.

25. John A. Loftus (Special Assistant to the Director of the Office of International Trade Policy, Department of State), "Petroleum in International Relations," *U.S. Department of State Bulletin*, vol. XIII, August 5, 1945, pp. 173–75; emphases added.

26. Jack Anderson, *Washington Exposé* (Washington, D.C.: Public Affairs Press, 1967), p. 202; emphases in original. For specific examples, see pp. 202–05.

27. Engler, note 11 above, pp. 247–48.

28. *Ibid.*, p. 190.

29. *Platt's Oilgram*, April 27, 1965. This statement is particularly significant in light of the following report:

U.S. Bolsters Oil Role in State Department: . . . Until now oil problems have been handled within the State Department by its Fuels and Energy Division. That division was only one of five reporting to the Office of International Resources. Now the fuels group has been upgraded from a division to a full-fledged office. Director of the new office of Fuels and Energy will be Andrew Ensor, who has been Chief of the Fuels Division.
Behind the move to strengthen the State Department agency handling oil affairs: A growing conviction that oil issues are a critical element in U.S. relations with a great number of countries. The re-organization is viewed in Washington circles as the handiwork of Thomas Mann, Undersecretary of State for Economic Affairs. Mann is known to believe that oil, already important in foreign affairs, is likely to become even more so in the days ahead.

It's expected that under the new setup oil problems will come to the attention of top policy makers more quickly than in the past and that the views of the Office of Fuels and Energy will be given a good deal more weight [*Petroleum Intelligence Weekly*, June 28, 1965].

30. From "Approach to the Problems in Maintaining and Expanding American Direct Investments Abroad," a talk by Leo D. Welch, Treasurer of Standard Oil Co., New Jersey, November 12, 1946, quoted in Engler, note 11 above, p. 267.

31. Engler, note 11 above, pp. 310–12. Engler goes on to state:

The sympathy and support enjoyed by oil cannot be explained simply by the number of its representatives within or advising the government. The engulfing character of oil operations brings a host of related business enterprises and leaders into its orbit. . . . This identity with which the men of power view one another, along with their frank admiration for their capacity to make deals, to get things done, and for efficiency, bigness, and power, and the frequent sharing of a common social outlook, are also fundamental factors in explaining oil's treatment by other businessmen in government [pp. 315–16].

32. Engler, note 11 above, p. 334, also notes:

A press inquiry about the propriety of having General White serve as director of the Defense Department's Division of Petroleum Logistics while on the Esso pay roll evoked the explanation from the President [Eisenhower] that White was a Reserve rather than a Regular officer and hence not required to divest himself of his economic interests: "It would be idle to employ as a consultant anyone who didn't know something about the petroleum business. He is bound to come from the petroleum industry."

33. *Ibid.*, p. 335.

CHAPTER SIX

1. For an idea of oil's potential as a sparkplug in economic development, and the lost opportunities, compare Chapters 22 and 23 on Mexico and Iraq, respectively.

2. David Hirst, *Oil and Public Opinion in the Middle East* (New York: Frederick A. Praeger, 1966), p. 81.

3. Hirst, note 2 above, page 29, states:

Saudi Arabia, it should be pointed out, is peculiarly vulnerable to hostile criticism. . . . Little imagination is needed to grasp the grotesque contrast between the backwardness and poverty of the people and the corrupt extravagance of the ruling family, and between the harsh puritanism of the Wahhabite tradition and the libertine tastes of the wealthy.

Hirst also notes, page 32:

Indeed, even the most impartial observers sometimes criticize Aramco for paying so much money to such irresponsible rulers. . . .

From the oil companies' profit maximization viewpoint, however, it is better for the rulers to squander oil revenues than to use them to go into competition with the companies themselves.

4. David Hirst, in his *Oil and Public Opinion in the Middle East* (New York: Frederick A. Praeger, 1966), pp. 40–41, says,

> In Iraq, for instance, when the Kassem government used to vie with Egypt in the fervour of its nationalism, nationalization would obviously have been a source of great prestige. But even if the government was interested, it was evidently not prepared to take the risk, fearing no doubt that the oil companies would mount an economic blockade as they did against Iran in 1951.

5. *Ibid.*, p. 11.

6. J. E. Hartshorn, in his *Politics and World Oil Economics* (New York: Frederick A. Praeger, 1962), pp. 176–179, says,

> In the United States, Britain and Europe, headquarters of the largest international companies, taxes paid abroad can be set off against tax liabilities at home . . .
> Following the Venezuelan example, Saudi Arabia in 1949 decreed that an income tax should be imposed: This would have entitled it to 50 per cent of Aramco's income *after* payment of United States tax. Negotiations between the government, the company and its corporate shareholders, and the United States Treasury ensued: eventually the law was drafted to give Saudi Arabia 50 per cent of profits *before* payment of American tax . . . But the sizeable increase in host government revenue did not bring about a corresponding reduction in the income of Aramco and its parent companies. Indeed, doubts have sometimes been expressed by non-American oilmen operating in the Middle East whether the acceptance of 50:50 cost the American companies operating in Arabia anything at all in net income —on the argument that the whole increase in Arabian tax, eligible for tax credit, was in practice offset by reductions in American tax on the same income [emphases added].

7. For a good summary discussion of these arrangements, see *ibid.*, pp. 160–165.

8. *Ibid.*, p. 163. Recent reports indicate that, owing to pressure from the Iranian government, the companies have agreed to liberalize their rules on "overlifting" one's quota, thereby improving the competitive position of Iranian oil. See *Petroleum Intelligence Weekly*, March 4 1968.

9. In Iran each company nominates a figure for total production in the following year. The highest total figure acceptable to an unstated percentage majority of the shareowners is then distributed among each company on the basis of their percent ownership. BP and Gulf's 47% bloc may well be sufficient to set the highest figure. See Hartshorn, note 6 above, p. 163.

10. Hirst, note 4 above, p. 42. Similar charges have been made by other oil-exporting countries:

> Producing companies operating in Venezuela are slowing their output of crude oil as a bargaining weapon in current negotiations with the government over back income tax claims and service contracts, according to views of former Oil Minister Juan Pablo Perez Alfonzo reported in the Caracas press [*Petroleum Intelligence Weekly*, May 9, 1966].

11. Hisham Nazer and Muhammed Joukhdar, "Oil Prices in the Middle East," *Middle East Economic Survey*, Beirut Weekly, 23rd September 1960; quoted in Hirst, note 4 above, p. 43.

12. The most prominent theoretician of this new philosophy is an American lawyer who has served as legal advisor to several Middle Eastern countries. His views have been summarized by Hirst, note 4 above, p. 63, as follows:

> . . . the "companies have managed to preserve their relationship with their host nations in an atmosphere of legal morality which had no legal basis either in creation or in continued existence." The original concession agreements were "surrounded with an assumed aura of untouchability and permeated with a self-created sanctity."
>   . . . the sovereign nation, acting in the ever-evolving interests of its people, should have the right to exercise its legislative authority to insure an equitable relationship with its concessionaire.

For the traditional view see George Lenczowski, *Oil and State in the Middle East* (Ithaca: Cornell University Press, 1960), pp. 94–101.

13. Hirst, note 4 above, pp. 57–58.

14. *Ibid.*, note 4 above, p. 58.

15. Speech at Ourgla, Algeria, September 28, 1964; quoted in *ibid.*, note 4 above, p. 58.

16. *Ibid.*, note 4 above, pp. 100–102.

17. See Hussein Abdallah Abdel-Barr, *The Market Structure of International Oil with Special Reference to the Organization of Petroleum Exporting Countries* (Ph.D. thesis, University of Wisconsin, 1966).

18. *Ibid.*, p. 56. The defection of Royal Dutch Shell from the other majors may be attributable to the fact that it is the weakest of them in terms of its ownership of Middle Eastern crude oil.

19. *Ibid.*, p. 49.

20. Hirst, note 4 above, pp. 114–115.

21. David Hirst, in his *Oil and Public Opinion in the Middle East* (New York: Frederick A. Praeger, 1966), pp. 112–113, says,

> Iran's membership in OPEC has always been potentially awkward for the Arab producers, both for political reasons and because, in strictly oil matters, Iran has always been lukewarm on some of OPEC's policies and prone to attitudes of embarrassing moderation . . . it has been suggested that Iran's interest in OPEC, to begin with at least, was largely the negative one of keeping Egypt out of the Organization and thereby preventing it establishing control over Middle East oil policies. . . . Little is known of the give-and-take of argument among OPEC's members, but it is more or less certain that Iran has always sought to dissuade the organization from asking its member governments to take legislative measures against the companies.

Again, it is reported that at the Sixth Arab Oil Congress, "The ideological split between the Socialist-style and monarchist regimes—or between 'nationalists' and 'moderates'—prevented the conferees from advancing any closer to a unified or coordinated oil policy . . ."—*Petroleum Intelligence Weekly*, March 20, 1967.

22. "On the whole, it is a moderate and realistic public opinion to which the technocrats have always sought to make their appeal. For to advocate extreme measures would bring them into conflict with the political executive in their home countries"—Hirst, note 4 above, pp. 103–104.

23. *Petroleum Intelligence Weekly*, March 20, 1967.

24. Hirst, note 4 above, p. 114.

25. OPEC, "OPEC and the Principle of Negotiation," paper presented to the Fifth Arab Petroleum Congress, Cairo, March, 1965, pp. 18, 19.

26. *The New York Times*, May 16, 1967.

27. *Petroleum Intelligence Weekly*, October 23, 1967.

28. *The Oil and Gas Journal*, November 6, 1967, p. 69.

29. *Ibid*.

30. *The New York Times*, December 13, 1967.

31. P. H. Frankel, *Mattei: Oil and Power Politics* (New York: Frederick A. Praeger, 1966), pp. 94–96.

CHAPTER SEVEN

1. This was not always the companies' position, however, particularly where a very large deal could be landed by barter; e.g., Jersey made a deal in 1963 with ENI to sell it crude oil partly in exchange for oil equipment. But of course, this was capital equipment that the company would have had to purchase anyway, rather than the diverse consumer goods the Soviets were willing to accept.

2. Harold Lubell, "The Soviet Oil Offensive," *Quarterly Review of Economics and Business*, November 1961, p. 12.

3. *Ibid.*, pp. 9–10.

4. Quote from "An article in an authoritative Soviet publication" reported in: National Petroleum Council, *Impact of Oil Exports from the Soviet Bloc, Vol. I* (Washington, D.C.: National Petroleum Council, 1962), p. 37, hereafter referred to as "NPC Report."

5. "On November 21, the Soviet Government began exerting direct pressure. It cancelled all orders with three major Finnish exporters . . . [One week later] the Soviets cut off shipments of crude oil to Finland for the rest of the year"—U.S. House of Representatives, Committee on Foreign Affairs, *Background Information on the Soviet Union in International Relations* (Washington, D.C.: Government Printing Office, September 27, 1961), p. 52.

6. NPC Report, note 4 above.

7. *Ibid.*, note 4 above, p. xiii.

8. *Ibid.*, note 4 above, p. 41.

9. Donald K. David, "A Plan for Waging the Economic War"—an address before the Business Advisory Council, Hot Springs, Virginia, Oct. 17, 1958 (Committee for Economic Development, n.d.) pp. 3–4, emphases in original.

10. *Ibid.*, p. 7, emphases in original.

11. Not only was the power of the United States government felt to be needed by the companies, but to the extent that coordinated action would be needed among the United States companies, this would require a special dis-

pensation from the Department of Justice which administers the United States antitrust laws. One ingenious rationale put forth for seeking such a Justice Department dispensation was the notion that special measures were needed to counteract a "Communist conspiracy in restraint of U.S. trade."

12. *World Petroleum*, April, 1962.

13. The National Petroleum Council on p. 28 of its NPC Report, note 4 above, claimed,

> . . . Insofar as Soviet Bloc oil exports reduced the volume of exports from Free World producing countries, Venezuela and the Middle East were the principal sufferers. . . . An estimate of the cumulative loss of direct income by Middle Eastern and Venezuelan governments over the period 1954–61, based on the above method, is shown below to have reached $US 490 million. This estimate is given without considering the depressing effect on prices of the dumping of Soviet oil in the free world.

14. Lubell, note 2 above, p. 16.

15. David Hirst, *Oil and Public Opinion in the Middle East* (New York: Frederick A. Praeger, 1966), pp. 43–44.

16. M. A. Adelman, "The World Oil Outlook," in Marion Clawson, ed., *Natural Resources and International Development* (Baltimore: Johns Hopkins Press, 1964), pp. 93–96.

17. "These exports, which had risen steadily until 1959, lost their momentum during 1960–62, but by 1963 had regained their 1959 level of $80 million. And in 1964, Soviet exports of petroleum products reached an all time high of roughly $100 million"—Carole A. Sawyer, *Communist Trade with Developing Countries: 1955–65* (New York: Frederick A. Praeger, 1966), p. 36.

18. Adelman, note 16 above, p. 94.

Again, a 1967 United States brokerage house survey of "overseas conditions" in the oil industry reported [Goodbody & Co., *Monthly Letter,* July 1967, p. 14]:

> On the brighter side, one of the important political problems of the past now seems to have been mitigated. Russia's crude oil exports are no longer growing at an inordinate rate . . . Moreover, Russian crude is no longer forcing its way into new markets via heavy price cutting. In fact, one Russian official recently complained of the price-cutting tactics employed by "monopolistic international oil companies."

CHAPTER EIGHT

1. International Bank for Reconstruction and Development, *Articles of Agreement.*

2. Abbas Ordoobadi, *The Loan Policy of the International Bank for Reconstruction and Development* (New School for Social Research, Ph.D. thesis, 1963), p. 188.

3. *Fortune*, March 1963, p. 210.

4. Andrew Shonfield, *The Attack on World Poverty* (London: Chatto and Windus, 1960), p. 110.

5. Black has been described as "a Southern-born aristocrat who is a

long-time employee of the Rockefeller interests." [G. William Domhoff, *Who Rules America?* (Englewood Cliffs, New Jersey: Prentice-Hall, 1967), p. 67.]

While the current Bank president, Robert McNamara, has a reputation for being a pragmatist, there is nothing in his background to suggest that he will ultimately be any less favorably disposed toward private enterprise.

6. Shonfield, note 4 above, pp. 135–37.

7. *Fortune*, March 1963, p. 212, says,

This brings us back to the case for multilateralism [in foreign aid] propounded by bankers Black, Champion [Chairman of Board of Chase Manhattan Bank], and Prochnow [President of First National Bank of Chicago] . . . It also means . . . a combined effort to organize the capacity of the underdeveloped countries to produce more and more primary commodities for export, the only path for those countries toward true "self-sustaining" growth and social stability.

8. IBRD, *Third Annual Report*, 1947–48, p. 17.

9. Shonfield, note 4 above, p. 136.

10. *Ibid.*, p. 130. A United Nations study of financing of petrochemical plants notes that: "The international agencies [including the World Bank] would rather encourage developing countries to appeal to the major international petrochemical companies to help them not only with transmission of know-how, but also with provision of funds"—United Nations, *Studies in Petrochemicals*, papers presented at the United Nations Interregional Conference on the Development of Petrochemical Industries in Developing Countries, Teheran, Iran, November 1964 (New York: United Nations, 1966), p. 931.

11. Eugene R. Black, *The Diplomacy of Economic Development* (Cambridge, Mass.: Harvard University Press, 1960), pp. 66–67.

12. *Ibid.*, p. 26.

13. *International Review Service; Energy Developments*, September 1966, p. 11.

14. Michael L. Hoffman (advisor to the editor of the official Bank and IMF publication), *The Fund and Bank Review: Finance and Development*, March 1967, p. 66.

15. Shonfield, note 4 above, p. 110.

16. Robert E. Asher, et al., *The United Nations and Promotion of the General Welfare* (Washington, D.C.: Brookings Institute, 1957), p. 1047.

17. See Francis Cassel, *Gold or Credit* (New York: Frederick A. Praeger, 1965), pp. 49–50.

18. Reported in Shonfield, note 4 above, p. 43.

19. *Ibid.*, pp. 42–43.

20. *Ibid.*, pp. 37–38.

21. *The New York Times*, February 5, 1967.

22. *The New York Times*, May 14, 1967.

23. *The New York Times*, June 1, 1967.

24. "Thoughts on Troubled Oil," *Far Eastern Economic Review*, December 19, 1963.

25. *International Review Service: Energy Developments*, May 1966.

26. *World Petroleum Report*, March 15, 1966.

27. *The Oil and Gas Journal,* January 9, 1967, p. 65.
28. *The New York Times,* May 8, 1966.
29. Robert Engler, *The Politics of Oil* (New York: The Macmillan Company, 1961), p. 265.
30. Alvin Z. Rubinstein, *The Soviets in International Organizations: Changing Policy Toward Developing Countries, 1953–1963* (Princeton, New Jersey: Princeton University Press, 1965), pp. 186, 190.

CHAPTER NINE

1. Pierre Maillet, in "Economic Aspects of the Importation of Energy," paper presented at the United Nations Interregional Seminar on Energy Policy in Developing Countries; Breau, France, May 1965, pp. 4–7, says,

> In order to compare the cost of imported products and that of national production, it is necessary to have a proper assessment of the cost of foreign currencies in the countries considered. . . .
> The problem of determining such a standard may be considered from two angles. On the one hand, the cost of marginal exports may be taken as a basis, that is to say the cost in national currency of goods which, when exported, will provide the foreign currencies required for the payment of energy. In practice, the cost of these marginal exports comprises two factors: On the one hand internal expenditure, such as cost of labour, on the other hand the cost of imported material required to achieve production (in some ˆcases, raw material and energy, in many cases, the mechanical equipment to manufacture the goods). The cost of foreign currencies, calculated in relation to marginal exports, can then be expressed as follows: value of exports (in foreign currencies) = cost of imports (in foreign currencies) + local costs (in national currency).
> The other method is to consider the marginal utility of imports other than energy. In most cases, these imports will consist of capital equipment. Insofar as they are necessary to ensure economic growth it is theoretically possible to relate a given amount of imports to a corresponding increase in national production of the country and thus to determine the marginal utility of the foreign currency.

2. National Council of Applied Economic Research, *Demand for Energy in India* (New Delhi: National Council of Applied Economic Research, 1966), p. 95.
3. Maillet, note 1 above, pp. 9–10.
4. *Ibid.,* p. 10.
5. United Nations, *Studies in Petrochemicals* (New York: United Nations, 1966), inside cover.
6. *Ibid.,* p. 3.
7. M. S. Williams and J. W. Couston, *Crop Production Levels and Fertilizer Use* (Rome: Food and Agriculture Organization, 1962), p. v.
8. See Arthur P. Lien, "Plastics as Construction Materials for Developing Countries," in UN, *Studies in Petrochemicals,* p. 553. See note 5 above.
9. Ralph Landau, "Technology and Obsolescence in the Petrochemical Industries for Developing Countries," p. 52. See note 5 above.
10. See note 5 above, p. 18.

11. See note 5 above, p. 389.

12. See note 5 above, p. 16.

13. See note 5 above, p. 927.

14. An extreme example of the impact of energy policy on the very life of a country can be seen in the case of North Vietnam in recent years. There a decision to import from the Soviet Union a large number of small-scale power plants, averaging less than 3,000 kw capacity, has helped "to keep North Vietnam's factories humming despite the repeated bombing of key power stations . . . it is believed that none of the small plants has been damaged by the raids"—*The New York Times,* November 6, 1966.

CHAPTER TEN

1. In order to forestall this the Middle Eastern governments in particular have been encouraging underdeveloped countries like India to explore for oil in the Middle East, on a joint venture basis with Middle Eastern governments. If such ventures were set up on a government to government basis, and oil were found, this would serve a dual purpose. First, it would essentially hold the Middle Eastern governments' revenues constant, with the saving in oil costs for the Indian government coming at the expense of the oil companies. Second, in the long run it might provide a relatively guaranteed outlet for the oil-exporting country's oil, and hence make the country less dependent on the international companies.

2. In fact, until 1962, the United States Government's Export-Import Bank was generally unwilling to provide financing and guaranties for United States exporters of equipment for oil exploration (and refining) to be sold to government-owned oil operations—*The New York Times,* March 25, 1962. Largely due to the pressure of competing Soviet offers, the government subsequently adopted a more flexible policy permitting such loans.

3. Walter J. Levy, *The Search for Oil in Developing Countries: A Problem of Scarce Resources and Its Implications for State and Private Enterprise* (for the International Bank for Reconstruction and Development, 1961).

4. While the Bank's general philosophy and actions clearly indicated opposition to governmental entry into the oil industry, the Bank could hardly explicitly endorse such opposition in a widely circulated document. This could be branded as an attempt to infringe on national sovereignty.

5. "Walter J. Levy . . . a petroleum consultant whose clients included Esso, Caltex, and Shell"—Robert Engler, *The Politics of Oil* (New York: The Macmillan Company, 1961), p. 311.

The New York Times reported (May 31, 1968):

Walter J. Levy, an internationally known petroleum consultant, has been presented a plaque by the State Department in recognition of his services. Secretary of State Dean Rusk presented the award.

The inscription on the plaque reads, "In grateful appreciation for the invaluable contribution to the welfare of the United States." Mr. Levy is a consultant to many of the nation's major oil companies as well as to the Government.

6. This is particularly true for a report like this, which was to deal with a controversial area of economic policy.

7. J. E. Hartshorn, in his *Politics and World Oil Economics* (New York: Frederick A. Praeger, 1962), p. 231, says,

> Mr. Walter Levy, the American oil consultant, in a study that he prepared for the World Bank in 1961, argued very powerfully that exploration and development, at any rate, are usually too expensive and risky an operation for the governments of underdeveloped countries to engage in on their own. With the limited capital resources they usually possess, he thought they might well concentrate on investments offering a more guaranteed return, since the international oil industry was prepared to put in the risk capital. This was wise counsel . . .

8. In the first group are Colombia, Nigeria, Canada, Guatemala, Turkey, and Libya; in the second group, Brazil and Mexico; in the third group, Argentina, Bolivia, and India.

9. W. J. Levy, "Basic Considerations for Oil Policies in Developing Countries," *Techniques of Petroleum Development: Proceedings of the United Nations Interregional Seminar on Techniques of Petroleum Development,* Jan.–Feb., 1962 (New York: United Nations, 1964), pp. 323–30.

10. *Ibid.,* pp. 325–26.

11. *Ibid.,* p. 326.

12. United Nations, Department of Economic and Social Affairs, *Petroleum Exploration: Capital Requirements and Methods of Financing* (New York: United Nations, 1962).

13. *Ibid.,* pp. 22, 29.

14. For example, the concept of the effect of an investment on the balance of payments suggests that the real value of foreign currency to the underdeveloped country is greater than the official exchange rate; if true, this can be incorporated in a rate of return on investment calculation by use of appropriate "shadow prices" which value foreign exchange at the assumed higher level. For discussion of this and other advantages of the rate of return on investment criterion see Wolfgang F. Stolper, *Planning Without Facts: Lessons in Resource Allocation from Nigeria's Development* (Cambridge, Massachusetts: Harvard University Press, 1966), particularly Chapter V, "Investment Criteria from a Planning Standpoint," pp. 138–218.

15. This can be seen from the following equation, where I equals the amount of investment, O the total output generated by the investment, and C the current cost of producing the output. Using this notation, by definition the rate of return on investment equals $(O - C)/I$. As Harrod-Domar have shown, the contribution of the investment project to economic growth is equal to the savings ratio from any given level of output divided by the capital-output ratio. In terms of our notation, the real savings ratio from a project may be defined as $(O - C)/O$, while the capital output ratio is I/O. Simple algebra will show that the growth rate caused by the investment equals $(O - C)/I$, or is equivalent to the rate of return on investment.

It might be noted that the capitol-output criterion, which implies minimizing the initial investment required for any given level of output, would be equivalent to the rate of return on investment criterion on the assumption that there were no real current costs associated with the investment.

16. Levy, note 9 above, p. 324.

17. United Nations, note 12 above, pp. 1–3.

18. P. Leicester, "The Risk and the Reward," *Techniques of Petroleum Development*, note 9 above, p. 135.

19. Royal Dutch Shell, *The Petroleum Handbook* (London: Shell International Petroleum Company Limited, 1959), pp. 71–72.

20. United Nations, note 12 above, p. 13.

21. E. N. Avery, "The Odds in Oil Exploration"—paper presented at the ECAFE Symposium on the Development of Petroleum Resources of Asia in the Far East, Third Session, Tokyo, Japan, November 1965, p. 2.

22. *Ibid.*, p. 13.

23. Levy, note 9 above, p. 324.

24. United Nations, note 12 above, p. 11. This report also presents many detailed specific figures and estimates, but in a way which renders them almost worthless for planning policy. For example, it provides a "simplified hypothetical seven-year program budget" for oil exploration. This program involves capital costs over a seven-year period of $9 million and operating costs of $31 million, for a grand total of $40 million. However, the reader has no clear idea how much drilling and testing for oil could be undertaken in such a program, nor whether this is a "minimum" or "average" or "maximum" size program.

25. Leicester, note 18 above, pp. 136–37.

26. Actual import parity would probably be between $1.75 and $2.00 per barrel, but we have deducted $0.25 per barrel as an upper limit estimate of "post development" production costs not included in the Levy estimate.

27. This ignores the lag (unknown for this example) between investment and production, which would reduce the real rate of return. The aim here is not precision, but simply to indicate that planners should view this kind of data with the proper perspective.

28. Data from *Petroleum Outlook*, June 1964, p. 98. The method used by this journal is to divide capital expenditures for finding and developing oil by the amount of oil produced during the period plus the increase in estimated oil reserves. Capital expenditures are defined as "costs of acquiring leases and concessions, dry holes, productive wells, and lease equipment, but excludes lease rentals, geological and geophysical expense."

29. Levy, note 9 above, p. 327.

30. United Nations, note 12 above, pp. 15, 23.

31. ENI, *Energy and Petroleum in 1964* (Italy: ENI, 1965), pp. 73–77.

32. The economic incentive for the international oil companies to explore for oil in underdeveloped countries in recent years has been further weakened by declining transport rates for international oil shipments, which have widened the area over which Middle Eastern oil can be competitive. In addition, the high discount factor used by the international oil companies for evaluating all investment projects militates against further exploration efforts. Combined with the uncertainty of success and the time lag between oil exploration and production, this high discount rate makes the calculated value of future oil, and hence present exploration efforts, even less than might appear: the present value of a

dollar's worth of oil found seven years from now when discounted at 20% per year is only 21¢.

33. In general it may be stated that for any established major, its willingness to invest in indigenous exploration will be positively correlated with: (a) the resolve of the government to embark on oil exploration if the company does not undertake it; (b) the chances of keeping the government out altogether, or its efforts minimal by going in itself; (c) the profitability of any crude oil found, as determined by agreements on pricing, tax rates, extent of preference over other oil, etc.; (d) the cost of importing its own oil from external sources. It will be negatively correlated with: (e) the market share of the company within the country; (f) the probable cost of finding oil; (g) the impact upon its other international operations, in terms of pressures from other governments for oil exploration.

34. Avery, note 21 above, p. 17.

35. While this in general would be true, as it is likely the "intruder" would be guaranteed that his crude oil would be processed within an indigenous refinery whatever its ownership, it should also be recognized that such an intruder has to be wary of retaliation in other countries where he may have a major established position. More important, the newcomers' problems frequently stem from the fact that they have found large amounts of crude oil but lack marketing and refining facilities, not the other way around.

36. Chase Manhattan Bank, Petroleum Department.

37. ENI, note 31 above, pp. 91–92.

38. The enormous expenditure of money on oil exploration in the United States, despite the fact that it consistently has the highest per barrel finding cost in the world, is eminently sensible from a profit maximizing viewpoint. The incentive is the sheltered high-price market combined with highly favorable tax treatment as explained by J. E. Hartshorn in his *Politics and World Oil Economics* (New York: Frederick A. Praeger, 1962), pp. 181, 200–201:

Income earned in oil production by American companies anywhere, though not on later operations, also qualifies for the "depletion allowance" mentioned above of 27½ per cent before computing taxable income. Moreover, it is possible to "expense"—i.e., to write off against income in the year it is incurred—all "intangible" costs of drilling as well as other expenses of operation. These "intangible costs"—i.e., those that do not remain embodied in physical assets on which depreciation can be charged—amount to about 75 per cent of drilling costs. . . .

There are cynical statisticians inside the American oil business who will point to the surge of drilling activities that seems regularly to take place in the second half of each financial year, and to link this with the point at which entrepreneurs begin to get fairly firm estimates of what their final income, and hence their prospective tax liability, may be. There is even a phrase for it—"drilling up one's tax."

39. See Annual Reports of Phillips Petroleum and ENI.

40. This unique aspect of oil exploration may be even more significant for a market economy than in a planned economy. While the planned economy can take into account the fact that some resources utilized in a steel mill may

be a "free good" because they would not be utilized otherwise, private entrepreneurs still have to pay money for these resources.

41. Once it is granted that the potential social return to the economy from crude oil exploration investments is so great, this weakens the rationale for joint ventures with foreign oil companies on a profit-sharing basis. Normally, there would be no great incentive to the companies to explore for crude oil in a given underdeveloped country unless the prospects for finding large amounts of crude oil were extremely promising. If that is the case, why should the underdeveloped country cut the companies in on a good thing? This is particularly true since the services of oil drillers can be purchased on a flat fee basis.

CHAPTER ELEVEN

1. Data from United Nations, *World Energy Supplies.* In 1950, Mexico and Argentina, where state oil entities dominated refining, accounted for 60 percent of this refined fuels production, as compared to only one third of the total consumption. Excluding these two countries, the aggregate self-sufficiency ratio was only one fourth.

2. C. H. Gamer, "A Modern Refinery," *Techniques of Petroleum Development: Proceedings of the United Nations Interregional Seminar on Techniques of Petroleum Development,* Jan.–Feb., 1962 (New York: United Nations, 1964), p. 192.

3. Part of the difference usually stems from differences in the degree of sophistication of the refinery, including the range of petroleum products which can be produced.

4. M. E. Hubbard, "The Economics of Oil Transport and Refining Operations," *Techniques of Petroleum Development,* note 2 above, pp. 215–24.

5. Calculated from *ibid.,* pp. 219, 222–23.

6. For Iran's product price data see United Nations, *Yearbook of International Trade Statistics: 1964* (New York: United Nations, 1966); for refinery yield data see Hubbard, note 4 above, p. 223.

7. Hubbard, note 4 above, p. 218.

8. Relatively more of the investment is in tankage in a simple refinery as contrasted to a large complex refinery.

9. This would be true whether depreciation was assumed to be in 12 or 18 years. This is because where the gross return on investment is high, depreciation costs typically are swamped by the true profit.

10. Based on Hubbard, note 4 above, p. 220.

11. In addition, with either the large or the small indigenous refinery, there would be further savings from the fact that it usually is cheaper to transport crude oil than refined products.

12. This certainly seems undeniable in the case of construction costs, since refinery construction is a specialized and relatively competitive business; the oil companies themselves frequently use outside contractors.

13. *Far Eastern Economic Review,* March 20, 1964.

14. R. E. Bittner, G. P. Baumann, A. R. Crosby (Esso Research and Engineering Company, Madison, New Jersey), "Recent Developments in the De-

sign of Small Refineries"—paper presented at United Nations Conference on the Application of Science and Technology for the Benefit of the Less Developed Areas (Geneva: November 1962), p. 6.

CHAPTER TWELVE

1. Shipping charges typically range from 10 to 30 percent of the total oil import bill.

2. Excluding that between the Gulf of Mexico and the U.S. East Coast, which is reserved for tankers flying American flags.

3. See J. Bes, *Tanker Shipping; Practical Guide to the Subject for all Connected with the Tanker Business* (Hilversum, Netherlands: C. de Boer, 1963), p. 83.

4. "The companies claim that they attempt to supplement their ownership with vessels chartered on a long-term basis, to a total of 90% of their expected requirements. For the remaining 10%, they choose to depend on the spot market"—Zenon S. Zannetos, *The Theory of Oil Tank Ship Rates: An Economic Analysis of Tank Ship Operations* (Cambridge, Massachusetts: MIT Press, 1966), p. 3.

5. Bes, note 3 above, p. 34.

6. Spot rates for tankers more than doubled from $2.00 per thousand ton-miles in mid-1956 to $4.50 per thousand ton-miles in February 1957.

7. See "Oil Abundance: Shortage Fears Raised by Mideast War Yield to Threat of New Glut," *The Wall Street Journal*, October 24, 1967.

8. *Ibid.*

9. The basic underlying reason for economies of scale in tanker ships is that construction cost varies primarily with the surface area while carrying capacity depends on the volume enclosed by the surface area; in general, for a cylindrical object like a tanker, doubling the surface area will result in a four-fold increase in the volume of carrying capacity, thereby tending to reduce unit construction costs by 50 percent. There are, at any point in time, limitations on how far one can carry this, but technological advance is continually raising the upper limits.

10. Bes, note 3 above, p. 23.

11. *The New York Times*, January 1, 1967.

12. See "Tanker Markets in the Sixties," *Petroleum Press Service*, May 1964, p. 172.

13. *Petroleum Intelligence Weekly*, May 2, 1966.

14. "Tanker Markets in the Sixties," note 12 above.

15. *Ibid.*, note 12 above, p. 173.

16. *Petroleum Intelligence Weekly*, July 11, 1966.

17. *The New York Times*, March 12, 1967.

18. J. Bes, in his *Tanker Shipping: Practical Guide to the Subject for all Connected with the Tanker Business* (Hilversum, Netherlands: C. de Boer, 1963), pp. 154–155, states

The "International Tanker Nominal Freight Scale"—code name: "Intascale"—which became effective on 15th May, 1962. This Scale replaced the London Market Tanker Nominal Freight Scale No. 3, known as London

Scale, which has served, with minor modifications, as the standard of reference in the tanker markets outside the U.S.A. since 1st November, 1952.

... It was solely intended as a standard of reference for various voyages against which the market level at any given time could be readily measured and by means of which rates for various voyages could be expressed comparatively. The London Scale set out rates of freight per ton of 2,240 lbs. ... "Intascale" ... in the main follow[s] the general principles adopted in Scale No. 3. ...

... in London Scale No. 3 a representative modern "general purpose" tanker was adopted as the standard vessel for all calculations viz. a tanker of 19,500 tons deadweight on summer loadline, having an average service speed of 14 knots ... [etc.]

An advantage of this system of quoting rates is that if the charterer of the vessel does not yet know his specific ports of departure or arrival, the rates can be set in terms of Scale, and when the ports are known the exact dollar amounts can be computed.

19. Zannetos, note 4 above, p. 246.

20. "... [AFRA] is used in industry for long-term agreements as an indicator of the cost of transportation ..."—*ibid.*, note 4 above, p. 194.

21. "... the company-owned vessels are multiplied by the average rate of the long-term contracts ..."—*ibid.*, note 4 above, p. 194.

22. *Ibid.*, note 4 above, p. 194.

23. "... the AFRA awards have ... imparted a valuable element of stability to oil prices by averaging widely fluctuating freight rates through time"—"Tanker Markets in the Sixties," note 12 above, p. 174.

24. *Ibid.*, note 12 above, pp. 171–73.

25. *Petroleum Intelligence Weekly*, September 25, 1967.

26. Another indication that AFRA largely serves the interests of the companies is the fact that immediately following the 1967 Suez crisis, when spot rates were skyrocketing, computation of AFRA was switched from a semiannual basis to a monthly basis—*ibid.*, August 28, 1967. This meant that the higher spot rates would more quickly be reflected in AFRA than otherwise. No such rush to more quickly reflect market rates was shown when spot prices were falling.

27. Peter Odell, *Oil: The New Commanding Height* (London: Fabian Society Research Series 251, 1965), p. 1.

28. "Tanker Markets in the Sixties," note 12 above, p. 174.

29. *Petroleum Intelligence Weekly*, April 6, 1964; emphasis added.

CHAPTER THIRTEEN

1. Economic data from India's Central Statistical Organization, reported in *The Eastern Economist: Annual Number 1967*, December 30, 1966, pp. 1372–73.

2. Commercial energy is defined here as including oil, natural gas, coal, and nuclear and hydroelectric power. We exclude wood, charcoal, vegetable wastes, and dung cakes which, while sometimes sold commercially, are not generally suitable for a modern economy.

3. For estimates of energy consumption by sector, see National Council of

Applied Economic Research, *Demand for Energy in India* (New Delhi: National Council of Applied Economic Research, 1966), pp. 160–73.

4. Different energy sources are converted here to a common denominator of "fuel oil equivalent" (FOE). Coal is statistically converted to fuel oil at one ton equals 3.91 barrels of fuel oil, based on the caloric values of Indian coal of about 11,000 BTU per pound, and of fuel oil, 6.2 million BTU per barrel. Hydroelectric power is converted at one million kwh equals 2,200 barrels FOE, based on assuming thermal efficiency of 25 percent.

CHAPTER FOURTEEN

1. Burmah Oil also owns 25 percent of the common stock of British Petroleum.

2. Such exclusion could come about from nationalization or by the government allowing other oil companies to build the refineries in exchange for guaranteed crude supply rights. This could ultimately result in complete or partial loss of the established majors' valuable marketing facilities in India.

3. Based on after-tax profits on Middle East crude oil of about $1.00 per barrel, estimated from posted prices which prevailed in the mid-1950s.

4. In later years, the refinery agreements came under strong criticism from the Indian Oil Minister, K. D. Malaviya:

> While the Minister said India would not take action, he gave the companies ample notice that the agreements would be subject to substantial revision when renewable in 1968.
> "It is a bad agreement," he said.
> Mr. Malaviya said the agreement was signed one year after independence when India knew little about such things. He said "even American experts had advised me it was a bad agreement."
> He said he had no personal quarrel with the companies, but criticized them for being slow in bringing about voluntary revisions. He said in some cases "they were not listening to advice of their own governments" [*New York Journal of Commerce,* July 12, 1960].

5. Source: Reserve Bank of India.

6. In hindsight this success appears to be due to the combination of relatively modest targets, unusually good agricultural weather and crops, and the fact that "the First Five Year Plan sought to rehabilitate the economy from the ravages of war, famine and partition . . ."—Government of India, Planning Commission, *Fourth Five Year Plan: A Draft Outline,* p. 1. As experience in war-torn countries has shown, rehabilitation is considerably easier than *de novo* growth.

7. *Foreign Commerce Weekly,* March 23, 1959.

8. United States Department of Commerce, Bureau of Foreign Commerce, *Investment in India,* January 1961, pp. 2–3.

9. Wolfgang G. Friedmann and George Kalmanoff (eds.) *Joint International Business Ventures* (New York: Columbia University Press, 1961), reprinted in Gerald M. Meier (ed.), *Leading Issues in Development Economics* (New York: Oxford University Press, 1964), p. 159.

10. There were rumors in India of such steps being taken. For example, in 1959 a New Delhi newspaper "reported" the following:

The Burmah Shell oil storage and distributing company of India, Limited, is likely to be converted into a rupee company shortly. A proposal to this effect is believed to be under consideration by the management of the company in London. Certain administrative and business aspects of the proposal are said to present some problems, but these are expected to be overcome soon [*The Hindustan Times*, July 5, 1959].

The implication of course was that conversion to a rupee (Indian currency) company would be followed by an offering of local equity.

11. *Esso Eastern Review*, June 1963.

12. As one Indian newspaper generally sympathetic to the oil companies noted:

The fact that there is no *Indian* private enterprise in oil makes it easier for the Government to resist suggestions for expansion from the private sector which is foreign. The unfortunate impression still prevails in the public mind that foreign oil companies (anywhere and not only in India) are meddlesome and rapacious [*The Hindu Weekly Review*, June 26, 1961; emphasis in original].

In 1962 the foreign oil companies announced a willingness to allow local common equity under certain conditions. By then, however, it was probably too late to undo the original damage, and, while discussions were held between the companies and the government, no agreement was ever reached.

CHAPTER FIFTEEN

1. *National Herald* (Lucknow, India), July 16, 1960.

2. See Harold Lubell, "The Soviet Oil Offensive," *Quarterly Review of Economics and Business*, November 1961, p. 11.

3. The offered price cut was originally believed to be about 30 percent on a CIF basis. The final deal worked out in 1960 was "apparently at about a 20 percent discount from f.o.b. posted prices"—*ibid*.

4. This government-owned company had been set up in 1959 as the government's distribution company for handling refined products produced by indigenous public sector refineries or imported on a government-to-government basis, particularly from Eastern Europe.

5. Data from Government of India, Department of Commercial Intelligence and Statistics (Calcutta), *Monthly Statistics of the Foreign Trade of India*.

6. Even as late as 1965–66, Soviet barter oil product imports amounted to only 27 percent of India's product imports (by value).

7. "About this period [1960] the cognoscenti were in the habit of referring knowingly to the Indian government as the NIL Government—Nehru-Indira-Lal (Shastri). . . . it suggested that there was no effective government of India (which was what the critics alleged) . . ."—Geoffrey Tyson, *Nehru: The Years of Power* (New York: Frederick A. Praeger, 1966), p. 158.

8. Reserve Bank of India.

9. See Lubell, note 2 above, p. 11.

10. *National Herald* (Lucknow, India), July 16, 1960.

11. *Ibid.*

12. *World Challenge: A Bulletin on the Communist Offensive As It Affects American Business in World Affairs*, Vol. I, No. 1, July 1960, pp. 5–6.

13. *Hindustan Times* (New Delhi, India), July 16, 1960.

14. *National Herald* (Lucknow, India), July 16, 1960.

15. *The International Petroleum Cartel, Staff Report to the Federal Trade Commission*, August 22, 1952, pp. 365–67, states:

The establishment of ECA early in 1948 and the allotment of substantial sums for the dollar financing of crude oil purchases by European refineries had made the question of Middle East oil prices a matter of public interest and scrutiny. More than 94 per cent of ECA-financed bulk oil shipments to Europe, for the year ending April 2, 1949, were made by six of the seven major international oil companies, nearly all of which took the form of shipments to their own affiliates or subsidiaries. . . .

As expressed by ECA administrator Hoffman in his statement before the Senate Foreign Relations Committee on February 17, 1949: "ECA has taken the position in respect of offshore procurement that . . . in general, prices charged on ECA-financed transactions should fully reflect competitive conditions affecting the market as the source of the commodity."

. . . As the result of this pressure, the companies, following the lead of Gulf Oil Corp. on its Kuwait crude oil, reduced their price for Middle East crude . . . [this reduction] appears to have been made for the sole purpose of improving the industry's public relations.

16. The companies knew they would not be pressed to match Soviet offers because of the following clause of the United States foreign aid act:

It is of paramount importance that long-range economic plans take cognizance of the need for a dependable supply of fuels, which is necessary to orderly and stable development and growth, and that dependence not be placed upon sources which are inherently hostile to free countries and the ultimate well-being of economically underdeveloped countries and which might exploit such dependence for ultimate political domination. The agencies of government in the United States are directed to work with other countries in developing plans for basing development programs on the use of the large and stable supply of relatively lowcost fuels available in the free world [Sec. 647, Public Law 87–195, Part Three].

17. It should also be noted that another provision of the United States foreign aid law also contributed to the reluctance of the oil companies to press for government financing of oil imports. This was a clause requiring that half of the tonnage involved in foreign aid be shipped in American-flag vessels. This clause would have required the oil companies to charter American vessels, and since most of their own tankerage flew the flags of other nations (because of the reduced cost) this would increase shipping costs by as much as 50% or more. If these increased costs were not somehow reflected in increased product prices, which is not likely, or the companies reimbursed (as in Pakistan where they were given "counterpart deposits" of blocked rupees), then the profitability of the refinery would be reduced.

18. This could occur, for example, if when such aid was given there was a price dispute between the Indian government and the oil companies; then the Indian government could officially declare to the Agency for International Development that it was not getting full benefit from U.S. assistance because of the oil companies' overpricing. Such a charge would undoubtedly have to be investigated by AID. This would then put the U.S. foreign aid program in the middle between the Indian government and the oil companies.

19. *National Herald* (Lucknow), July 16, 1960.

20. Periodically in India there are "demands for nationalisation of oil industry made by the left wing in the Indian Parliament"—*The Statist*, March 9, 1962.

21. *New York Journal of Commerce*, July 12, 1960.

22. *Petroleum*, November 1962.

CHAPTER SIXTEEN

1. *The Statist*, March 9, 1962.

2. See Government of India, Ministry of Steel, Mines and Fuel, Department of Mines and Fuels, *Report of the Oil Price Enquiry Committee* (New Delhi: Government of India, 1961); hereafter cited as "Damle."

3. Damle, note 2 above, pp. 16–24.

4. Government of India, Ministry of Petroleum and Chemicals, Department of Petroleum, *Report of the Working Group on Oil Prices* (New Delhi: Government of India, 1965), p. 39; hereafter referred to as "Talukdar."

5. One of the companies also "stated frankly" that it had an overall investment in refining and marketing to consider, which investment the newcomer price-cutters did not as yet have to take into account in their pricing decisions—*ibid.*, p. 40. Objectively speaking, the fact that the established majors had significant marketing and refining investments would seem irrelevant, since the price of crude oil would not affect in any direct way the profitability of either the marketing or refining investments, viewed separately. As the companies did not share their crude oil profits with India, it hardly seems consistent for them to assume that India has a stake in maintaining the worldwide profitability of the companies.

6. Alirio A. Parra, "Oil and Stability"—paper presented by the Venezuelan Delegation to the Third Arab Petroleum Congress, October 1961, pp. 7–12.

7. See *Petroleum Intelligence Weekly*, July 19, 1965, for evidence of higher crude prices in India at that time. In 1968 M. A. Adelman estimated that independent Japanese refiners were paying 10 percent less than India for the same oil from Saudi Arabia and Iran (*Petroleum Intelligence Weekly*, April 8, 1968).

8. Damle, note 2 above, p. 34.

9. *Ibid.*, pp. 24, 32–34.

10. *Ibid.*, p. 32.

11. If an independent marketer were foolish enough to try and use his discounts on imported products to reduce his prices and increase his market share at the expense of an established major, the latter could view its greater refinery profits as a subsidy to meet or beat the competition.

12. In Japan in 1962 foreign-tied companies (either affiliates or joint-venture companies) controlled 57 percent of Japan's refining capacity and 62 percent of the domestic market; thus, about two-fifths of the industry was controlled by independent companies, or a far higher share than anywhere else in Asia. See Ching Chih Chen, "Capital Movements in the Japanese Oil Industry—a Study of Direct Investment and Tied Loans—1949–64"—Massachusetts Institute of Technology, MSc thesis, 1965, pp. 17–18.

13. A further indication of product price discounting by the international majors in later years also can be seen from the fact that until September 1966:

> In the case of Saudi Arabia, discounts had been accepted for tax purposes on sales made by Aramco of crude and products to third parties. Such discounts were close to 10% and 25% of the posted prices for crude and petroleum products respectively—OPEC, "Collective Influence in the Recent Trend Towards the Stabilization of International Crude and Product Prices," paper presented to the Sixth Arab Petroleum Congress, Baghdad, March 1967, p. 8.

The fact that the discounts were disallowed in September 1966 does not mean that they had disappeared in the market place; rather this was a way for the government to increase its revenues and put pressure on the companies toward eliminating such market discounts.

14. *National Herald* (Lucknow), July 16, 1960. This makes a sharp contrast with the acumen demonstrated by the Japanese government in dealing with the problem of affiliate pricing. In the 1950s the Japanese government developed an ingenious method for stimulating competition and putting pressure on crude oil prices. The mechanism used by the Japanese government was based on the fact that, as in India, the government controls both the level of refining capacity and the amount of foreign exchange allotted to the refineries for importing crude oil. Moreover, in Japan as in most underdeveloped countries including India, with rising oil demand all companies were anxious to increase their refinery capacity. The Japanese government took advantage of this by linking each company's allowable refinery expansion to the quantity of crude oil imported by it with its allocation of foreign exchange. It was thus tempting for each company to work its hardest to get a lower price for the crude oil imported by it, so that it could import more crude oil than otherwise, and hence possibly in the future get a greater foreign exchange allocation for oil imports and a greater amount of refining capacity. See K. Oshima, *The Present International Petroleum Situation and the Petroleum Policy of Our Country* (Tokyo: Economic Planning Conference of Japan, mimeographed, April 25, 1964), p. 45. Of course, an important condition for this competitive mechanism to work effectively was the existence of a fairly large number of independent refining companies, rather than three major ones as in India.

15. Damle, note 2 above, pp. 13–15.

16. *Ibid.*, p. 63.

17. One point should be noted. It seems clearly incorrect for the companies to claim that in a regulated industry which would be allowed a specified return on all capital employed, e.g., 12 percent, interest payments to long-term

debt holders should be treated as a deductible expense (rather than coming out of the after-tax profits of the company). That is to say, the 12 percent return (before tax) is presumably designed to pay both the equity capital and the debt capital, and allowing an interest deduction would in effect be allowing for paying capital twice; moreover, it would greatly encourage the companies to use Indian debt capital rather than their own money.

18. Damle, note 2 above, p. 34.

19. *Ibid.*, pp. 27, 28, 34.

20. See note 4.

21. *Petroleum Intelligence Weekly*, January 18, 1965.

22. *Ibid.*, July 26, 1965.

23. *Ibid.*, July 19, 1965.

24. *Ibid.*, July 19, 1965.

25. *The New York Times*, July 10, 1965.

26. *Petroleum Intelligence Weekly*, April 5, 1965.

27. *Ibid.*, June 7, 1965.

28. *Ibid.*, June 14, 1965.

29. *Ibid.*, June 28, 1965.

30. *Ibid.*, June 21, 1965.

31. *Ibid.*, July 5, 1965.

32. Talukdar, note 4 above, pp. 28, 42.

33. *Petroleum Intelligence Weekly*, October 25, 1965.

34. *Ibid.*

35. *Ibid.*, December 20, 1965.

36. In the latest round of price cuts for crude oil coming into India—3 cents per barrel in June 1968—it was reported that one causal factor was an Indian government threat to reduce the foreign exchange allocation for crude oil (*Petroleum Intelligence Weekly*, July 1, 1968). However, it was also reported that the Indian government was considering a Burmah-Shell proposal for a major expansion in its refinery (*Ibid.*, June 17, 1968).

37. Talukdar, note 4 above, p. 29.

38. It is interesting to note that if the Russians had charged AFRA rates for their 1960 crude oil offer to India, because of the much longer haul from the Black Sea to Bombay their quoted price for oil delivered in Bombay would have been just about the same as the oil companies' delivered price at that time. Hence, a realistic use of spot prices helped the Russians to underbid the Western oil companies.

39. It should be noted that, since the nautical distance from the Persian Gulf to Madras is much greater than to Bombay, use of AFRA for pricing transport costs into the newly scheduled Madras refinery would increase the level of overcharges on the order of 50 percent (and for the Cochin Refinery by some 20 percent).

40. Damle, note 2 above, p. 39.

41. Both Committees seemed more concerned about getting the oil companies to place their marine insurance with Indian insurance companies rather than foreign ones. (See Damle, note 2 above, p. 41 and Talukdar, note 4 above, p. 47.) Any loss in foreign exchange to India from marine insurance could only be a fraction of the loss from AFRA transport pricing.

CHAPTER SEVENTEEN

1. Data from Petroleum Information Service, Statistics Division, *The Indian Petroleum Handbook: 1966* (New Delhi: Petroleum Information Service, 1966).

2. See M. Ramabrahmam, General Manager, Gauhati Refinery, "One Year of Refining," *Noonmati News* [Gauhati, India], January 1963.

3. See, for example, *Petroleum Intelligence Weekly*, April 12, 1965, and May 9, 1966. The Barauni refinery was completed in 1964 and the Koyali refinery in 1965.

4. *Financial Express,* January 2, 1962.

5. *The Statist,* March 9, 1962.

6. For example, Krishna Menon of the "Menon-Malaviya" wing was forced out from his post as Defense Minister.

7. Government of India, Planning Commission, *Third Five Year Plan* (New Delhi: Government of India, 1961), p. 482.

8. *Petroleum Intelligence Weekly,* February 24, 1964.

9. See *Petroleum Intelligence Weekly,* March 23, 1964.

10. "A blow—but not a mortal one—has been dealt established refiners' hopes for major expansion in India. That's the initial reaction to Oil Minister Malaviya's statement . . ."—*Petroleum Intelligence Weekly,* January 14, 1963.

11. *The Eastern Economist,* June 28, 1963.

12. *Far Eastern Economic Review,* October 17, 1963, p. 146.

13. *World Petroleum,* October 1963.

14. *World Petroleum,* November 1963.

15. *Petroleum Intelligence Weekly,* December 2, 1963.

16. *Ibid.,* February 10, 1964.

17. *Ibid.,* May 11, 1964. Jersey got Standard Vacuum's business in India.

18. *Ibid.,* May 4, 1964.

19. *The Eastern Economist,* August 11, 1961 states as follows:

The draft agreement with ENI provides for a 12 year credit, at 6 per cent rate of interest, for an amount in lira equivalent to $120 million. A sum of $33 million out of this can be utilized for setting up a refinery, another $33 million for financing the laying of two products pipe lines—one from Barauni to Calcutta and the other from Barauni to Delhi via Kanpur or Lucknow—$20 million for oil prospecting in collaboration with the Oil and Natural Gas Commission and the remaining $34 million for putting up a gas separation plant in Assam, a naphtha cracker project to manufacture petrochemicals and a lubricating oil unit.

20. In the interim ENI might still make a good profit on purchases of crude oil for the projected Indian refinery:

"ENI makes a big part of its profits on the crude oil side of the business, even though it must buy a large proportion of its oil from third parties, company executives reveal . . ."—*Petroleum Intelligence Weekly,* May 25, 1964.

21. *Ibid.*

22. *Ibid.,* January 11, 1965.

403

23. *Ibid.*

24. *Petroleum Intelligence Weekly*, May 18, 1964 states,

The Indian government is also said to be insisting on only short-term contracts for oil supply. This demand is believed to stem from a growing conviction that the 15-year commitment for crude supply signed with Phillips Petroleum was unwise. India's reasoning: the price of crude oil may decline further over the next decade and a half.

25. See *Petroleum Intelligence Weekly*, May 11, 1964, and July 27, 1964. In fact, subsequently the Indian government through its ONGC was allowed into offshore exploration in Iran, in combination with ENI, Phillips, and NIOC.

26. *Petroleum Intelligence Weekly*, April 26, 1965.

27. *The Oil and Gas Journal*, August 28, 1967, p. 52.

28. *Ibid.*, p. 52.

29. *Indian Express*, September 14, 1962.

30. *Petroleum Intelligence Weekly*, June 29, 1964.

31. *Carta Semanal* (Journal of the Venezuelan Ministry of Mines and Hydrocarbons, English translation edition), July 3, 1965.

CHAPTER EIGHTEEN

1. See V. S. Swaminathan, "Oil Exploration: A Search That Never Ends," *Commerce*, September 30, 1961, p. 646.

2. *Ibid.*

3. Reported by K. K. Sahni, "Progress of the Oil Industry," *The Financial Express*, September 20, 1961.

4. Of each $100 spent fruitlessly, Standard Vacuum would directly provide $75 and the government $25. However, if oil were not found, Standard Vacuum could then deduct $50 ($75 minus $25) from its marketing income, which with a 50 percent tax rate would mean a $25 reduction in its tax bill to the government.

5. Government of India, Press Information Bureau, "Towards Self-Sufficiency in Oil and Oil Products in India"—press release, mimeographed, August 15, 1966, p. 4. Punctuation changed from original; no semicolon after "help" but a comma after "reason" in original.

6. The USSR allotted $70 million in loans for oil exploration in India, or more than was spent by the ONGC in 1956–61. In 1961 Malaviya negotiated smaller credits for oil exploration—$8 million from France and $20 million from ENI—*Commerce*, August 26, 1961.

7. Until completion of the government-owned refineries in that area, all the crude oil produced had to be processed by the established majors' refineries in Bombay.

8. *The Oil and Gas Journal*, November 6, 1967, p. 67.

9. Government of India, Ministry of Petroleum and Chemicals, Department of Petroleum, *Report of the Working Group on Oil Prices* (New Delhi: Government of India, 1961), p. 32.

10. This makes it difficult to estimate the exact amount of current oil production which is attributable to previous expenditures as opposed to current

outlays; the latter are both for current production and for developing future production. Hence use of this data overestimates the true cost of oil exploration and therefore understates the rate of return.

11. The upper limit on value would be set by the price of imported crude oil (CIF), which in recent years was at least $1.75 per barrel, less the real social cost to the economy of producing the discovered oil, which would probably be no more than $0.25 per barrel. The lower limit is based on an estimate that CIF prices of crude oil over the relevant future period will not average less than $1.25.

12. *The Eastern Economist*, August 11, 1961.

13. *The Eastern Economist*, June 28, 1963.

14. *Far Eastern Economic Review: Year Book, 1966*, p. 173.

15. *The Economic Times*, May 7, 1962, p. 8.

16. Shri Raj Narain Gupta, "India's Oil Policy—An Appraisal," *AICC Economic Review*, June 22, 1961, p. 25.

17. *The Eastern Economist*, June 28, 1963.

18. *The Eastern Economist, 1966 Year Book*, December 31, 1965, pp. 1295–96.

19. Walter J. Levy, *The Search for Oil in Developing Countries: A Problem of Scarce Resources and Its Implications for State and Private Enterprise* (for the International Bank for Reconstruction and Development, 1961).

20. *Petroleum Intelligence Weekly*, September 2, 1963; emphases added.

21. *Ibid.*, July 15, 1963. Ultimately, however, the Indian government was able to obtain a stake in Iranian oil exploration after granting part ownership of the Madras refinery to the Iranian state oil company (NIOC). The one-fourth share of an off-shore Iranian oil exploration company (Iminico) given to the ONGC appears now to be very valuable since reports indicate that production from a field discovered in 1966 should reach 100,000 barrels a day by 1970—*The Wall Street Journal*, February 8, 1968; on that basis India's annual share would equal 9 million barrels, or almost one-fourth of the ONGC's estimated domestic production in that year.

22. *The Oil and Gas Journal*, November 6, 1967, pp. 65–67.

23. The new petroleum rules enunciated in November 1959 contained some standard provisions for oil exploration licenses and leases, but left considerable room open for negotiation. The most important of the standard provisions were: (1) Exploration rights would run from four to six years. (2) Exploitation rights would run for twenty years. (3) Surface and royalty fees would be paid to state governments, with the royalty fee to equal 10 percent of the gross value at the wellhead of all oil and gas produced—"The Petroleum and Natural Gas Rule, 1959" in Government of India, Ministry of Steel, Mines and Fuel, Department of Mines and Fuel, *Oil Prospects in India* (New Delhi: 1959).

24. ENI, *Energy and Petroleum in 1963* (Italy: ENI, 1964) pp. 65–66. Computed by dividing 1961 reserves by 1961 production.

25. Assuming the same effective tax rate in India as in the Middle East.

26. United Nations, Department of Economic and Social Affairs, *Petroleum Exploration: Capital Requirements and Methods of Financing* (New York: United Nations, 1962), p. 28.

27. *Ibid.*
28. *The Statist,* March 9, 1962.
29. *The Hindustan Times,* January 26, 1962, p. 111.

CHAPTER NINETEEN

1. 1960–61 data are from Food and Agricultural Organization, *Fertilizers: An Annual Review of World Production, Consumption and Trade, 1962* (Rome: FAO, UN, 1963), p. 18.

2. Based on FAO data presented in: Jung-chao Liu, "Fertilizer Application in Communist China," *The China Quarterly,* October–December 1965, p. 43. Moreover, the Indian fertilizer price appears to be on the high side—*ibid.,* p. 43.

3. Fertilizer consumption data from Government of India, Planning Commission, *Fourth Five Year Plan: A Draft Outline (1966),* p. 63.

4. Fertilizer production data from *Eastern Economist: Annual Number, 1967,* p. 1385.

5. *Ibid.,* p. 1410.

6. *Quarterly Bulletin of the Eastern Economist: Records and Statistics,* May 1966, p. 132. Again, on March 11, 1966, it was reported that India was to receive a West German loan of $3 million for buying fertilizers—*ibid.*

7. Government of India, Ministry of Information and Broadcasting, Publications Division, *India—A Reference Annual: 1965* (Delhi: Government of India, 1965), p. 279.

8. Sindri at Bihar, Fertilizers and Chemicals, Travancore Ltd. (FACT) at Kerala, and Nangal in the Punjab.

9. Government of India, Planning Commission, *Third Five Year Plan* (New Delhi: Government of India, 1961).

10. *Eastern Economist: Annual Number, 1967,* p. 1385.

11. Thus, of the seven fertilizer factories actually in production as of March 1967, only two small ones accounting for 3 percent of total capacity were in the private sector. See Government of India, Ministry of Petroleum and Chemicals, *Report: 1966–67* (New Delhi: Government of India, 1967), p. 10.

12. While there is some room for changing refinery design so as to change the proportions of different products which can be derived from a given barrel of crude oil, there are limitations on this degree of flexibility. Moreover, most of the research in this area has been developed for the United States market, which has sought to maximize the amount of motor gasoline which can be derived from a barrel of crude oil, rather than minimize it.

13. As a result of the limited refinery expansions of the established majors in the early 1960s, India's exports of petroleum products, about three-fourths of which were gasolines, rose from $7 million in 1961 to $17 million in 1964; exports of motor gasoline rose in this period from 1.4 million barrels to 3.5 million barrels.

14. The established majors' belief that they would not be able to solve any naphtha problems by exporting gasolines from India to the rest of the area was borne out by the fact that product exports from the Abadan refinery, the major refinery of the international consortium in Iran, began to fall sharply:

406

Iran—only 67% of capacity, that's what the exports of the world's second largest refinery have slipped to. Product exports in April averaged 278,000 barrels daily, down 11% from the 312,000 b/d shipped a year earlier and way below Abadan's rated capacity of 412,000 b/d.

The chief reason for the slide, of course, is the proliferation of new refineries around the world which are making it harder to find markets [*Petroleum Intelligence Weekly,* June 14, 1965].

15. The foreign exchange economics of the situation make this virtually a foregone conclusion. An investment in fertilizer facilities often pays for itself in foreign exchange savings in two years or less.

16. Jung-chao Liu, note 2 above, p. 43.

17. "India Opens Wider to Foreign Funds," *Business Week,* April 11, 1964.

18. *Ibid.*

19. *Discussion Reports: Investment Conference, New Delhi, India, April 13–18, 1964.*

20. *Ibid.,* p. 47.

21. *Ibid.,* pp. 44–45.

22. *Ibid.,* p. 32. It is interesting to note that one reason the oil companies were so willing to invest in refineries in India was that repatriation for the great bulk of their profits was never a problem. This was because most of the profits lay in the sale of crude oil which was immediately paid for in foreign exchange by the Indian affiliate. The same would not be true of petrochemical projects which did not provide a very large crude oil outlet.

23. *Ibid.,* pp. 28–30.

24. *Ibid.,* p. 25.

25. After all, the reported rate of return on American manufacturing investment in India was 20 percent per year. To the extent that the oil companies make less on their indigenous operations, of course, this is more than made up by their high profits on crude oil.

26. *Discussion Reports:* . . . , note 19 above, p. 10.

27. *The New York Times,* January 10, 1965.

28. *Far Eastern Economic Review: 1966 Yearbook,* p. 175.

29. *Petroleum Intelligence Weekly,* October 4, 1965.

30. *The New York Times,* December 18, 1965.

31. *The New York Times,* May 15, 1966. American International is a subsidiary of Standard Oil of Indiana.

32. *The New York Times,* April 28, 1966; emphases added.

33. *The New York Times,* January 15, 1967. Further evidence of strong opposition to this rightward shift in petrochemical policy was not long in forthcoming:

An offer of concessions on [fertilizer] markets and pricing made to potential investors in December, 1965, is due to expire at the end of this year. The fact that it has already been extended once increases the political price Prime Minister Indira Gandhi might have to pay if she decides to opt for a second extension. . . .

Of all the projects now under negotiation, the largest would be one proposed by Allied Chemical and the Tata group of industries here for Okha in Gujarat. Indeed, it would be the largest fertilizer complex in the world, achieving a production of 1.15 million tons by 1976.

The whole concept of the project is startlingly imaginative. In its final phase it would involve the construction of a nuclear desalination plant capable of producing ammonia from seawater and air.

The snag is that it would involve imports of ammonia for the first decade. This is something the Government has been chary of approving because of the possibility of producing ammonia from naphtha, a petroleum byproduct available here.

Last month the Cabinet killed a proposal by the Kuwait Chemicals and Fertilizer Corporation to erect a plant because it depended on the import of ammonia from Persian Gulf producers. The proposal had been approved by the Ministry of Petroleum and Chemicals and encouraged by the International Bank for Reconstruction and Development and Western aid officials.

There is strong sentiment in the governing Congress party that further concessions would represent a "sellout" to Western business interests. It is too early to say whether the Cabinet will be able to withstand that pressure and accept the attractive Tata-Allied Chemical proposal.

—*The New York Times,* October 13, 1967

34. *I. F. Stone's Weekly,* December 12, 1966.

CHAPTER TWENTY

1. See Chapter 13, note 4, for method of converting different energy sources to fuel oil equivalency.

2. Thus, industrial production in the 1965/66 to 1970/71 period is targeted to increase by 69 percent, compared to a projected 31 percent increase in agricultural production.

It might also be noted that the Indian government appears unwilling to attempt to significantly restrain the growth of energy consumption in the home. Since kerosene consumption is a major component of household energy consumption, this increases the pressure for building oil refineries, particularly since kerosene at present must be imported into India. The fact that after the mid-1967 devaluation of the Indian rupee (the aim of which was to reduce India's trade deficit by making imports more expensive), the government moved to offset any impact on kerosene consumption by keeping internal kerosene prices at the old pre-devaluation level is indicative of the government's approach.

3. See *Fourth Five Year Plan: A Draft Outline,* (1966), pp. 267–68.

4. Under the "Industrial Policy Resolution" of April 30, 1956, which in theory still governs Indian policy, the whole energy sector was placed in the "Schedule A" category:

A fresh statement of industrial policy, necessitated by the acceptance of a socialist pattern of society as the national objective, was announced on April 30, 1956. Under this, industries specified in Schedule A will be the exclusive responsibility of the State . . . [*India, A Reference Annual: 1965* (Delhi: Government of India, 1965), p. 279].

5. The average per annum growth rate in India's industrial production index from 1964 to 1967 was close to 4% compared to 7.5% in the 1956–64 period.

6. *Far Eastern Economic Review,* February 2, 1967, p. 165.

7. *Far Eastern Economic Review: 1967 Yearbook,* pp. 192–93.

8. Geoffrey Tyson, *Nehru: The Years of Power* (New York: Frederick A. Praeger, 1966), p. 64, emphasis added.

9. Tyson, *Ibid.,* p. 64, quoting from John Kenneth Galbraith, *Economic Development* (Cambridge, Massachusetts: Harvard University Press, 1964), p. 23.

10. For a discussion and critique of the Harrod-Domar type model see: Gerald M. Meier, *Leading Issues in Development Economics* (New York: Oxford University Press, 1964), pp. 466–69 and 101–104.

11. Moreover, this completely omits the foreign exchange requirements for investments by indigenous groups, which, of course, would be at a distinct disadvantage vis-à-vis the foreign investors in terms of access to foreign exchange.

12. Based on data from Reserve Bank of India, *Financial Statistics of Joint Stock Companies in India, 1950–51—1962–63* (Bombay: Reserve Bank of India, 1967). The data used here for foreign branches were increased by 28% and for foreign-controlled rupee companies by 33% in line with the Reserve Bank's estimate of the published data's percentage of actual coverage.

13. See *Reserve Bank of India Bulletin,* January 1963, pp. 8–19.

14. Data on reinvested earnings from K. M. Kurian, *Impact of Foreign Capital on Indian Economy,* cited in R. L. Goel, "Social and Economic Costs of Private Foreign Investments," *All India Congress Committee Economic Review,* September 15, 1967, p. 24; data on investments from Reserve Bank of India.

15. Goel, note 14 above, pp. 23–24.

16. See K. M. Upadhyay, "Foreign Investments in India after 1956: A Study of Their Industry-wise Direction," *All India Congress Committee Economic Review,* September 1, 1966, p. 16. The author also states on p. 15:

> It is important to note that foreign business delegations visiting India in recent years have been found interested in petrochemical industries, pharmaceuticals, electronics, steel forgings and castings, hotel industry, aircraft and agricultural implements, machine tools, foundry machinery, heavy steel construction and roller bearings.

Many of these industries would fit into the monopolistic class in India. On the other hand, the same study notes that there has been very little foreign interest in the coal mining sector—*ibid.,* p. 15. Coal, as contrasted to oil, would seem clearly to fit in the category of industries in our model in which, barring artificially high prices set by the government, the ratio of profit to output would be relatively low, thereby discouraging foreign investment.

17. Meier, note 10 above, pp. 150–52.

18. World Bank loans, which are based on financing "top priority" projects in the underdeveloped countries, may give the Bank even more control over the total economy.

19. In terms of our numerical model, raising the savings ratio from 0.10 to 0.20, assuming the same capital-output ratios, would raise the average growth rate of the economy as a whole to over 7% per year, while even an increase to 0.15 would raise the per annum growth rate to 5.5%.

20. W. Arthur Lewis, "Some Reflections on Economic Development," *Economic Digest,* Institute of Development Economics, Karachi, Vol. 3, No. 4, Winter 1960; quoted in Meier, note 10 above, p. 98.

21. Goel, note 14 above, p. 26.

CHAPTER TWENTY-ONE

1. Dwight H. Perkins, *Market Control and Planning in Communist China* (Cambridge, Massachusetts: Harvard University Press, 1966), p. 215.

2. *Ibid.,* p. 225.

3. Yuan-li Wu, *Economic Development and the Use of Energy Resources in Communist China* (New York: Frederick A. Praeger, 1963). It is interesting to note how crucial Wu viewed the energy sector in Communist China. In his preface, pp. vii, viii, he states:

> . . . a number of sectoral studies should be made so that we shall gain a better understanding of Communist China's economic structure and be in a position to make continual revisions and refinements of the aggregative studies on the basis of sectoral findings. . . .
>
> Our interest in sectoral studies presented us with an immediate problem of priority. We selected the energy resources sector for our initial effort on the ground that it represents a pivotal segment of the economy which has far-reaching relationships with every other economic sector. The use of energy resources has an especially intimate relation to the process of industrialization—not to mention the development of nuclear capability —and it is the working of this process since 1958 that aroused our initial curiosity. It was hoped that the energy resources sector would reveal to us certain data that would not be obvious otherwise; and in this we have not been disappointed.

4. *Ibid.,* p. 42.

5. 1952 oil import data not available.

6. Wu, note 3 above, p. 192.

7. K. P. Wang, "The Mineral Resource Base of Communist China," in *An Economic Profile of Mainland China,* studies prepared for the Joint Economic Committee, Congress of the United States (Washington: Government Printing Office, February 1967), p. 174.

8. 1965 estimate from John Ashton, "Development of Electric Energy Resources in Communist China," *An Economic Profile of Mainland China,* note 7 above, p. 309.

9. Wang, note 7 above, p. 186. In *World Petroleum,* December 1966, p. 12, it was reported:

> According to a group of European industrialists recently returned from China, the output of crude today might well have passed the 13 million-ton/yr mark. . . .
>
> Many new refineries have been built throughout the country, which should explain the surprising declaration made by the party's newspaper "Remnin Riboa" (The Red Flag) that China's refinery production increased by 82.8% during the first nine months of this year.

10. Wang, note 7 above, p. 194.

11. Wu, note 3 above, pp. 32–34.

12. Wu, note 3 above, p. 175.

13. This is particularly significant not only because pig iron is a basic material for industrialization, but also because it is a heavy user of energy. For example, close to one-fifth of all the raw coal produced in China in 1960 was consumed in producing pig iron—Wu, note 3 above, p. 111.

14. Ashton, note 8 above, p. 311.

15. The situation is analogous to the ease with which China and other countries quickly raised production in recovering from wartime dislocations.

16. For data on China between 1952 and 1962 see Jung-chao Liu, "Fertilizer Supply and Grain Production in Communist China," *Journal of Farm Economics*, Vol. 47, No. 4, November 1965, p. 925.

17. *Far Eastern Economic Review*, July 14, 1966, p. 68.

18. For 1963 estimate see Jung-chao Liu, "Fertilizer Application in Communist China," *The China Quarterly*, October–December 1965, p. 32.

19. Wang, note 7 above, p. 182.

20. Liu, "Fertilizer Supply . . . ," note 16 above, pp. 920–21.

21. *Ibid.*, pp. 917–18.

22. *Far Eastern Economic Review*, July 14, 1966, p. 68.

23. Liu, "Fertilizer Supply . . . ," note 16 above, p. 922.

24. *Ibid.*, pp. 922–23.

25. Leo A. Orleans, in his "Research and Development in Communist China: Mood, Management and Measurement," in *An Economic Profile of Mainland China*, note 7 above, pp. 557–58, states,

What little science and technology activity there was in China prior to 1949 was mainly Western oriented. . . . the conversion from English, German, and French to the Russian language posed great problems both for established professionals and aspiring students. Thousands of Russian texts and reference works were translated into Chinese; thousands of scientists, engineers, and students spent years studying the Russian language—a tongue that does not come easily to the Chinese—and many of the institutions of higher learning adopted Russian materials in their courses.

This laborious effort of reorientation toward the U.S.S.R. came to a sudden halt in 1960 with the major political schism between the two countries. China was left with Russian-made factories and Russian blueprints for industrial construction projects, but without the necessary spare parts and without scientific and technical advisors. After some 10 years of intensive Russian influence in science, industry, and education, China once again found herself in the throes of a painful and expensive readjustment. . . .

The loss of Soviet support, particularly at a time when China's economy was already experiencing a sharp downward trend, had conspicuous economic consequences.

With respect to oil specifically we have the following report from *World Petroleum*, March 1961, p. 70:

An acute shortage of petroleum and products is reported from Red China, believed by Western observers to be a result of calculated Soviet economic pressure. Most of China's petroleum is supplied by the Soviet Union.

The shortage appears to be growing ever more acute as the scale of Chinese industrialization increases. Strenuous measures are now being taken to meet the situation. They include cuts in public transport services. There has also been noticed a marked reduction recently in the activity of Chinese jet fighters when trying to intercept Nationalist Chinese reconnaissance planes. China's jet and aviation fuel is provided entirely by the Soviet Union.

This may be contrasted with a report in *The Oil and Gas Journal*, February 20, 1967, pp. 33–36:

Russian experts are credited with introducing into China many of the newer methods and skills, including seismic and aerial-magnetic techniques. In production, they brought in gas repressuring and waterflooding.

But in July 1960 the situation changed radically due to the deepening ideological differences between the two countries. Russia abruptly stopped much of her technical and material aid. . . .

Nevertheless, Chinese sources indicate that the nation's petroleum industry made much progress in the 1960's after recovering from the initial setback.

It has been claimed that the Chinese machine-building industry is now able to make relatively-sophisticated equipment. The tools range from drilling rigs to many precision instruments used in exploration work—including gravity meters, seismometers, and various types of logging instruments. . . .

The Chinese also claim that their engineering industry can now turn out complete refineries in the 20,000-b/d size range. . . .

. . . Information on Red China's refining, while incomplete, is sufficient to make it clear that it has developed into a major industry. . . .

[China's estimated refining capacity] is small potatoes by western standards. But the Red Chinese have crossed the highest hump—establishing a technological base. And expansion from now on should accelerate.

26. *The New York Times*, September 17, 1967.

CHAPTER TWENTY-TWO

1. An additional advantage of this method is that it avoids the necessity of evaluating the "fairness" of the compensation for nationalization. The cost side omits the compensation but the revenue side omits the oil discovered prior to nationalization.

2. The actual social value of crude oil discovered in Mexico would be the "import parity" price of crude oil less the real social costs to the Mexican economy of producing the oil discovered. The maximum value is based on historical import prices and the minimum on the lowest conceivable future import prices.

3. *The Wall Street Journal*, January 26, 1967.

4. Looked at in a simplified way, in Mexico the average exploration cost for finding crude oil over the years has been less than 10 cents per barrel. Even if, owing to the existence of previously discovered reserves, it takes 15 years to realize the $2 worth of oil from 10 cents worth of investment, the payoff is still so enormous as to be worth the wait. It is akin to a dam which

takes 15 years to build, but can then yield over $20 worth of increased output for every $1 originally invested.

5. From Antonio J. Bermúdez, *The Mexican National Petroleum Industry: A Case Study in Nationalization* (Stanford University: Institute of Hispanic American and Luso-Brazilian Studies, 1963).

6. These estimates are based on the following assumptions. First, we treat all of Pemex's nonexploration investments in this period, $615 million, as if they went into refining, which is clearly an overestimate; in estimating the return from this investment, no credit is given for distribution profits. Second, the quantity of refined oil attributed to this investment was the number of barrels that would be refined indigenously above and beyond what could have been processed by refinery capacity existing at the end of 1938. It was further assumed that the average longevity of the post-1938 refining capacity existing in 1958 would be about 12 years (through 1970). On this basis it is estimated that 1.3 billion additional barrels of refined products would be generated by the new investment. This quantity was then valued at a rough average world market import parity price (including transport costs) for refined products produced in Mexico, $3 per barrel, less the $1 to $2 per barrel value of crude oil, leaving $1 to $2 per barrel.

7. *The Wall Street Journal*, January 26, 1967.

8. Bermúdez, note 5 above, p. 52. Again, in 1966 a later Director General (Reyes Heroles) of Pemex stated: "But we prefer to confine our estimates of what we have available to proved reserves, making it plain at the same time that in determining these we have applied extremely conservative standards"—*Petroleum Policy: Report of the [Pemex] Director General*, 18 March 1966, p. 5.

9. Miguel S. Wionczek, "Electric Power: The Uneasy Partnership," in Raymond Vernon, ed., *Public Policy and Private Enterprise in Mexico* (Cambridge: Harvard University Press, 1964), p. 77.

10. *The Wall Street Journal*, January 26, 1967.

11. *Ibid.*

12. Bermúdez, note 5 above, p. 120.

13. *Ibid.*, pp. 120–22.

14. Data from William E. Cole, *Steel and Economic Growth in Mexico* (Austin, Texas: University of Texas Press, 1967), pp. 67, 86.

15. Bermúdez, note 5 above, p. 122.

16. *The Wall Street Journal*, January 26, 1967.

17. Petróleos Mexicanos, "Natural gas reserves in Mexico as a factor of the social and economic development of the country by means of nitrogenous compounds," in United Nations, *Studies in Petrochemicals* (New York: United Nations, 1966), p. 427.

18. H. R. Shawk and D. L. Campbell, "Application of Advanced Technology to Developing Countries for Basic Petrochemical Intermediates," *ibid.*, p. 283.

19. *Ibid.*, pp. 923–25. The lack of coincidence is underscored by the United Nations' note, pp. 921–22:

Despite the availability of large amounts of natural gas and petroleum in a number of countries and despite the interest of governments and entrepreneurs, there are still relatively few plants in operation or under construction in Latin America. One reason for this is that the national markets for most of the products are even smaller than the minimum economic capacities of the plants; another reason is the large investments required for the establishment of these industries which in some cases exceed the financial capacity of national entrepreneurs. In addition, the technical skills needed for the industrial processes are in short supply and, as a general rule, the latter are controlled by the international enterprises that have traditionally supplied the Latin American market.

20. *The Wall Street Journal,* January 26, 1967.

21. *Petroleum Policy: Report of the Director General of Petróleos Mexicanos, Lic. Jesus Reyes Heroles, March 18, 1967,* p. 11.

22. *The Wall Street Journal,* January 26, 1967.

CHAPTER TWENTY-THREE

1. Abbas Alnasrawi, *Financing Economic Development in Iraq: The Role of Oil in a Middle Eastern Economy* (New York: Frederick A. Praeger, 1967), p. 3. Unless otherwise noted, factual materials presented in this chapter are taken from this study.

2. *Ibid.,* p. 1.

3. BP, Shell, and CFP each own, and Jersey and Mobil share, a 23.75 percent interest in IPC; the Gulbenkian interests, the original entrepreneurs in the deal, own the remaining 5 percent.

4. On a current price basis, income in the oil sector grew an average of 19 percent per year compared to income growth in the nonoil sector of 8 percent per year.

5. Alnasrawi, note 1 above, p. 87.

6. *Ibid.,* p. 88.

7. See Alnasrawi, note 1 above, pp. 53–54.

8. Alnasrawi, *ibid.,* pp. 63–65, observes:

Iraq is basically an agricultural country with about 78 percent of its gainfully occupied workers absorbed by the agricultural sector. This sector's contribution to the net national product during the period under consideration ranged from 16 percent to 29 percent, depending on a number of factors, including the weather conditions. Iraq is no different from other underdeveloped countries in that this sector suffers from chronic underemployment and seasonal unemployment reaching as high as 75 to 80 percent . . .
It is important, therefore, that agriculture should be of major concern in all efforts to increase the rate of economic growth in Iraq.
The development of the agricultural sector is vital to the process of economic development because: (1) it releases agricultural labor for nonfarm employment, (2) it helps to increase exports or reduce imports, (3) it is necessary to increase agricultural output in order to meet the expected rise in demand for food as a result of the increase in income which the process of economic development will create, and, (4) it increases the savings which are needed for investment in the agricultural and the nonagricultural sectors of the economy.
Any development policy, therefore, should have as its first and foremost target the reduction, if not the eradication, of the high rate of unem-

ployment. But to achieve this, the industrial sector must grow so as to provide employment opportunities for the absorption of the surplus labor on the farm, and for the growth in the labor force. . . . It may be safe to say that without the simultaneous increase in the rate of investment in both sectors, the twin goals of a higher standard of living and a higher rate of employment may not be attained.

9. *Ibid.,* p. 134. These big holdings ranged in size from 600 to 600,000 acres.
10. *Ibid.,* pp. 66–68.
11. International Bank for Reconstruction and Development, *The Economic Development of Iraq* (Baltimore: The Johns Hopkins Press, 1952), p. 33—IBRD Report.
12. IBRD Report, pp. 38, 40–41.
13. The following is from Alnasrawi, note 1 above, pp. 77–78:

. . . it may be worth noting that other advisors recommended a similar policy. Professor Carl Iversen states that though in the long run Iraq will have to industrialize, "it seems to be in the best interest of Iraq not to force this development but to concentrate on raising the efficiency of agricultural production" . . .
Lord Salter followed the same line of thinking when he said that the expenditure of money and effort on developing industries which could not compete with foreign manufacturers except in a highly protective domestic market, could be wasteful and damaging to Iraq's general progress and prosperity.
Finally, and along the same line of thinking, Arthur D. Little, Inc., which was commissioned by Iraq to make a survey of potential industries, states that: "forced industrialization with the aid of tariff protection or government subsidy might lead to inefficiency and waste of economic resources, and no industry has been recommended unless it is anticipated that it will produce at a cost below the landed price of comparable imported goods or raw material before import duty has been levied."

14. *Ibid.,* pp. 78–81.
15. If historically such antiprotectionist advice had been followed by underdeveloped countries like the United States in the 18th and 19th centuries, it is quite possible that today Great Britain would still be the only industrialized country in the world.
16. Alnasrawi, note 1 above, p. 82.
17. *Ibid.,* pp. 72–74, 82.
18. See *ibid.,* pp. 73, 83.
19. Harvey O'Connor, *World Crisis in Oil* (New York: Monthly Review Press, 1962), p. 312.
20. Alnasrawi, note 1 above, p. 88.
21. This inflation factor of 4 percent per year in Iraq's nonoil sector is conservatively based on the implicit average figure for 1953–61 of 4.4 percent per year, derived from K. Haseeb, *The National Income of Iraq: 1953–1961* (London: Oxford University Press, 1964), pp. 17, 18.
22. The general formula for the annual income derivable in year N from fully investing government oil revenues in year $T = (1/\text{capital-output ratio})$ times (government oil revenues in year T) times $(1 + \text{savings-income ratio}/\text{capital-output ratio})^{N-T}$.

23. To anticipate objections, in our examples we have assumed (as is usually done) no timelag between the original capital investment and increased output. While this is clearly unrealistic, appropriate adjustment can be made for greater realism; if, for example, there is a 3-year average timelag between initial capital investment and its stream of output, then a $300 or $700 level of per capita income would probably have been reached by 1969 instead of 1964. This is clearly inconsequential compared to the decisive fact that in one generation Iraqi oil revenues could have decisively broken the bonds of underdevelopment.

24. O'Connor, note 19 above, p. 317.

25. See Alnasrawi, note 1 above, pp. 73–75.

26. Venezuela is another good example as described in the following from Edwin Lieuwen's *Petroleum in Venezuela: A History* (Berkeley: University of California Press, 1954), pp. 119–20:

> On no point is the record so black as on the question of the government's investment of its huge petroleum revenues. Though tremendous tax wealth has come in, the program of "sowing the petroleum" has accomplished surprisingly little. The mass of the people in Venezuela today are poor, unhealthy, illiterate and live in the most primitive surroundings. The economy is still shockingly backward. . . .
>
> The state becomes more and more opulent; the populace continues to live in misery. Meanwhile the nation becomes more and more dependent on a single extractive industry, more and more sensitive to events abroad.

Fifteen years and many billion dollars later this description still appears basically true.

27. Agrarian reform could lead to an energized electorate which might also end Iraq's exemption of agricultural income from taxation.

28. Alnasrawi, note 1 above, p. 62.

29. *Ibid.*, pp. 142–43.

1. Any oil boycott may conveniently be analyzed on the basis of six characteristics: who originates it, what is the cause, what are the aims, what are the methods used, what kind of country is affected, and what are the results. Of the dozen realized or almost realized oil boycotts in history, about half have been originated by the international oil companies and the other half by governments. In the case of boycotts originated by the international oil companies, the causes have either been nationalization by the country of oil company properties, or fear of "unfair" competition. Boycotts originated by governments have been triggered by a wide variety of political acts by the boycotted countries. Similarly, the aims of boycotts originated by the international oil companies have been narrowly restricted to the oil business, while the aims of boycott-launching governments have been to bring about widespread political changes. In some cases, however, the two aims are closely related, with the companies and government consequently working hand in hand. While nominally the methods utilized by the company boycotts are "nonviolent," as we shall see it has sometimes been the case that govern-

mental power has been brought to bear to help make the boycott successful.

2. Harvey O'Connor, *World Crisis in Oil* (New York: Monthly Review Press, 1962), pp. 82–83.

3. *Ibid.,* pp. 82–90.

4. "In a final interview with President Cárdenas, the companies' representatives asked who would guarantee that the award would be fulfilled. 'I, the President of the Republic,' answered Cárdenas. 'You,' retorted a British official, mockingly. Cárdenas rose to his feet and answered, drily, 'Gentlemen, we have finished.' The expropriation decree followed . . ."—*ibid.,* p. 112.

5. For a discussion of these factors see O'Connor, *The Empire of Oil* (New York: Monthly Review Press, 1955), pp. 315–16, and Bermúdez, chapter 22, note 5, p. 117.

6. "In 1950 . . . Iran had received $45 million and the British Treasury $140 million. In that year Iran realized more revenue from the state tobacco monopoly than from petroleum"—O'Connor, *World Crisis in Oil*, note 2 above, p. 290.

7. See Hossein Sheikh-Hosseini Noori, *A Study of the Nationalization of the Oil Industry in Iran* (Colorado State College, Graduate Division of Education: University Microfilms, Inc., 1965), pp. 1–162.

8. *Ibid.,* p. 182.

9. Reported in *The Times* (London), October 7, 1951; quoted in Noori, note 7 above, p. 201.

10. Henry F. Grady, "What Went Wrong in Iran?" *Saturday Evening Post*, January 5, 1952, p. 57.

11. Noori, note 7 above, pp. 236–37.

12. Noori, note 7 above, pp. 248–53.

13. Jerrold L. Walden, "The International Petroleum Cartel in Iran—Private Power and the Public Interest," *Journal of Public Law*, Vol. 2, No. 1, Spring 1962, p. 17; quoted in Noori, note 7 above, p. 245.

14. David Wise and Thomas B. Ross, *The Invisible Government* (New York: Random House, 1964), p. 111.

15. *Ibid.,* p. 110.

16. Hossein Sheikh-Hosseini Noori, in his *A Study of the Nationalization of the Oil Industry in Iran* (Colorado State College, Graduate Division of Education: University Microfilms, Inc., 1965), pp. 249–50, states:

. . . Dr. Mossadeq sought the assistance of W. Alton Jones, Chairman of the Cities Service Corporation of New York, with the hope of renewing oil operations in Iran. This was the most serious move of all, because Jones was believed capable not only of setting up an adequate marketing organization, but also of providing a fleet of tankers to carry Iranian oil to foreign ports. On August 16, 1952, Jones arrived at Teheran. After a tour of works, including the refinery at Abadan, he was brimming with enthusiasm. "The free world knows the Iranians can manage their own oil industry and operate the refinery," he was prompt to declaim. He found the refinery at Abadan upon inspection to be in good repair and estimated that full scale production could be resumed with a minimum expenditure of some $10,000,000. Jones also volunteered the services of Cities Service in enlisting European and American technicians to assist in starting up the industry again, and indicated that tankers could be secured within a matter of months to make resumption of full scale production in Iran. Upon leaving Iran, he publicly announced that "there was interest in buying Iranian oil and indicated strongly, although he would not enlarge upon it, that his company shared this interest." As for the blocked [sic] and the possibility of extended litigation over oil purchases, Jones proclaimed that "he might buy oil from Iran . . . irrespective of whether Britain or the Anglo-Iranian Oil Co. would take legal action against his Company for handling Iranian oil products."

With these auspicious beginnings, the ultimate failure of the Jones mission to revive the Iranian oil industry is a matter for major conjecture. As to what actually happened one can only surmise. "It is known that," contends Walden, "two concerns with which Jones was affiliated, Cities Service and Sinclair, each received handsome long-term contracts for Middle Eastern crude [oil] at substantial discounts from the posted price of the international cartel." "It has been publicly charged that," continued Walden, "one of these contracts was offered to Cities Service by cartel members in order to dissuade it from executing its reported agreement to market Iranian oil. We at least know that companies with which Jones was identified were subsequently given a share in the consortium arrangement which ultimately compromised the Iranian controversy."

17. Wise and Ross, note 14 above, p. 110.

18. For an informed discussion of this oil boycott see Herbert Feis, "Oil for Italy: A Study" in his book, *Seen from E.A.: Three International Episodes* (New York: Alfred A. Knopf, 1947).

19. *Ibid.*, pp. 305–6; emphases in original.

20. *Ibid.*, pp. 306–8.

21. *Ibid.*, p. 308.

22. There has been considerable dispute within the United States as to what and who were ultimately "to blame" for the deteriorating relations between the United States and Cuba. For two contrasting interpretations see Theodore Draper, *Castro's Revolution: Myths and Realities* (New York: Frederick A. Praeger, 1962) versus William A. Williams, *The United States, Cuba and Castro* (New York: Monthly Review Press, 1962).

23. O'Connor, *World Crisis in Oil*, note 2 above, p. 259.

24. Maurice Zeitlin and Robert Scheer, *Cuba: Tragedy in Our Hemisphere* (New York: Grove Press, 1963), pp. 175–76, emphases and brackets in original.

25. O'Connor, *World Crisis in Oil*, note 2 above, p. 258.

26. *Boston Daily Globe*, February 23, 1960. The economics of international oil strongly suggest that this belief of the majority of Cubans, common in other countries, is a myth. Once discovered, oil is usually so cheap to produce that a company finding it within a country can make more profits by producing it than by continuing to import its own external crude oil. Moreover, if a company can increase its market share because of governmental preference for indigenous crude, it can be highly profitable to produce the oil discovered. Thus, the logic of international oil is not that the majors will not produce oil once discovered in an oil-importing country; rather, it is that they will not look very hard in the first place.

27. O'Connor, *World Crisis in Oil*, note 2 above, p. 259.

28. From *The Times* (London), July 1, carried in *The Statesman* (Calcutta), July 6, 1960.

29. Zeitlin and Scheer, note 24 above, pp. 176–77.

30. J. P. Morray, *The Second Revolution in Cuba* (New York: Monthly Review Press, 1962), p. 102.

31. 1960 per capita energy consumption data from United Nations, Department of Economic and Social Affairs, *World Energy Supplies: 1959–62* (New York: United Nations, 1964).

32. *The Wall Street Journal*, September 27, 1960.

33. Zeitlin and Scheer, note 24 above, p. 176.

34. "The oil companies were diverting their tankers, allowing stocks to dwindle . . . Cuban workers in the refineries kept the government informed on the decline in stocks and the departure of American personnel, signs of an intent to shut down"—Morray, note 30 above, p. 102. That this choice of country and government over foreign companies is generally likely to be the case can be seen from the similar responses of the Mexican and Iranian petroleum workers when boycotts were attempted against their respective petroleum industries. A more recent example was the previously discussed threatened strike in 1965 of refinery employees of the established majors in India. The cause was the workers' belief that the oil companies were causing petroleum shortages by their boycott of Soviet refined products—see *Petroleum Intelligence Weekly*, July 5, 1965.

35. ". . . the Cubans took over the management of the refineries without incident. The crews had been nearly all Cuban, and so about the only formality was the hoisting of the Cuban Petroleum Institute's emblem on the properties" —O'Connor, *World Crisis in Oil*, note 2 above, p. 261.

36. *Ibid.*, p. 266.

37. This lesson should have been learned by the oil companies years before the Cuban events. As J. E. Hartshorn says on p. 295 of his *Politics and World Oil Economics* (New York: Frederick A. Praeger, 1962):

And the sorry affair of the Suez Canal pilots, demonstrated that some of the claims of the West to a monopoly of technical competence were arrant nonsense. That point, politically unimportant in itself, might in the medium run be as important as any other implication of Suez for the oil industry in the Middle East.

38. *The Wall Street Journal*, September 27, 1960, stated:

Fidel Castro's aids in Houston have been trying to buy more than $1,400,000 of U.S. industrial equipment but so far haven't been able to make a single purchase. . . .
Some parts the Cubans are seeking are essential to the operation of oil refineries they have seized from Shell Oil Company, Standard Oil Company (New Jersey) and Texaco, Inc. . . . some sources are convinced that if the Cubans can't buy the parts in the U.S. they will have to seek them elsewhere, possibly in Russia.

39. Quoted in O'Connor, *World Crisis in Oil*, note 2 above, p. 261; brackets in O'Connor. The enormous scope of this declaration led the same author to conclude: "Reminiscent of its ancient wars against Russia and later Mexico, Standard Oil issued a warning to tanker companies. . . . As Cuba was not

mentioned specifically, it seemed like another world-wide declaration of war by Standard on Russia . . ."—*ibid.*, p. 261.

40. Quoted *ibid.*, pp. 261–62.

41. *The New York Times*, November 3, 1960.

42. O'Connor, *World Crisis in Oil*, note 2 above, p. 260.

43. *The New York Times*, July 3, 1960.

44. *The Wall Street Journal*, July 25, 1960.

45. O'Connor, *World Crisis in Oil*, note 2 above, pp. 260–61. It is interesting to note that Superior Oil was solely a crude oil producer, with large amounts of surplus oil in Venezuela; hence, it was less likely to worry about the worldwide implications of refinery seizures than were the integrated international oil companies. In any event, possible future problems for the international majors from Superior's oil were largely eliminated by Texaco's purchase of Superior's Venezuelan properties in 1964.

46. Assuming conservatively that the round trip from the Soviet Union to Cuba and back would take about two months.

47. "Approximately 15% of the world total capacity is operating in the spot market. . . . the range of the tonnage operating in the spot market varies from 20% under depressed market conditions to 8% during periods of strong demand"—Zenon S. Zannetos, *The Theory of Oil Tank Ship Rates: An Economic Analysis of Tank Ship Operations* (Cambridge, Massachusetts: The M.I.T. Press, 1966), p. 154.

48. "As of January 1, 1959, there were over 600 tanker owners with none owning more than 7 per cent of the total capacity available . . ."—*ibid.*, p. 174.

49. *The Free Press Bulletin* (India), July 19, 1960.

50. *Ibid.*

51. O'Connor, *World Crisis in Oil*, note 2 above, pp. 261–62.

52. *The New York Times*, October 2, 1960.

53. O'Connor, *World Crisis in Oil*, note 2 above, p. 262.

54. National Petroleum Council, *Impact of Oil Exports from the Soviet Bloc, Vol. 1* (Washington, D.C.: National Petroleum Council, 1962), p. 110.

55. Cuban oil data from United Nations, *World Energy Supplies* (various issues).

56. "Reportedly, the exchange of products now being discussed is the purchase by Canada and Britain of sugar and gasoline and their sale to Cuba of machinery, spare parts, chemicals and other materials"—*The New York Times*, November 6, 1960.

57. During the period 1959 through 1965, gasoline consumption increased by 39 percent, or somewhat faster than all oil fuels consumption. This rise in gasoline consumption may have been partly fostered by the government. Thus, a 1962 study noted that in Cuba:

> . . . the transport administration is charged with conducting its operations as economically as possible which means that the purchase and use of diesel trucks and tractors are to be preferred if the balance sheet at the end of the year is to turn up with optimum figures. But the petroleum administration has a surplus of gasoline, and on balance it would probably be better for the national economy at this stage if more gasoline trucks and tractors were bought and used [O'Connor, *World Crisis in Oil*, note 2 above, p. 266].

Finally, success in changing the gasoline refinery yield may be seen from the fact that while gasoline's share of total refined fuels production jumped from 19 percent in 1959 to 25 percent in 1961 (the first year during which only Soviet crude oil was used) it was then reduced to 22 percent in 1962.

58. O'Connor, *World Crisis in Oil*, note 2 above, pp. 266–67.

59. *The Wall Street Journal*, July 25, 1960.

60. *Look*, November 8, 1960.

61. To date no significant supplies of indigenous crude oil have been found in Cuba. In this connection, the Cuban experience tends to refute a common myth in underdeveloped countries—that the international oil companies discover large amounts of oil within the country and then "sit on it": "Many Cubans suspected that big reserves had been found which the companies were hiding for future exploitation, in case of war. Subsequent examination of surveys did not sustain that contention"—O'Connor, note 2 above, p. 258.

On the other hand, the Cuban experience also tends to support our thesis that the established majors have little interest in exploring for crude oil in an underdeveloped country when they have surpluses of oil in proximity (in this case, Venezuela), at least not without enormous incentives:

The Batista government had been most liberal to oil-searchers. The oil laws had been overhauled in 1954 to increase exploration areas available, to provide government loans equalling 60 percent of foreign investment (cancellable if no oil were discovered) and to guarantee that there would be no changes in legislation for 15 years. A flurry of exploration followed, both by the cartel companies and by many independents operating on a shoestring. When a 250 b/d well was discovered at Jatibonico in 1954 Batista cried: "What we have long suspected has been confirmed—Cuba has oil!" But nothing came of it. . . .

The [Cuban Petroleum] Institute's technicians picked up some curious facts. Promising surveys had in some cases not been followed by drilling . . . [*ibid.*, pp. 257–58]. (See note 26 above.)

62. *Monthly Review*, February 1967, p. 7.

63. *The New York Times*, August 10, 1967.

64. O'Connor, *World Crisis in Oil*, note 2 above, p. 262.

65. *The Indian Express*, July 14, 1960.

66. The problem is further complicated by the fact that 1960 was an election year in the United States, and the incumbent president, Eisenhower, could not succeed himself. The behavior of a "lame duck" leader in a crisis is always more difficult to predict.

67. Testimony of General C. P. Cabell, Deputy Director, United States Central Intelligence Agency, in *Hearings, Committee on the Judiciary, United States Senate, 86th Congress, 1st Session*, November 5, 1959, pp. 162–64.

68. *The Boston Globe*, February 25, 1960.

69. J. P. Morray, *The Second Revolution in Cuba* (New York: Monthly Review Press, 1962), p. 102.

70. *Far Eastern Economic Review*, February 2, 1967, p. 165.

71. *Commerce* (Bombay), July 23, 1960.

72. *Far Eastern Economic Review*, February 2, 1967, p. 197.

73. Terence McCarthy in "The Garrison Economy," in *The Columbia University Forum*, Fall 1966, p. 31, says:

> For all today's trouble spots—from Algeria to Cuba, from the Dominican Republic to Vietnam, India and Pakistan, and all the nations bordering on Mainland China—are countries of appalling backwardness. It is not their attempts to escape backwardness but their insistence that progress be based on transformation of social institutions which brings the U.S. garrison economy, in country after country, into open conflict with the local populace. For the United States, it seems, has determined that while aid shall be extended, no attempts at basic change in social relations shall be tolerated.
>
> These social relations cluster around the single factor of land ownership.

A recent study of Vietnam seems to support this United States approach:

> The study was written by Edward J. Mitchell, an economist at the Rand Corporation, a largely United States Government-financed research center in Santa Monica, Calif. . . . His analysis, he said, suggested that, during wartime, Government control was more extensive in those Vietnamese provinces that are "essentially feudal in social structure."
>
> "The ideal province from the point of view of government control," Mr. Mitchell wrote recently, "is one in which population density is high, cross-country mobility is low, few peasants operate their own land, the distribution of farms by size is unequal, large estates (formerly French-owned and now primarily Government-run) exist, and no land distribution has taken place." [*The New York Times*, October 15, 1967].

74. *Far Eastern Economic Review: 1967 Yearbook*, p. 197.

CHAPTER TWENTY-FIVE

1. J. Fred Rippy, "Latin America's Postwar Golpes De Estado," *Inter-American Economic Affairs*, Vol. 19, No. 3, Winter 1965, p. 80.

2. These are Bolivia, Yacimientos Petrolíferos Fiscales Bolivianos; Chile, Empresa Nacional del Petróleo; Colombia, Empresa Colombiana de Petróleos; Ecuador, Empresa Petrolera Ecuatoriana; Peru, Empresa Petrolera Fiscal; Uruguay, Administración Nacional de Combustibles Alcohol y Refinería; and Venezuela, Corporación Venezolana del Petróleo.

3. The reasons for establishing each of these state oil entities in Latin America varied from country to country. For a survey background see Harvey O'Connor, *World Crisis in Oil*, Part IV (New York: Monthly Review Press, 1962). Undoubtedly, however, a necessary condition which allowed some of these countries to have a government oil sector before World War II (as compared to the total absence in Afro-Asian underdeveloped countries) was that the Latin American countries were legally independent (while almost all of the other underdeveloped countries were outright colonies). In addition, under Spanish law mineral rights belong to the state rather than the owners of the land.

4. O'Connor, *World Crisis in Oil*, note 3 above, p. 189.

5. Data from U.S. Bureau of Mines.

6. Quoted in O'Connor, *World Crisis in Oil,* note 3 above, pp. 195–96.

7. Julian Justo Gomez, "A Review of Argentina's Petroleum Policy: 1958–1962" unpublished Master's dissertation, Faculty of Political Science, Columbia University [circa 1963], p. 26.

8. From "The Anti-imperialist Struggle," a preface to Frondizi's book, *Petróleo y Política [Petroleum and Politics],* 1954, which "was issued as a pamphlet and used during the Presidential campaign"; quoted in O'Connor, note 3 above, p. 194.

9. Simon G. Hanson, "The End of the Good-Partner Policy," *Inter-American Economic Affairs,* Vol. 14, No. 1, Summer 1960, pp. 79–80.

10. *Ibid.,* pp. 81–83, emphases in original. In the light of this scathing published indictment of the Argentine oil contracts, it is worth noting that the Levy report for the World Bank concluded in its analysis of the Frondizi program that the participation of private enterprise should help greatly to achieve self-sufficiency in oil; in a footnote the report added that no conclusions were implied about specific individual contracts. In his later 1962 paper for the United Nations, Levy stated: "While arrangements such as in Argentina have proved effective, we do not suggest that they are in any sense models that should be copied elsewhere"—Levy, "Basic Considerations . . . ," chapter 10, note 3 above, p. 328.

Finally, it is interesting to note that the previously analyzed 1962 UN study on financing petroleum exploration concluded a brief analysis of Argentina with the statement: "The fact that Argentine production has doubled in two years has meant a saving of nearly \$300 million a year in the country's oil imports"—United Nations, *Petroleum Exploration . . . ,* chapter 18, note 26 above, p. 21. One study of Argentina's oil policy in this period notes that this is "A quite exaggerated estimation"—Gomez, note 7 above, p. 58.

11. *The New York Times,* November 11, 1963. As part of the maneuvering, Senator Hickenlooper introduced an amendment to the United States foreign aid bill requiring the cutoff of aid to any country that "repudiated or nullified" existing contracts with United States citizens or companies; it was explicitly aimed at Argentina (and also Peru, "which has indicated she may take similar action")—*The New York Times,* November 16, 1963.

12. *Platt's Oilgram News Service,* December 4, 1963, p. 2.

13. *World Petroleum,* July 1964, p. 7.

14. *Petroleum Intelligence Weekly,* November 2, 1964.

15. *The New York Times,* February 10, 1965.

16. *World Petroleum,* April 1966, p. 56.

17. *Petroleum Intelligence Weekly,* February 14, 1966.

18. *Petroleum Intelligence Weekly,* March 7, 1966.

19. *Ibid.,* February 14, 1966.

20. As Simon G. Hanson noted earlier in his "The End of the Good-Partner Policy," *Inter-American Economic Affairs,* Vol. 14, No. 1, Summer 1960, pp. 80–81:

It was the [State Department] practice to stress that Frondizi had come in as the first completely freely-elected president in thirty years. The fact was that the dominant party in the country had been banned and the army had determined which groups might be permitted to participate in the election. And the elections after the Frondizi regime had come in had shown him distinctly a minority leader. He was, indeed, in office for as long as the military were interested in keeping him there.

21. This was somewhat ironic since in 1963, as noted by *The New York Times,* October 13, 1963:

Dr. Illia was accompanied in the inaugural parade by Gen. Juan Carlos Ongania, the army Commander in Chief. General Ongania is credited with steering Argentina to elections last July despite conflicts within the military ranks.
A military group ousted President Frondizi on March 29, 1961. Dr. Illia was elected President last June, replacing Dr. Frondizi's military-appointed successor, Dr. José Maria Guido.

22. See *Petroleum Intelligence Weekly,* September 5, 26, November 14, December 9, 1966.

23. *The Oil and Gas Journal,* June 19, 1967, p. 92. Perhaps the most ironic aspect of the 1966–67 developments was that the proposed new law was attacked by former President Frondizi. He particularly objected to granting outright concessions (rather than the "service contracts" of his era) and argued against the new oil policies because:

They again open the door to ideological debate—this over the idea of concessions. They also give priority to juridical formalities rather than to the concrete need to obtain oil. And based on the belief that there are insufficient reserves, they lead to an excessively cautious production policy. . . .
I believe that, as soon as the government calls for proposals from foreign companies, many will be in the running, and particularly the larger ones. Here it should be considered that, to the extent that it proves of greater interest to these companies to sell to Argentina oil from other areas, so will they be disinclined to produce oil from our own subsoil. . . .
. . . I do know of the unfavorable reaction in military circles caused by the suspicion that the law has been elaborated with the direct participation of representatives of international monopolistic companies [*Petroleum Intelligence Weekly,* May 29, 1967].

24. *The Oil and Gas Journal,* July 3, 1967, p. 51. The provision barring "companies owned by foreign governments" was presumably aimed at excluding either the Soviet Union or state oil entities like Petrobras or Pemex.

25. Levy, "Basic Considerations. . . ." note 10 above, p. 328.

26. O'Connor, *World Crisis* . . . , note 3 above, p. 175.

27. *The Oil and Gas Journal,* various issues.

28. *Ibid.*

29. Hanson, note 9 above, pp. 74–77.

30. U.S. Bureau of Mines.

31. *World Petroleum,* March 1961, p. 38.

32. Yet, there was a curiously contradictory note to the views espoused in their favor. The petroleum press chided Petrobras for wasting the Brazilian people's money in a fruitless hunt for oil. Thus, in February 1961, *World Petroleum,* self-styled "Management Publication of the International Petroleum

Industry," published an editorial entitled "Disaster in Brazil," which said in part:

> There seems no room for doubt that the country's oil potential is not even a fraction of what was so fondly hoped five years ago. The Brazilian press has published in exact detail the confidential reports submitted by Walter Link, Petrobras exploration chief since 1954 [and former chief geologist for Jersey]. His final evaluation of the industry's present state is the result of all Petrobras's effort, since he has been in charge of that company's exploratory effort virtually the entire period of Petrobras's existence. In his educated opinion, outside the small coastal basins of Bahia and Tucano there are no good prospects for oil discovery in Brazil.

Again, *World Petroleum*, in April 1963 seemed to imply that given Link's report the companies did not really think there were good oil exploration prospects in Brazil:

> A few years ago international oil companies were ready and willing to spend many millions in Brazil to find oil and help make the nation less dependent on foreign resources. Had permission been granted at that time the nation would have been spared the cost of wildcatting and had the benefit of skills available in the experienced geological and geophysical departments of the oil companies. The nation would have gained whether much oil was found or not.

This suggests then the possibility that the real aim of the companies in the 1960s might not have been to explore for crude oil in Brazil, which was what the country badly needed, but to get permission for refinery expansion; the latter of course could allow the companies outlets for their own external low-cost crude oil.

33. *World Petroleum*, May 1964, p. 29.
34. *Petroleum Intelligence Weekly*, July 12, 1965.
35. *Ibid.*, September 13, 1965.
36. *Ibid.*, October 11, 1965.
37. *The Wall Street Journal*, November 23, 1966.
38. Soon after the overthrow of the Goulart regime, Petrobras announced a major find which, it was said, "may be big enough to bring the country close to self-sufficiency in oil production"—*Petroleum Intelligence Weekly*, December 21, 1964.
39. *World Petroleum*, January 1967.
40. *Ibid.*, p. 28.
41. *Petroleum Intelligence Weekly*, November 8, 1965.
42. *I. F. Stone's Weekly*, April 1, 1963.
43. With all these swings in Latin America toward the private sector and away from the state oil entities (including purges or mass resignations in the latter) it is hardly surprising that in 1967:

> In contrast to the often vituperative tone of the Arab Oil Congresses, an almost total absence of attacks on private oil companies pervaded the recent United Nations' sponsored meetings of government oil officials from 10 Latin American countries. . . . All sources agreed that they were a bit surprised at the absence of criticism of private oil companies.

In fact, some of the government participants from Venezuela, Argentina and Bolivia, among others, even had varying degrees of praise for the role of private foreign oil in their respective countries . . .

Argentina, especially, argued for the vigorous participation of private companies to help t ie state achieve its oil objectives [*Petroleum Intelligence Weekly*, April 3, 1967].

44. *Petroleum Intelligence Weekly*, April 10, 1967. These events are probably not unrelated to the following early 1967 report about Ghana:

Help for the new Government in overcoming the lack of foreign exchange has been provided through credits from the International Monetary Fund, which has thus helped to restore confidence in Ghana's currency. It has also endorsed the council's recovery program. This action, together with new government policies, has restored domestic and foreign business confidence in the Ghanaian economy. . . .

. . . the council has announced plans to turn a number of state-owned enterprises over to private operations and seek private participation in others [*The New York Times*, January 27, 1967].

45. *The Wall Street Journal*, April 18, 1967.

46. *Petroleum Intelligence Weekly*, February 20, 1967.

47. *Ibid.*, May 1, 1967.

48. *The New York Times*, November 20, 1967.

49. *Ibid.*, emphasis added.

Index

Adelman, M.A.: on production costs, 12; on price structure, 13; on Soviet pricing, 86–87; on Soviet oil, 88

Affiliates: 382; and prices, 14–16; and underdeveloped countries, 15; and balance of payments, 46; and Japan, 401

AFRA. *See* Average freight rate assessment

Africa. *See* North Africa; Union of South Africa

Agency for International Development: 400

Agriculture: and underdeveloped countries, 108; and Iraq, 414–415

Alagesan, O.V.: 217

Aldrich, Winthrop W.: 55

Alfonzo, Juan Pablo Perez: 384

Algeria: 23, 59, 343

All-India Congress Committee: 269

Alnasrawi, Abbas: 313; on Iraq, 304–305; on oil revenues, 306–307; on land tenure, 308–309; on industrial development, 309–310

Altos Hornos de Mexico: 301

American Management Association: and Soviet competition, 84

AMOCO–NIOC: 218–219

Anderson, Robert B.: 56

Anglo-Iranian: 321

Anglo-Persian Oil Company. *See* British Petroleum

Ankleshwar field: 225

Arab Oil Congress, 6th, Baghdad, 1967: 73, 385

Argentina: 120, 132, 350–357; percent of oil imports, 19; and foreign oil companies, 356–357; and oil contracts, 423

Arthur D. Little, Inc.: survey for Iraq, 415

Assam, India: 208

Atlantic Refining: 30

Australia: 246

Average freight rate assessment (AFRA): 153–159, 207, 396; relation between AFRA and market rates in period of declining market rates, 157; and underdeveloped countries, 158

Avery, E.W.: on exploration, 126–127; on exploration by companies, 131

Balance of payments: and oil industry, 45–48; and investment, 391

Balance of Payments of the Petroleum Industry: quoted, 381

Bandung Conference, 1955: 80

Barauni: refinery at, 208

Bechtel Corporation: 253

Ben Bella, Ahmed, President of Algeria (1962–1965): 69

Bermúdez, Director General: quoted on Pemex, 300

Black, Eugene: 92; on World Bank, 94

Bolivia: 120

Bosch, Juan, President of Dominican Republic (1963): 362, 363

Boycott(s): 6, 77, 319–348; of Mexico, 320–321; of Iran, 321–326; of Italy, 326–327; of Cuba, 327–348; characteristics of, 416

Brazil: 19, 94, 120, 122, 132, 350, 357–362, 424–425

Bridgeman, Sir Maurice: 222

British Petroleum: 21, 30, 43; purchase of, 49; and Iraq, 60, 305; and Kuwait, 60, 63–64; and Iranian boycott, 326

Buckley, John: quoted, 76–77

Bunker, Ellsworth: 186

Burmah Oil Company: 223, 238, 239; joint venture agreement in India, 225

427

Hurley, Patrick J.: 56
Hydroelectric power: 110

Illia, Arturo: 353, 355
Income: in Iraq, 314–315
Independence: and energy, 6; and foreign investment, 271
India: 26, 120, 374; and lack of energy, 4; percent of oil imports, 19; and World Bank, 95, 189; petroleum struggle in, 163–167; energy consumption pattern in, 164; political parties in, 165–166; refineries in, 168–170, 208–222; evolution of oil industry in 1950s, 168–177; oil companies and local equity, 172–177; and Soviet Union, 178–193, 208; and nationalization, 183; and U.S. government financing of oil imports, 188; its government and oil prices in 1960s, 194–207; and Japan, 199; exploration in, 223–242; government expenditures and returns from oil exploration investment, 1956–1966, 226; DCF rate of return on ONGC oil exploration investment, 1956–1966, 229; DCF rate of return on Indian government share of Oil India, Ltd., investment in oil exploration, 1958–1966, 231; fertilizers in, 243–256; U.S. investment in, 250–256; future of, 257–274; and ownership of oil industry, 260; and foreign investment, 264, 267–274, 409; and China, 282–283, 286–287; and Cuba and Soviet oil, 344–348; government and land reform, 347–348; and Iranian exploration, 405; and petrochemical policy, 407–408; energy consumption in homes, 408
Indian Congress Party: 346
Indian Oil and Natural Gas Commission (ONGC): 224–225, 228
Indian Oil Company: 179
Indonesia: 59, 349, 363–364
Indo-Stanvac: 223–224
Intascale: 153, 395–396
International Bank for Reconstruction and Development. *See* World Bank
International Monetary Fund (IMF): 26–27, 29, 95–100, 271
International oil companies. *See* Oil companies

International organizations: and oil, 26–27
International Tanker Nominal Freight Scale. *See* Intascale
"Interregional Seminar on Techniques of Petroleum Development": quoted, 120–121
Investment. *See* Foreign investment
"Investment in India": quoted, 172
The Invisible Government: quoted on Iran, 325
Iran: 18, 23, 59, 64, 374, 379–380, 384, 417; and oil companies, 60; and ENI, 75; boycott of, 320, 321–326
Iraq: 18, 23, 59, 64, 67, 289, 304–318, 415; and British Petroleum, 60; and Mobil, 60; and Royal Dutch Shell, 60; and ENI, 75; growth of oil production in, 305; oil revenues in, 306; World Bank mission to, 1952, 309; industrial development in, 309–310; politics in, 311–312; revolution, July 14, 1958, 311–312; actual and hypothetical income growth, 1950–1964, 314; actual and hypothetical annual income expressed in 1964 prices, 315; agriculture in, 316, 414–415; Detailed Economic Plan of 1962, 317; and oil companies, 318
Iraq Petroleum Company (IPC): 305
Israel: 81, 111; refinery in, 147; and boycotts, 320
Italy: 23, 111; and boycotts, 320; boycott of, 326–327
Iversen, Carl: 415

Jacobsson, Per: 96–97
Japan: 23, 63; and India, 199; and affiliate pricing, 401
Japanese Arabian Oil Company: 23, 30
Jersey. *See* Standard Oil of New Jersey
Johnson, Lyndon: 57, 58
Jones, W. Alton: 417
Jung-chao Liu: quoted, 285–286

Kabir, Prof. Humayun: 254
Kalol field: 225
Kassem regime: 60–61
Kassim, Abdul Karim al: 311
Kennedy, John F.: 57
Kerosene: 377, 408

Index